T0189844

DYNAMICS OF
COASTAL SYSTEMS

Advanced Series on Ocean Engineering — Volume 25

DYNAMICS OF COASTAL SYSTEMS

Job Dronkers

Rijkswaterstaat, The Netherlands

World Scientific

NEW JERSEY · LONDON · SINGAPORE · BEIJING · SHANGHAI · HONG KONG · TAIPEI · CHENNAI

Published by

World Scientific Publishing Co. Pte. Ltd.

5 Toh Tuck Link, Singapore 596224

USA office: 27 Warren Street, Suite 401-402, Hackensack, NJ 07601

UK office: 57 Shelton Street, Covent Garden, London WC2H 9HE

British Library Cataloguing-in-Publication Data
A catalogue record for this book is available from the British Library.

ISBN-13 978-981-256-207-4
ISBN-10 981-256-207-9
ISBN-13 978-981-256-349-1 (pbk)
ISBN-10 981-256-349-0 (pbk)

Editor: Tjan Kwang Wei

Typeset by Stallion Press
Email: enquiries@stallionpress.com

Printed in Singapore

Preface

'There, twice in every twenty-four hours, the oceans's vast tide sweeps in a flood over a large stretch of land and hides Nature's everlasting controversy about whether this region belongs to the land or to the sea. There these wretched peoples occupy high ground, or manmade platforms constructed above the level of the highest tide they experience; they live in huts built on the site so chosen and are like sailors in ships when the waters cover the surrounding land, but when the tide has receded they are like shipwrecked victims.'

So wrote the Roman historian Pliny the Elder (23–79) after his visit to the northern Netherlands.

Last year I spoke with a French TV-producer who was shooting a documentary on flood-prevention policy in The Netherlands. He was a bit disappointed: 'The sea does not seem to worry the Dutch; everywhere they have hidden it behind dikes; the sea is out of sight and out of mind'. This documentary perpetuates the myth that is expressed in the popular saying: 'God created the earth but the Dutch created Holland'.

One may also look from a different perspective. When the sea filled the Southern Bight of the North Sea 8000 years ago, central Holland was on average 15 metres below the present mean sea level. Most of Holland was drowned by the sea. In the Roman era, 6000 years later, tidal basins had been filled up with (mainly marine) sediment, and the greater part of Holland had grown close to the mean sea level. From that time on, human activities have caused soil subsidence of several metres and large tidal basins have formed in the southwestern and northern parts of the country. Maybe we should say: 'God created the earth but the sea created Holland'.

Is it possible that the sea creates land? Most often the sea is viewed as a destructive, land-swallowing force. It is the way the Dutch looked at the sea for a long time. And they had good reasons for that.

v

When I was five years old The Netherlands was hit by a devastating storm surge, which destroyed many sea defences. My father worked at Rijkswaterstaat, the governmental public works department responsible for safety against flooding. He developed new tidal computation techniques for the Delta project [115], designed after the disaster of 1953 to raise coastal defences to resist almost any storm surge. He died before the completion of the Delta project. I entered Rijkswaterstaat a few years later.

The Delta project gave a major impetus to coastal research in The Netherlands. The scale of the interventions in the Rhine–Meuse–Scheldt delta required the broadening of the coastal engineering perspective to long-term and large-scale aspects of coastal morphodynamics. Geologists and physical geographers joined the research teams and this interdisciplinary cooperation gave birth to the Netherlands Centre for Coastal Research. This organisation has contributed since its establishment in 1992 to the foundations of sustainable coastal management through interdisciplinary education and research programmes.

Great progress in understanding the basic nature of sea-land interaction has been achieved worldwide in the past decades. Insight in the nonlinear feedback processes inherent to this interaction has been a major breakthrough. It has modified the old perception of the sea as a destructive and dispersive force, by revealing its creative and structuring power. Progress in this area has greatly benefitted from the Marine Science and Technology Programme of the European Commission and the international conferences of the American Society of Civil Engineers [329] and the International Association of Hydraulic Research.

Our conception of the role of the sea has fundamentally changed. This is expressed in the present coastal defence strategy of the Netherlands. This strategy is based on working with the sea, instead of working against it, by stimulating the capacity of the sea to create and maintain new land [291].

Can we understand the creative and structuring power of the sea? That is what this book is about. It is an extended version of a course on the Physics of Coastal Systems, which I have given during the past ten years at the universities of Utrecht and Delft.

Acknowledgements

The partial closure of the Eastern Scheldt estuary drew me in the mid-seventies of the past century into the study of coastal dynamics. The challenge of this project was to protect the Netherlands against storm surges without sacrificing the ecological value of this largest tidal basin of the Dutch delta. My fortune was that no experts and no models were capable to predict the impact of partial closure on the morphologic response of the basin and that a large research programme was initiated in which I was engaged.

My first teacher in the dynamics of estuaries and coastal seas was Co van de Kreeke. From him I learned how little we basically know about the physics of these systems; he also taught me how to design measurement campaigns and how to improve our insight by using idealised models to interpret the observations. During the field campaigns I often got very useful information from the survey staff on board, who impressed me with their skills and their experience of the processes occurring in the field. Many of the ideas presented in this book are based on field data gathered by the survey departments of Rijks-waterstaat, which have not been published before. I value the discussions with colleagues from RIKZ, Jan Mulder, Ruud Spanhoff, Jean-Marie Stam, Cornelis Israël and Jelmer Cleveringa, on the interpretation of the extensive monitoring programmes of Rijkswaterstaat and on the morphodynamic behaviour of the Dutch coastal systems.

A breakthrough in my conception of coastal morphodynamics was due to the book 'Exploring Complexity' [325] by G. Nicolis and I. Prygogine, which I bought by chance in the library of MIT in Boston. My way of analysing physical phenomena was greatly inspired by talking and working with my colleagues of the Netherlands Centre for Coastal Research, Henk Postma, Sjef Zimmerman, Marcel Stive and Huib de Swart. The eagerness of my students, their critical questions and their misunderstanding of my explanations have been very important to sharpen my own understanding.

When writing this book I experienced how difficult it is to be clear, precise and concise at the same time, and to find a right balance between context and detail. Therefore I am very grateful to all those who have read and criticised the voluminous drafts I have sent to them: Wilfried ten Brinke, Rolf Deigaard, Keith Dyer, Albert Falqués, Frans Gerritsen, Co van de Kreeke, Ashish Mehta, Leo van Rijn, Giovanni Seminara, Richard Soulsby and Sjef Zimmerman. They have helped me clarify several questions and provided me with constructive comments and suggestions that greatly improved many parts of this work.

The former director of the National Institute for Coastal and Marine Management (RIKZ) of Rijkswaterstaat, Martin Beljaars, encouraged me to write this book and allowed me to retire after twelve years from my position as the head of the research department and to take a sabbatical. All the figures of the book have been drawn by Jules Modde during the last year before his retirement. The photographs of Fig. 5.27 were taken by my friend Maarten Michel, with whom I visited several beaches in Spain and who was intrigued as much as me by the beach cusp phenomenon.

List of Symbols

a	tidal amplitude *or* wave amplitude (half the root mean square wave height H_{rms}) [m]
A	total channel cross-sectional area [m^2] *or* coefficient
A_C	channel cross-section (flow carrying part) [m^2]
b	instantaneous width *or* time-averaged width [m]
b_C	channel (conveyance) width [m]
b_S	surface (storage) width [m]
c	wave propagation speed [m/s] *or* concentration of suspended sediment [kg/m^3]
c_D	friction (drag) coefficient $\tau_b/\rho \overline{u}^2$
C	sediment load (volumetric) [m] *or* coefficient
d	grain diameter of sediment particles [m]
d_*	dimensionless grain diameter $(g\Delta\rho/\rho\nu^2)^{1/3}d$
D	total instantaneous water depth $(Z_s - Z_b)$ [m]
D_L	longitudinal dispersion coefficient [m^2/s]
D_S	propagation depth A_C/b_S [m]
De	deposition rate (volumetric) [m/s]
e	2.7183
E	wave energy *or* tidal energy [J/m^2]
Er	erosion rate (volumetric) [m/s]
f	Coriolis parameter [s^{-1}] *or* a function (amplitude)
F	Froude number $u/\sqrt{gh}$,
g	gravitational acceleration 9.8 [ms^{-2}]
G	scaling coefficient
h	time-averaged water depth (channel depth) [m]
h_S	time-averaged propagation depth [m]
h_{cl}	closure depth [m]

ix

H	wave height $H_{rms} = \sqrt{8E/g\rho} = 2a$ [m]
HW	time of high water (also indicated by superscript +)
HSW	time of high slack tide
i	$\sqrt{-1}$ *or* index
j	index
k	wave number (in x-direction) [m^{-1}]
K	diffusion coefficient *or* eddy diffusivity [m^2/s] *or* absolute value of complex wave number [m^{-1}]
I	time-averaged water surface slope
l	length [m] (in particular basin length) [m]
L	wavelength of tide *or* wave [m]
L_A	length of barrier tidal inlet [m]
L_b	convergence length of river tidal inlet [m]
LW	time of low water (also indicated by superscript $-$)
LSW	time of low slack tide
m	metre
n	velocity exponent in sediment flux formula *or* ratio of wave-group and wave propagation speed
N	turbulent viscosity (eddy-viscosity) [m^2/s]
$O[..]$	order of magnitude estimate of . . .
p	pressure [Nm^{-2}] *or* seabed porosity
p_n	probability that the net displacement of a water parcel during n tidal periods exceeds a certain distance
P	tidal prism [m^3]
P_n	cross-sectional average of p_n
q	volumetric sediment flux per unit width [m^2/s] *or* total [m^3/s]
q_b	bed load transport [m^2/s]
q_s	suspended load transport [m^2/s]
Q	water discharge per unit width [m^2/s] *or* total [m^3/s]
Q_R	river discharge [m^3/s]
Q_T	tidal discharge [m^3/s]
Q_a	half the difference of maximum tidal flood and ebb discharges $(Q_T^+ - Q_T^-)/2$
Q_m	amplitude of tidal discharge [m^3/s]
r	coefficient for linearised bottom friction [m/s]
r_A	earth radius [m]
R	curvature radius of channel-bend [m]

Ri	Richardson-number
s	second
S	salinity [parts per thousand]
$S^{(xx)}$, $S^{(xy)}$	radiation stresses [N/m]
t	time [s]
T	wave-period *or* tidal period [s],
$\vec{u}$	flow velocity vector [m/s]
u	flow velocity in x-direction [m/s]
u_{cr}	critical flow velocity for incipient sediment motion [m/s]
u_*	shear velocity $\sqrt{\tau_b/\rho}$ [m/s]
U	amplitude of tidal velocity *or* amplitude of wave-orbital velocity [m/s]
v	flow velocity in y-direction [m/s]
V	longshore current [m/s] *or* sand volume ebb-tidal delta [m^3]
w	flow velocity in z-direction [m/s]
w_s	settling velocity [m/s]
x	longitudinal coordinate [m]
y	lateral coordinate (x, y, z right-turning coordinate system) [m]
z	vertical coordinate (z-axis upward) [m]
z_b	seabed height relative to unperturbed seabed [m]
Z_b	seabed height in a fixed reference frame [m]
Z_s	water surface height in a fixed reference frame [m]
α	coefficient sediment flux [m^{2-n}s^{n-1}]
β	bed slope/beach slope/shoreface slope *or* ratio of tidal acceleration to tidal friction $h\omega/r$
γ	bed-slope coefficient in sediment transport formula
γ_{br}	criterion for wave breaking H_{br}/h_{br}
δ	boundary-layer depth [m]
Δ	difference
Δ_S	average time delay between HW/LW and corresponding slack tides [s]
Δ_{EF}	difference in the duration of ebb and flood [s]
Δ_{FR}	difference in the duration of falling tide and rising tide [s]
$\Delta\rho$	density difference between sediment and water [kg/m^3]
∂	partial derivative

ϵ	infinitesimal quantity
ε_b	efficiency coefficient of bed-load transport
ε_s	efficiency coefficient of suspended-load transport
ζ	vorticity $(\overline{v}_x - \overline{u}_y)$ $[\text{s}^{-1}]$
η	departure from mean water level *or* water level with respect to horizontal reference [m]
θ	angle of wave incidence *or* angle between flow and bathymetry [rad]
κ	complex wavenumber $k + i\mu$ $[\text{m}^{-1}]$
λ	wavelength of rhythmic seabed pattern [m]
μ	imaginary part wave number *or* coefficient exponential damping $[\text{m}^{-1}]$
ν	kinematic viscosity $[\text{m}^2/\text{s}]$
Ξ	tidal diffusion coefficient $[\text{m}^2/\text{s}]$
π	3.14
ρ	density of (sea)water $[\text{kg/m}^3]$
ρ_{sed}	sediment density $[\text{kg/m}^3]$
σ	complex radial frequency of a rhythmic seabed perturbation $[\text{s}^{-1}]$
σ_r	radial frequency of a rhythmic seabed perturbation $\Re\sigma$ $[\text{s}^{-1}]$
σ_i	growth rate (positive or negative) of a seabed perturbation $\Im\sigma$ $[\text{s}^{-1}]$
τ	shear stress $[\text{Nm}^{-2}]$
τ_b	bottom shear stress $[\text{Nm}^{-2}]$
τ_{cr}	critical bottom shear stress for erosion $[\text{Nm}^{-2}]$
ϕ	phase angle between perturbed and unperturbed flow [rad]
φ	phase angle between tidal variation of velocity and water level [rad]
φ_r	angle of repose
Φ	flow potential $\Phi_x = u$, $\Phi_z = w$ $[\text{m}^2/\text{s}]$ *or* tide-averaged salt flux $[\text{mkg}^3/\text{s}]$
ψ	stream function $\psi_x = -w$, $\psi_z = u$ $[\text{m}^2/\text{s}]$
ω	radial frequency waves *or* radial frequency (semi-diurnal) tide [rad/s]
Ω	radial frequency earth rotation [rad/s]

Subscripts

$i = 0; 1; 2; ..$	residual component; basic frequency component; first higher harmonic component; ...
or	unperturbed state; 1st; 2nd order perturbation; ...,
$x; y; z; t$	partial derivative $f_x = \partial f/\partial x$; $f_t = \partial f/\partial t$; $f_{xx} = \partial^2 f/\partial x^2$; etc.
$i =, 1; , 2; ..$	partial derivative $f_{,1} = \partial f/\partial x$; $f_{,2} = \partial f/\partial y$; etc.
eq	equilibrium value f_{eq}
br	value at the breaker line f_{br}
C	refers to the tidal channel f_C
R	refers to river flow f_R
T	refers to tidal flow f_T

Superscripts

$(x; y; z)$	$x; y; z$-component, $\vec{u} = (u^{(x)}, u^{(y)}, u^{(z)})$
$+ ; -$	value at HW *or* at maximum flood (f^+); LW *or* maximum ebb (f^-)
$'$	perturbation from equilibrium f' *or* deviation from mean
$*$	dimensionless variable f^*

Averaging

$\langle f \rangle$	time-average $\frac{1}{T} \int_t^{t+T} f\, dt$
$\bar{f}$	depth-average $\frac{1}{D} \int_{-h}^{\eta} f\, dz$
$\bar{\bar{f}}$	cross-sectional average $\frac{1}{A} \int \int_A f\, dy\, dz$
$[f]$	order-of-magnitude estimate of f

Contents

Chapter 1

Introduction

1.1. What is this Book About?

Why are coasts the way they are?

Why are coasts the way they are: Multiform, infinitely complex, quasi-fractal, always changing and unpredictable in many aspects. But also: Why have coasts so many features in common? Any answer to these questions should be based on the universal nature of fundamental physical laws. This approach is known as 'coastal morphodynamics'.

This book is about the physical processes that shape sedimentary coastal landscapes. It introduces the reader to the physical-mathematical concepts developed during the past decades to explain the underlying basic principles. Most of the coastal landscape is under water, so landscape is a confusing word; we will use the term morphology instead.

Coastal morphology is shaped essentially by the action of waves and currents, with often an important role of the tide. This looks as an obvious statement, but it raises a puzzling question. Coastal morphology exhibits in general a great richness of structures at many scales, from very small to very large. This strongly contrasts with the comparatively poor spatial variability of waves, tides and wind-driven currents. One of the central themes of this book is to provide clues for explaining this apparent contradiction.

Sediment in motion

Coastal zones are among the most dynamic and energetic environments on earth, as stated in many articles or textbooks on coastal processes. Waves, currents and tides are the very visible expression of this dynamic nature. Acting on the shore and the seabed, the motion of the sea causes erosion and transport of seabed material. In the coastal zone, large quantities of sedimentary particles are perpetually in motion. In the Dutch coastal waters, for instance, the average quantity of sediment in motion at any moment is comparable to the net annual volume of coastal erosion (a few million cubic metres). The magnitude and direction of this sediment transport depend on waves, currents and tides, which are driven by external forces such as incoming waves, incoming tides and wind.

The coastal zone is characterised by the interdependence of water motion and seabed morphology

A crucial notion is that waves, currents, tides and sediment transport do not depend only on external forces, but also on the local topography and composition of the seabed. Hence, the magnitude and direction of sediment transport is not the same at different places in the coastal zone. At some places there will be erosion and at other places there will be deposition of sediment. As a result, seabed and shoreline are continuously changing: Changing position, changing form and changing composition. This change, in turn, affects waves, currents and tides. In other words: Coastal morphology and water motion evolve in an interdependent way. Sedimentary coastal environments are characterised by the continuous mutual adaptation of coastal morphology and water motion. Coastal systems in this book are defined by:

- water motion is substantially influenced by oceanic conditions;
- water motion and seabed topography are interdependent.

Typical coastal systems corresponding to this definition are sketched in Figs. 1.1 and 1.3.

Analogy with traffic

The interaction between water motion and seabed topography bears some analogy with the mutual dependency of car speed and car density on a highway. Variations in the traffic density along a highway are caused by speed differences; the traffic density increases where fast cars catch up slower cars. But there is also an inverse relationship because the traffic density influences how fast you drive your car. Variations in traffic density along the highway cause variations

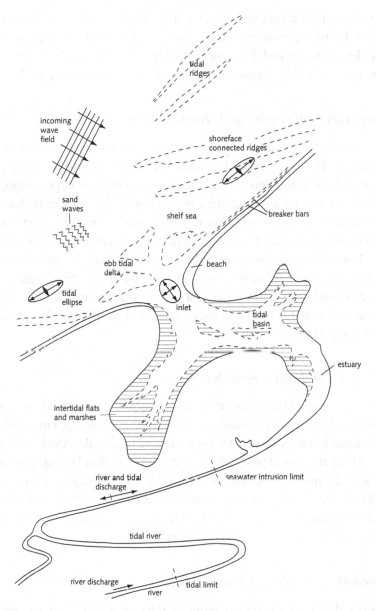

Fig. 1.1. The coastline may ingress far inland. The systems sketched in the figure are characteristic for low-lying coastal plains. The morphology of these coastal systems and the offshore seabed morphology mainly result from nonlinear interaction with water motion. The figure presents several features of the coastal environment and the corresponding terms used in this book.

in car speed and variations in car speed along the highway cause variations in traffic density. Hence, both evolve continuously in a mutually dependent way. If traffic density and speed differences are high, this mutual dependency leads to traffic jams caused by 'spontaneous' amplification of density peaks.

Scale dependency of erosion and sedimentation

At small spatial scales seabed morphology and water motion adapt to each other in a short delay, but at large spatial scales the adaptation period can be very long. If erosion and sedimentation are in balance averaged over large temporal and spatial scales it may happen that there is an imbalance at smaller scales or vice versa. In fact, the phenomena erosion, sedimentation and sediment transport always have to be defined with respect to particular spatial and temporal scales. In general we will choose these scales larger than the scales of turbulent motion and smaller than the time scale of sea-level rise. This last time scale is in the order of thousand years; during the past thousand years the sea level has risen about one metre. If necessary, a more precise definition of the spatial and temporal scales is given in the text.

Stability of a morphologic equilibrium

Suppose a situation where erosion and sedimentation are in balance; such a situation is known as a morphologic equilibrium. What happens if this morphologic equilibrium is disturbed, for instance, by an additional sedimentary deposit? There are several possibilities. The sediment may be dispersed and the deposit may disappear after some time; in that case the equilibrium is stable. The additional sedimentary deposit may also remain unaffected; in that case the equilibrium is called marginally stable. The third possibility is that the deposit starts to grow. In this last case the equilibrium is unstable.

Formation of morphologic features

The last possibility seems counter-intuitive if we agree that the added sediment does not possess any special attractive force. What makes the deposit grow? There is no unique answer to this question; it will be shown later that different processes may play a role. All these processes have in common that the flow perturbation produced by the sedimentary deposit affects the existing sediment motion in such a way that sediment converges at the deposit. This phenomenon is contained mathematically in the nonlinearity of the equations describing water motion and sediment transport. However, the underlying principles are more general and also play a role in other systems with nonlinear feedback processes, such as the earlier mentioned occurrence of traffic jams on a highway. In Chapter 2 it will be shown that these basic principles can be captured in the concept of symmetry breaking, which is inherent to the nonlinear nature of the interaction between seabed morphology and water motion [325]. The concept of symmetry breaking applies to the emergence and evolution of most morphologic structures in sedimentary coastal environments, from ripples to sandbanks, from creeks to tidal inlets.

Time scale of coastal morphodynamics

We have mentioned above that the physics of sedimentary coastal environments is related to temporal and spatial scales. The physical processes that determine coastal morphology span a range of temporal scales covering more than ten

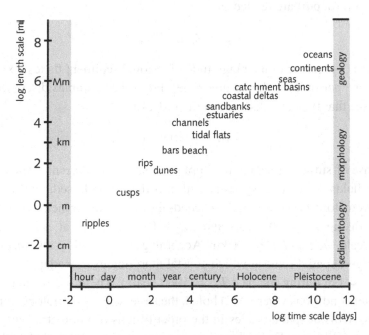

Fig. 1.2. Geomorphologic patterns in coastal systems span a very large range of space and time scales. Space and time scales are closely related. Adapted from [208].

orders of magnitude, see Fig. 1.2. At the lower end of this range we will not discuss processes at time scales smaller than the time scale of turbulence; this time
scale is typically between one second and a few hours, depending on the type
of flow. For these processes we will adopt empirical parametric descriptions.
At the upper end we will limit the discussion to the time scale of substantial sea
level change, which is in the order of ten thousand years. This excludes long-
term geological processes, such as plate tectonics and glacial cycles, which
are responsible for the global distribution of coastal environments. We will
not try to answer the question why certain types of coastal systems are where
they are; we accept inherited large-scale characteristics, such as the width of
the continental shelf, the seabed composition and the external hydrodynamic
forcing by waves, wind and tides.

Spatial scales of coastal morphodynamics

The restriction on the range of time scales is equivalent to a restriction on the
range of morphologic scales. The temporal scale T_m and the spatial scales L_m
(length), Z_m (depth) are related by

$$q_m = Z_m L_m / T_m,\tag{1.1}$$

where q_m is a measure of the magnitude of residual sediment fluxes (expressed
as a volume per unit width and unit time). From the examples discussed later
we will see that typical values are in the order of

$$q_m \approx 10^{-6} - 10^{-5}\ \mathrm{m^2/s},\tag{1.2}$$

in the case of sufficient sediment supply and sufficient current strength. This
estimate holds for coastal systems with a bed of mobile sediment and with
near-bottom flow that substantially exceeds the threshold for incipient sediment
motion. Taking $T_m \leq 10^4$ years and $Z_m \approx 10\,\mathrm{m}$ as a typical estimate for the
vertical scale, we find $L_m \leq 100\,\mathrm{km}$. According to (1.1), this is also an estimate
of the largest spatial scale at which coastal morphology can adapt to sea-level
rise ($Z_m \approx 1\,\mathrm{m}$ during the past $T_m = 1000$ years). Therefore we may expect
that, under conditions where (1.2) holds, the large-scale morphology of natural
coastal systems at spatial scales in the order of tens of kilometres will not be
far from equilibrium. Examples of such coastal systems are sketched in the
Figs. 1.1 and 1.3.

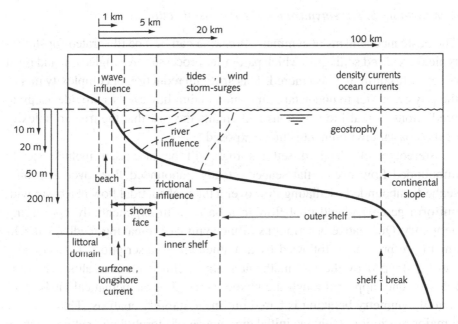

Fig. 1.3. The coastal zone is a continuum from beach to shelf break. Different zones can be distinguished related to prevailing hydrodynamic processes and external forces; these zones correspond to characteristic depth ranges, but the transitions are gradual. The intensity of bed-flow interaction increases from the outer shelf to the beach. At the time scale of sea-level rise the concept of morphodynamic equilibrium only applies to the beach-shoreface zone.

1.2. Why this Book?

Breakthrough in understanding

Our understanding of coastal morphology has made spectacular progress during the past decades. The productivity of coastal research has benefited from techni-cal innovation: Refined observation techniques and increased computing power. But a major key to this progress is conceptual innovation: Relating coastal morphology to symmetry-breaking properties of the nonlinear feedback with water motion. Time-symmetric sinusoidal forcing of the water motion yields a morphology-dependent time-asymmetric response; a uniform equilibrium morphology may under certain conditions become unstable and evolve spon-taneously into a spatially modulated morphology. This approach allows a uni-fying description of land-sea interaction and therefore runs as a leading thread through this book.

A *laboratory of stereotypical idealised coastal settings*

The basic mechanisms of symmetry breaking can best be illustrated for stereo-typical idealised settings in which particular processes are singled out and most of real-life complexity is ignored. This artificial reduction of complexity makes the physics easier to understand; it cannot claim, however, to provide accurate predictions in real-life situations and entails the risk that the true complexity of the coastal environment is underexposed.

Stereotypical idealised settings that will be discussed include steady unbounded flow over a flat seabed, steady unbounded flow over a sloping seabed, unbounded oscillating flow over a flat seabed, tidal flow in a basin with uniform geometry, and tidal flow in a basin with exponentially converging geometry. Qualitative descriptions of the symmetry-breaking feedback mechanism are presented, followed by a mathematical description. This last part can be skipped by the less mathematically inclined readers, although some features only appear through the mathematics. The mathematical analysis of spatial symmetry breaking is based on linear stability analysis. This imposes a major restriction: Only the initial emergence of morphologic patterns can be described in this way and not the evolution towards fully developed patterns.

Important contributions to the discipline

The great progress in physical coastal science of the past fifty years is related to the economic expansion after WWII, which led to major investments in coastal and harbour development. It was soon recognised that the coastal zone is a dynamic environment which may respond to interventions in an unexpected way. In many countries research laboratories have been established or expanded in order to develop better knowledge of coastal behaviour. This book highlights essential discoveries made by these research groups and puts them within a coherent framework. It does not attempt, however, to review all scientific developments; more complete reviews can be found in other recent textbooks [259, 496, 102, 304]. The approach of this book is closer to the symposium volume 'River, Coastal and Estuarine Morphodynamics' [399], edited by Seminara and Blondeaux (2001), with many excellent review articles. The editors rightly mention R. A. Bagnold [19] and F. Engelund [138] as the leading pioneers of sediment transport mechanics. Their work has led to important breakthroughs in understanding sedimentary processes in both the fluvial and the coastal environment. Due to the many similarities between the two fields, river research has strongly stimulated progress in coastal morphodynamics. Important generic insight has been gained from research on the formation of ripples, dunes, bars and meanders in rivers, with leading contributors such as L. B. Leopold [274], J. F. Kennedy [252], M. S. Yalin [501], W. H. Graf [177], G. Parker [343] and S. Ikeda [228].

Sediment transport mechanics

The Danish school of river and coastal morphodynamics founded by F. Engelund, with leading scientists as J. Fredsøe and R. Deigaard [157], has greatly contributed to the development of a theoretical framework for sediment transport mechanics and bed-flow interaction. Important insight in coastal zone wave dynamics and its influence on sediment transport is due to M. S. Longuet-Higgins (1953) [287] and later works of J. Battjes [24] on wave-breaking processes. Major contributions to the practical description of sediment transport in the coastal zone were made by E. Bijker (1967) [35], L. van Rijn [472], D. Huntley [224], R. Soulsby [418], P. Nielsen [326] and others. Our knowledge of the behaviour of cohesive sediments was greatly advanced by the pioneering experimental work of R. B. Krone [262] and A. J. Mehta [312].

Beach processes

The introduction of the equilibrium profile concept (Fig. 1.3) was a crucial step in coastal morphodynamics. It was first proposed a century ago [85, 86, 243], but in 1954 P. Bruun highlighted the implications of this concept for predicting long-term coastal evolution [56]; the theoretical foundation of this concept was advanced further by R. G. Dean (1973) [98] and A. J. Bowen (1980) [49]. The first comprehensive description of beach processes was made by L. D. Wright and A. D. Short (1984) [499], members of a very productive Australian coastal science community including other leading scientists such as R. W. G. Carter [67] and C. D. Woodroffe [495]. After first exploratory work by M. Hino in 1974 [201], the Catalan research group of A. Falqués and coworkers managed to unravel the basic mechanisms responsible for longshore surf zone instability [143]. The observational basis of this work was laid by R. Holman [208], who developed the ARGUS-camera technique and initiated the establishment of a worldwide network of observation sites.

Seabed structures

The coastal environment is characterised by a broad spectrum of seabed structures; sand ripples are at the small-scale end of this spectrum and tidal sandbanks at the large-scale end. The mechanics of ripple formation has long been one of the most challenging coastal phenomena: the basic mechanisms were first described in detail by J. R. L. Allen [2] and J. F. A. Sleath [410] and later simulated with process-based analytical models developed by the Genua morphodynamics school of P. Blondeaux and coworkers [44]. Important field investigations of tidal and nontidal sandbanks have been conducted by the research groups of T. Off [336], J. J. H. C. Houbolt [212], J. H. J. Terwindt [438], I. N. McCave [306], D. J. P. Swift [433], B. A. O'Connor [331] and others. The mechanics of tidal sandbanks were first described by J. D. Smith [414], J. T. F. Zimmerman [510] and J. M. Huthnance [226, 227].

Tidal morphodynamics

The first morphodynamic description of tidal basins was presented by J. van Veen (1950) [476]. This description was corroborated and extended with observational evidence on depositional processes in the estuarine environment by the research groups of G. P. Allen [5], R. W. Dalrymple [91], J. S. Pethick [348], C. L. Amos [8] and others. The essential role of tidal asymmetry and its morphological consequences was first recognised in 1954 by H. Postma [354]. Generation mechanisms of tidal asymmetry were first studied by P. H. LeBlond [271], D. Prandle [358] and D. G. Aubrey [16]; the relationship with estuarine morphology was further explored by C. T. Friedrichs [161], D. A. Jay [238], G. Seminara [401], S. Lanzoni [269] and others. Important steps towards process-based models of the full coastal system, including tidal basins and the adjacent shoreline, were realised within the Netherlands Centre for Coastal Research, with major contributions by H. De Vriend, M. J. F. Stive and D. J. A. Roelvink [424].

A complementary approach

This short overview of important contributions to our present understanding of sea-land interaction is far from complete. No existing textbook covers the entire field. The classical textbook on coastal sedimentary processes is K. Dyer's 'Coastal and Estuarine Sediment Dynamics' (1986) [123]. Recent important textbooks are 'Beach Processes and Sedimentation' by P. D. Komar (1998) [259], 'Coasts' by C. D. Woodroffe (2002) [496], 'Coastal Processes' by R. G. Dean and A. Dalrymple [102] and 'Introduction to Coastal Processes and Geomorphology' by G. Masselink and M. Hughes (2002) [304]. The present book complements these recent overviews by its focus on the basic physical principles underlying sea-land interaction.

1.3. Who is this Book Intended for?

Sustainable coastal management

Coastal zones directly support a growing part of the world population [412]. Highest densities of urbanisation are found in sedimentary coastal plains, which have been shaped by land-sea interaction. This interaction has not ceased, and coastal morphology continues to evolve. Coastal evolution can be frozen, locally and temporarily, by engineering interventions. But the large-scale evolution can hardly be stopped. Artificial constraints may even speed up coastal evolution instead of slowing it down; this will occur if the coastal system is brought further away from equilibrium. Sustainable coastal planning aims to avoid conflicts with natural coastal evolution; therefore it has to rely on a solid understanding of the mechanisms of land-sea interaction.

Continuing efforts to maintain the coastline

The ancient village of Egmond along the Dutch North Sea coast disappeared into the sea in the past centuries (Fig. 1.4). It could have been protected against the natural retreat of the coastline by the construction of sea walls, but for how long? At present, the coast around Egmond is nourished with sand taken from far offshore, in order to bring the coast locally closer to a morphologic equilibrium; it is expected that sand nourishment will have to continue for centuries before a situation close to equilibrium is reached. In the meantime the Dutch coast will continue to adapt to sea-level rise and to sand loss to the dunes. Hence, coastline maintenance can only be sustained by using the North Sea bottom as a permanent sand source for coastal nourishment [261].

Cost-effectiveness of coastal policies

The example of Egmond illustrates the dilemmas related to coastal development. It implies striking a balance between benefits and costs; this balance is

Fig. 1.4. The church of Egmond was swallowed by the sea in the 18th century. The picture was drawn shortly before.

more easily evaluated for the short term than for the long term, but the latter is often more significant. Several policies for responding to coastline retreat are possible, ranging from hard protection structures to abandonment of settlements, from supratidal beach nourishment to subtidal shoreface nourishment. The effectiveness of different policies depends to a large degree on the long-term response of the coastal system to the intervention. A coast eroded by tidal currents may require other measures than a wave-dominated coast. Estimating the cost-effectiveness of measures is possible only if the natural dynamics of the coastline is well understood.

Coastal observation is essential

Any attempt to understand the coast should start with observing the coast. This is a major investment, because the time scale of coastal evolution is long and trends are masked by short-term fluctuations. In The Netherlands a coastal monitoring programme started since the 1960s. Each year, the shoreface-profile is measured over a cross-shore distance of 800 m along the entire coastline, with a longshore spacing of 200 m. The coastal data set already covers almost 40 years; this information guides the annual coastal nourishment planning. Interpreting this data set is essential to the planning efforts. Each year large fluctuations occur in the coastline position, but many of these fluctuations do not correspond to long-term trends. Effective coastline management requires a thorough interpretation of observed coastline behaviour.

Management of river mouths, tidal basins and coastal wetlands

Sustainable management of river mouths, estuaries, tidal basins and coastal wetlands poses dilemmas similar to those in shoreline development and maintenance. Such dilemmas arise from the wish to protect existing features or from new claims on the coastal environment, which are mutually conflicting or conflicting with natural coastal evolution. Decision-makers need to know whether an intervention may conflict with natural coastal evolution and at which scale, in space and time. How can such conflicts be avoided or mitigated? At which costs? Does the intervention influence coastal evolution? Will this affect existing uses and opportunities? How can competing claims and interests be reconciled? Credible policy choices should make the best possible use of the progress achieved during the past decades in understanding coastline dynamics; this understanding has increased substantially the capability to produce more reliable predictions of the natural evolution of the coastal environment and its response to intervention.

Models, rules and analogies

Present knowledge is not yet sufficiently advanced, however, for delivering reliable standard tools for predicting large-scale and long-term coastal evolution. Certain (semi-)empirical relationships are often successful, for instance, the 'Bruun rule' for coastal retreat (Sec. 5.5.2), the 'Dean rule' for the shoreface slope (Sec. 5.2.1), the 'O'Brien rule' for the cross-sectional area of tidal inlets (Sec. 4.2.6) and the 'Walton rule' for the ebb-delta volume (Sec. 4.2.2). In this book some additional relationships are derived from simple models, which help predict coastal evolution and the coastal response to intervention. It must be emphasised that these relationships have a restricted validity; they are applicable only under certain conditions. Understanding these conditions is more important than knowing the relationship. This requires for each case knowledge of the specific field situation, which can only be obtained through observation programmes. The most important benefit of the morphologic relationships is the help they offer for interpreting the observations and for drawing the right conclusions. Models and rules can easily be misused. The emphasis in this book is therefore primarily on the processes governing land-sea interaction, rather than on practical field cases that may give rise to misleading analogies.

Coastal diversity

The land-sea interaction mechanisms discussed in this book are often illustrated with field evidence from the coastal zone of the Netherlands. This is a deliberate choice; many of these data were collected for special coastal management purposes by the Dutch Ministry of Public Works and Water Management and are not available in the open scientific literature. It must be realised that these data are not necessarily representative of other coastal environments; the main characteristics of the Dutch coast are the relative minor influence of ocean swell, the modest fluvial sediment supply and the geological setting of a steadily subsiding basin. The basic land-sea interaction mechanisms are the same in most sedimentary coastal environments; however, the relative importance of these processes may differ greatly from one coastal zone to another, even over short distances.

Students and coastal professionals

This book is written primarily for learning purposes; it is based on a lecture course on coastal morphodynamics given at the Universities of Utrecht and Delft in The Netherlands for graduate students who are familiar with the basic concepts of coastal hydrodynamics. The material has been extended to provide an overview of the main physical principles of land-sea interaction. It enables coastal engineers to complete their background knowledge and to facilitate access to cutting-edge scientific literature on specific topics. The book may also serve to familiarise professionals in other coastal disciplines with modern concepts of land-sea interaction. The mathematical subsections, often presented at the end of a section under the heading 'A simple model', can often be skipped without affecting the understanding of the essential concepts.

1.4. How is this Book Organised?

Principles of symmetry breaking

Chapter 2 introduces the concepts of spatial and temporal symmetry breaking; these concepts provide a unifying framework for the multiple morphodynamic feedback processes discussed in the following chapters. The tidal bore in the Qiantang estuary (China) is discussed as an illustration of time symmetry breaking; spatial symmetry breaking is illustrated by the instability of a system of parallel channels. The history of the Rhine delta (at a timescale of centuries) and the history of the Ameland reef (at a timescale of years) are discussed as examples.

Current-topography interaction

Chapter 3 starts with a brief overview of some general properties of water flow and sediment transport over a rough, mobile seabed. The main body of the chapter is devoted to seabed instability inherent to the interaction of currents with topography, resulting in spatial symmetry breaking and the generation of rhythmic seabed structures. Spatial symmetry breaking is associated with both steady currents and oscillating currents; the basic principle is the same in both cases. Different symmetry-breaking processes are discussed, which produce seabed structures at very different scales, from ripples (wavelength of tens of centimetres) to tidal sandbanks (wavelength of kilometres). Symmetry breaking in flow bounded by channel banks leads to channel meandering and building of tidal flats. Understanding these processes is a great help in interpreting observations of seabed and basin topography and in extracting information on morphologic processes and coastal evolution.

Tide-topography interaction

Chapter 4 deals with the morphodynamic feedback responsible for the topography of tidal basins. Tide-topography interaction affects the time symmetry of the tidal wave, producing differences in the strength of flood and ebb currents and in the duration of high-water slack tide and low-water slack tide. This asymmetry changes the topography of tidal basins due to net erosion or sedimentation, and results in feedback to tidal-wave asymmetry. We distinguish two types of tidal inlets: River tidal inlets and barrier tidal inlets. The stability criteria of both types of inlets are discussed with reference to a large number of tidal basins which are documented in the literature. The results are relevant for answering questions like: 'How do tidal basins respond to human interventions (dredging, sand mining, tidal flat reclamation) and to sea-level rise?'

The chapter ends with a discussion of fine-sediment dynamics in tidal basins. The fine-sediment fraction responds to tidal asymmetry in a different way than the coarse fraction and it is also more sensitive to other transport and sedimentation processes, including biotic activity. No in-depth treatment of eco-morphologic processes is presented, in spite of their potential importance for long-term morphologic evolution of tidal basins.

Wave-topography interaction

Chapter 5 deals with the nearshore zone where sediment transport is primarily influenced by wave activity. Waves interact with coastal topography in several different ways, involving both time-symmetry and space-symmetry breaking. The increasing wave asymmetry in shallow water, finally leading to wave breaking and to the production of radiation stresses, is at the root of net sediment fluxes and topographic adjustment, which result in feedback to wave asymmetry, wave breaking and radiation stresses. In this way a broad spectrum of seabed structures is generated, for instance, longshore bars, transverse bars, rip cells and beach cusps. The shape of the coastal profile itself is also the result of wave-topography interaction. Understanding these processes is essential for answering questions like: 'How can the sea help us maintain the coastline?'

Basic mechanisms

Throughout this book much emphasis is placed on explaining the mechanisms responsible for the generation of coastal morphology. A rough indication of the morphologic spatial and temporal scales can already be derived from these qualitative descriptions. These descriptions also provide an understanding of the conditions under which different topographies develop. For each interaction mechanism, a qualitative discussion is followed by an analytical model which is solved for a particular idealised stereotypical situation. The solution repeats the qualitative discussion, but sharpens the underlying assumptions. The predictive value of these models for real-life situations is very limited; they should be considered primarily as tools for analysis and understanding. The appendix provides an introduction to the basic equations and a mathematical derivation of certain results for which only a qualitative justification is given in the main text.

Chapter 2

Morphodynamic Feedback

2.1. Pattern Generation

A remarkable coastal feature

Once I went for a walk with friends along the beach of northern Brittany (France), near Morlaix. The sight of this coastal landscape is fascinating, with its granite rocks and pocket beaches, nested in between high cliffs protruding far into the sea. The lower part of the beaches is sandy and smooth, with a half-moon shaped shoreline. Higher up, the beach becomes steeper and it is covered with shingle. The upper part of the beach is made of nicely rounded and coloured boulders, a few tens of centimetres in size. It was low water, about 3 m below mean tidal level, and we could walk from one pocket-beach to the next by climbing the rocky promontories. One of the pocket-beaches looked different from the others, but it took some time before we realised the cause of this. The boulders at the top of the beach were arranged in a sequence of bows, terminating in sharp cusps of about one metre high, pointing seaward. The spacing between the cusps was the same everywhere; about 15 metres. We counted more than ten cusps along the upper rim of the beach, over a distance of a few hundred metres. This regular pattern looked quite artificial. My friends asked me whether the boulders could have been placed in cusps for coastal defence reasons. I answered that such coastal defence structures are most unusual, and that I could not see any good reason for coastal defence structures in this place. But if the boulder pattern had not been built by humans, what could be the explanation? And why did the cusp pattern not appear on other pocket beaches? My friends proposed that it might be the work of extraterrestrials.

A symmetry-breaking event

Talking about extraterrestrials, let us imagine that we can travel back in time. We go back to the time that the boulders are still uniformly scattered on the beach; the beach now has a perfectly uniform appearance. But suddenly it gets dark; clouds cover the sky and the wind starts blowing very strongly. A heavy storm arises and huge ocean waves beat against the coast. Breaking waves lift the boulders, which are entrained by a violent swash flow and hurled higher up the beach with a loud roar. After a few terrible hours the storm calms down, the clouds move away and it becomes clear enough to see the beach. The lower sandy beach strip has become much steeper, but the most striking is the regular pattern of the boulders, which are now placed in equally spaced cusps. The beach has lost its original uniformity (see Fig. 2.1). The boulder pattern introduces a new spatial scale that did not exist before. But, even stranger, a similar scale did not appear in the waves hitting the coast, which had a much broader front and a far greater wavelength.

Is natural symmetry breaking possible?

In this story waves and boulders together played a game with a remarkable outcome. Could this really happen as described above? Could the transformation from a uniform to a non-uniform coast be achieved without human intervention? Would physical laws allow that a perfect symmetry is broken? How can

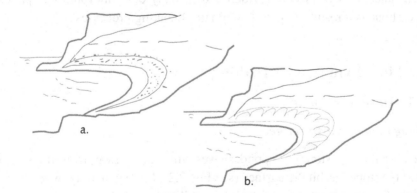

Fig. 2.1. Pocket-beach enclosed by rocky headlands on an exposed, macrotidal coast, at low water. The lower part of the beach is sandy, the higher part is covered with cobbles and boulders. (a) Longshore uniform beach. (b) Rhythmic pattern of beach cusps on the higher part of the beach. See also the beach cusp pictures in Fig. 5.27.

the cusp pattern be so regular and what can determine the spatial scale? We will leave these questions till later as far as this particular example is concerned. We notice, however, that many other phenomena in nature raise the same type of questions [325]. For instance, how can a fluid change into crystals? How does a regular wave field arise from a flat ocean surface? Symmetry breaking and pattern formation are fundamental properties of the physical laws postulated by Newton.

Temporal and spatial symmetry breaking

We can distinguish between two kinds of symmetry breaking. The first kind of symmetry breaking affects the temporal pattern of currents induced by waves or tides, and the second affects the spatial pattern of currents. The first kind is related to the interaction of waves with topography (wind waves and tidal waves) and the second kind to the interaction of currents with topography. In this chapter the nature of these two types of interaction will be discussed. Time symmetry breaking is a mutual adaptation process between topography and wave propagation, through the generation of residual sediment fluxes. It plays a major role in long-term coastal evolution, as discussed in Sec. 2.2. Spatial symmetry breaking is a self-exciting process that can be triggered by small disturbances of an initially uniform topography, as discussed in Sec. 2.3. It may explain, for instance, the formation of cusps on an initially uniform beach; later we will provide evidence that many other morphologic patterns come about as a result of spatial symmetry breaking processes.

2.2. Time-Symmetry Breaking

2.2.1. *Wave Asymmetry*

Propagation of a disturbance

When we throw a stone in a pond, waves will radiate away from the location where the stone has hit the surface (see Fig. 2.2). More generally, when a water body is disturbed locally, the disturbance will propagate through the water body away from the initial location. With time, an increasing part of the water body will be affected, depending on the propagation speed of the disturbance and on the rate at which the disturbance dissipates. For water bodies with a free

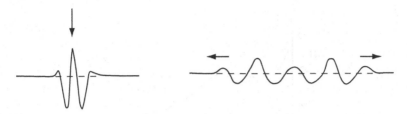

Fig. 2.2. A local impact on the water surface produces a disturbance which propagates away from the initial impact location.

surface, for instance the ocean, a shelf sea or a tidal river, the propagation speed depends mainly on water depth if the disturbance length scale is much larger than the water depth. The water depth, however, is not an intrinsic property of the water body; it is influenced by the disturbance, at least to some degree. If the water surface disturbance is only a very small fraction of water depth, the influence of the disturbance on the propagation speed may be ignored. If, in addition, the water depth is almost uniform over distances comparable to the length of the disturbance, then the disturbance will propagate throughout the water body without being distorted. Swell waves generated far offshore by local wind fields propagate shoreward without much distortion until they reach the coastal zone, where the wave length and water depth are no longer very large with respect to wave height.

Onshore wave propagation

The asymmetry of wind waves and swell in shallow water is clearly visible from the beach. The crest of the incoming waves becomes increasingly steep as the wave front approaches the shore. At a certain moment the wave crest catches up with the wave trough; then the wave overtops and breaks. Wave steepening in shallow water goes along with an increasing asymmetry between onshore and offshore orbital motion. The onshore motion is of shorter duration than the offshore motion, but also stronger (Fig. 2.3). The wave-orbital motion moves bottom sediments back and forth, but onshore sediment transport is higher than offshore transport. Sand is piled up against the coast by this mechanism; this goes on until a balance is reached between net onshore transport due to wave asymmetry and net offshore transport due to the effect of gravity on the offshore sloping coastal profile. In reality an equilibrium profile is never attained; wave-asymmetry mainly depends on the relative wave height (ratio

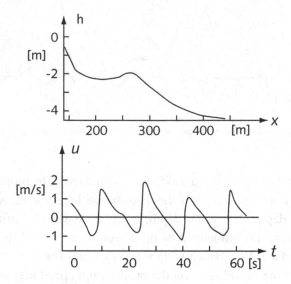

Fig. 2.3. The coastal profile near Duck (North Carolina, USA Atlantic coast) from the low-water line to 400 m offshore, representing the surf zone (zone where waves break). Below: observed near-bottom wave-orbital velocity at 1.5 m water depth in the above profile, just before wave breaking. Notice the dominance of onshore wave-orbital velocities and also the strong acceleration from offshore to onshore velocity. Redrawn after [137].

of wave height and depth) and this quantity changes continuously with tide and wind conditions. Moreover, the asymmetry of the wave-orbital motion is affected by wave breaking; the resulting net sand transport pattern produces a more complex, barred coastal profile.

Concave coastal profile

Simple models are insufficient for predicting the instantaneous coastal profile in detail. Coastal profiles have a complex shape and each profile can be quite different from other profiles at neighbouring locations along the coast. However, many coastal profiles share the common characteristic of concavity, i.e., the tendency of increased shoaling in the onshore direction [99]. A typical coastal profile is shown in Fig. 2.3; the shape is concave at the seaward edge of the surf zone and close to the shore, but not in between, where a bar is present. Observed near-bottom wave-orbital velocities, shown in the same figure, illustrate wave asymmetry; onshore orbital velocities dominate over offshore orbital velocities. The very fast reversion of offshore to onshore orbital velocity indicates that the wave crest is almost overtaking the wave trough. Flow acceleration is very strong at this moment; it has been shown that this acceleration asymmetry contributes to net onshore sediment transport, in addition to the orbital-velocity asymmetry [137, 189, 235]. The concavity of the coastal profile is related to the feedback between seabed slope and wave asymmetry. The seabed slope increases shoreward in order to produce sufficient gravity-induced offshore transport for compensating the onshore sand transport produced by shoreward increasing wave asymmetry. This shoreward increasing wave asymmetry is related in turn to the concavity of the coastal profile. This illustrates that the shape of the coast is a direct result of mutual feedback with wave propagation. However, it should be noted that the equilibrium between gravity-induced offshore transport and wave-induced onshore transport is not necessarily stable, due to the return current induced by wave breaking (undertow); coast-parallel breaker bars may originate from this instability. This, and other phenomena related to wave-topography interaction are further discussed in Chapter 5.

2.2.2. Tidal Asymmetry

Ocean tides are sinusoidal in time

Tidal waves in the ocean are not local disturbances; their wavelength spans an entire ocean basin and the water depth cannot be considered uniform over such distances. The tidal wave may assume more complex forms than a simple sinewave. The tidal amplitude is much smaller than the ocean depth and the ocean depth is much smaller than the tidal wavelength. The tidal propagation speed then depends only on the average water depth; it is independent of the tide and independent of time. Tides are generated by the forces of attraction exerted by the moon and sun on the ocean waters; earth rotation and the relative motions of sun and moon impose a sinusoidal time variation on these tide-generating forces. As tidal propagation does not modify the temporal variation, ocean tides assume the same sinusoidal time dependence as the tide-generating forces.

Symmetry of the ocean tides

The sinusoidal time dependence of ocean tidal waves implies symmetry of flood and ebb currents. Because tidal motion is the sum of many tidal wave constituents with different period and phase, ebb-flood symmetry does not apply to successive ebb and flood periods. But on average, the flood and ebb periods in the ocean have the same duration and flood and ebb currents have equal strength (there are exceptions, discussed later). If sand grains on the seabed are displaced by flood currents over a certain average distance, then on average they will be transported back by ebb currents over the same distance. If the initial seabed is a flat horizontal plane, ocean tides will not produce any net sand displacement. In other words, a sand grain can make no distinction between ebb and flood.

It should be noticed that this only holds for the hypothetical case that water motion is only due to tides. In the ocean, tidal currents are often superimposed on strong non-tidal currents. Because sediment transport is not a linear function of velocity (the sediment flux may increase by a factor 10 or even more if the flow velocity doubles) the superposition of tidal and non-tidal currents affects the uptake of sediment during flood and ebb in different ways and produces an asymmetry between flood and ebb sediment fluxes.

Asymmetric ocean tides

Symmetry of ocean tides applies to the different astronomic tidal components separately. However, irrespective of the averaging period, certain combinations of astronomic components yield asymmetric tides. The reason is that most astronomic tidal constituents result from a superposition of a limited number of basic periods, with constituent periods corresponding to sums and differences of the basic periods [114]. Therefore, different tidal constituents may interfere in such a way that flood and ebb are modulated in a systematic, asymmetric way. Such asymmetries may become significant in the case of strong diurnal tides; for dominant semidiurnal tides this effect is small [204].

Asymmetry of flood and ebb in shallow basins

The situation changes fundamentally when the tidal wave enters a shallow basin (coastal sea, tidal river) where the tidal amplitude is no longer very small compared to the water depth. The propagation speed now depends on the tidal wave and is no longer constant. The water depth at high water is significantly different from the water depth at low water; the high water crest of the tidal wave will thus propagate at a different speed than the low water trough. As a

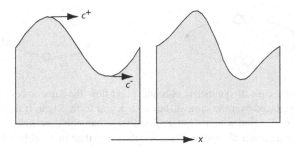

Fig. 2.4. A symmetric tidal wave becomes asymmetric in shallow water, where the tidal range to water depth ratio is larger. The case sketched in the figure supposes that HW propagates faster than LW, due to a greater water depth at HW compared to LW.

result the tidal wave becomes asymmetric (Fig. 2.4). The strength and duration of the flood current are not identical to the strength and duration of the ebb current. Due to the nonlinear variation of the sediment flux with flow velocity the uptake of sediment will be different during flood and during ebb. The average distance sediment particles travel during flood will also be different from the average distance they travel during ebb, see Fig. 2.5. The asymmetry between ebb and flood currents therefore causes a net tidal transport of sediment, although no net water transport takes place during the tidal period.

Feedback between tidal asymmetry and basin morphology

If the total sediment fluxes during flood and ebb in a tidal basin are not equal, sediment will be redistributed within the basin. The depth and width of the tidal basin will change. Such a morphologic change affects the propagation of the tide within the basin and therefore affects the asymmetry of the tidal wave; a change of tidal asymmetry produces, in turn, a change in net tidal sediment transport. Hence, the morphology of the basin and the tidal wave both evolve over time as a result of their mutual interaction. Tidal basins may nevertheless reach an equilibrium morphology, but only certain topographic configurations are possible, as shown in Chapter 4.

Breaking tidal waves

The breaking of wind waves is a familiar (though complex) phenomenon. One might wonder if tidal asymmetry may also result in tidal wave breaking. Tidal

Fig. 2.5. Left: In the case of symmetric ebb and flood flow the same average amount of bed material will be displaced over the same distance back and forth. Right: If the flood duration is shorter than the ebb duration with a stronger flood flow than ebb flow, more bed material will be transported over a greater distance in the flood direction than in the ebb direction.

wave breaking is far less common, but it is observed in some tidal inlets at spring tide, when the tidal range is at maximum. Tidal wave breaking produces a tidal bore; the flood advances upriver as a wall of water. Well developed bores occur only with spring tidal ranges which exceed about 6 m at the inlet [296]. The most famous example of tidal wave breaking in Europe was the tidal bore (mascaret) in the Seine, which could exceed five metres at spring tide, see Fig. 2.6. The bore was produced some 40 km from the inlet, where the tidal wave crest (high water) overtook the tidal wave trough (low water). Dredging works at the Seine inlet in the 1960s reduced the bore by increasing the propagation speed of the low tide compared to the high tide; as a result, high water could not catch up low water any more and the mascaret belonged to the past. The largest tidal bore in the world occurs in the Hangzhou Bay/Qiantang river in China at spring tide; this bore was reported by Chinese authors over a thousand years ago [68]. Breaking tidal waves are spectacular phenomena, but in fact they are nothing else but the ultimate stage of tidal wave asymmetry.

The Qiantang bore

Breaking tidal waves develop at tidal inlets under particular morphologic conditions. These conditions are very persistent, as shown, for instance, by the long history of the Qiantang bore. This points to the existence of a stabilising morphodynamic feedback between the bore and tidal inlet morphology. Characteristic morphologic features are the funnel shape of the inlet and the presence of a large bar (or shoal) near the inlet. In the Qiantang estuary the centre of the bar is located inshore of the inlet, see Fig. 2.7. The bar is 10 m high, has a length of 100 km and is mainly composed of silt (grain diameter mainly between 20 and 40 μm). Most other bore-developing inlets, for instance the Amazon and Ganges-Brahmaputra, have a similar morphology; in the last two

Fig. 2.6. Tidal bore at spring-tide in the Seine at Caudebec in 1963. The bore is also visible in tidal records, see Fig. 4.20. Photograph by J. Tricker, reproduced with permission of Scientific American.

cases the centre of the bar is located just seaward of the inlet. The large inlet bar slows down low water propagation compared to high water propagation; the water depth over the bar at low water spring tide is so small that the tide hardly advances, while the water depth at the subsequent high water is sufficient for fast tidal propagation. The bar also causes frictional energy dissipation, which decreases the tidal amplitude. This amplitude decrease is, at least partly, compensated by the convergence of the inlet geometry. Concentration of the landward propagating tidal energy flux in a converging cross-section produces an increase in the tidal amplitude; at the location where high water overtakes low water the tidal range is still large enough for producing a considerable bore.

Morphodynamic feedback to tidal asymmetry

Tidal asymmetry resulting from inlet morphology also contributes to maintaining this morphology. Flood dominance over the bar at the inlet (see Fig. 2.7) prevents the seaward escape of sediment and therefore promotes bar growth by trapping both fluvial and marine sediment. The funnel shape of the river mouth prevents upstream migration of the inlet bar due to upstream increasing river-related downstream sediment transport. In the Qiantang estuary the inlet bar and the funnel shape developed during the past millennia, possibly triggered

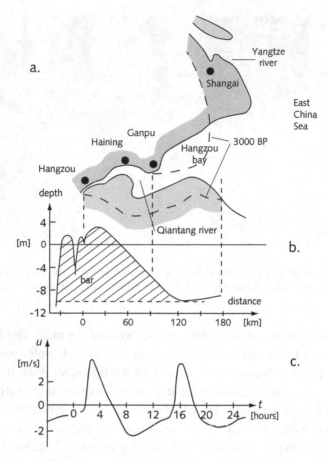

Fig. 2.7. The Qiantang tidal bore, from [242]. (a) Map of Hangzou Bay and Qiantang river, with the present shoreline and the shoreline in 3000 BP. (b) Longitudinal depth profile along the estuarine axis. (c) Tidal variation in the current velocity at Haining, near the centre of the bar, displaying a strong flood velocity dominance. The bar and the funnel-shaped geometry of the bay are not only the cause but also the consequence of tidal asymmetry.

by the sediment supply from the Yangtze river, just north of Hangzhou Bay, see Fig. 2.7. However, an analysis of sediment characteristics suggests that erosion of the outer Hangzou bay is the major sediment source for bar formation and for infill of the funnel geometry [242]. The example of the Qiantang bore illustrates that morphologic evolution has been affected more by tide-topography inter-action than by fluvial sediment supply (either from Qiantang or Yangtze) or by relative sea level change. In Chapter 4 it will be shown that similar dynamic principles govern the morphologic evolution of most river tidal inlet systems in the world.

Similarities in coastal basin morphology

The feedback between tidal propagation and basin morphology has important consequences. In the absence of such feedback, tidal basins evolve passively under the influence of tectonic motion, sea-level rise and sediment supply. Tidal basins would differ greatly because of differences in original bathymetry and differences in sediment transport; one might thus expect little similarity between tidal basins. Tidal feedback processes regulate sediment infill as a function of morphology and lead the basin towards equilibrium. For instance, tidal feedback stimulates net tidally-induced sediment import in response to sea-level rise and a net tidally-induced sediment export in response to large sediment supply, see Sec. 4.6.3. Tidal basin evolution is therefore much less dependent on original bathymetry and on sediment supply rates; there are likely to be strong similarities between tidal basins throughout the world. These similarities will be further investigated in Chapter 4. Figure 2.8 gives an impression of the different types of coastal systems occurring in nature. Fluvial sediment supply clearly plays a role, but the morphology of coastal basins also strongly depends on tides and waves. Tide-dominated coastal systems exhibit typical common morphologic characteristics and the same goes for wave-dominated coastal systems, even though the detailed morphologic structure of these coastal systems is very complex. There is no strong randomness in coastal typology; this indicates that most coastal basins are largely shaped by morphodynamic feedback processes between tides, waves and basin morphology.

2.3. Spatial Symmetry Breaking

2.3.1. *Morphodynamic Feedback to Seabed Perturbation*

Complexity of the submarine landscape

Looking at a geographical map we see that most landscapes possess an intricate morphology. This is also true for coastal sedimentary landscapes. Sometimes this morphology reflects the structure of erosion-resistant layers in the underground and sometimes it is structured by human intervention. But also in the absence of such constraints the coastal landscape morphology does not exhibit much uniformity. This conflicts with intuition, because (a) the characteristic spatial scales of external conditions, like wind and wave fields and tidal waves,

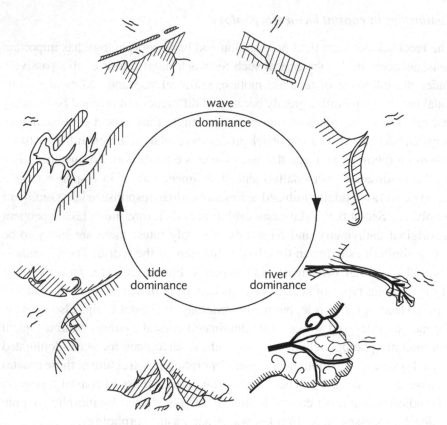

Fig. 2.8. The characteristic morphology of coastal systems differs greatly depending on the dominant external driving force; the figure shows characteristic coastal morphologies developing in response to the dominance of fluvial sediment supply, tide action or wave action. A few intermediate morphologies for combined river-tide dominance, river-wave dominance and tide-wave dominance are also shown. Other factors (tectonic, sedimentary, . . .) may also play a role and many coastal classifications have been proposed accordingly [347]. Coastal classification makes sense because morphodynamic feedback limits the diversity of coastal evolution. Every coastal system is unique though, due to the great complexity at sub-system scales. In this book we will only discuss tidal inlets represented in the left half of the figure. The tidal inlets in the upper left half will be designated Barrier Tidal Inlets and the inlets in the lower left half River Tidal Inlets. Adapted from [26].

are fairly large, (b) marine transport processes generally include a strong diffusion component and (c) seabed structures tend to be smoothed by erosion. The almost complete absence of uniformity in coastal morphology and the high complexity of spatial patterns is a puzzling feature!

A perfect morphodynamic balance does not exist

Complex patterns exist both in the underwater landscape and in the above-water landscape; we will focus on the former. The submarine landscape is subject to strong erosional forces due to currents; it evolves over time as a result of sediment being taken away or being deposited. The landscape will change unless a perfect balance exists at every location between incoming and outgoing sediment fluxes. In practice a perfect balance does not exist, at least not at every location and at every time scale; the submarine landscape never reaches a perfect equilibrium. In this section we will show that the feedback between morphology and water motion provides a mechanism for departure from a simple symmetric morphology towards asymmetric and complex morphologies. This mechanism will be illustrated through a few experiments in a highly idealised environment. However, the principle of symmetry breaking can also be experienced in many real life situations. Let us look at some examples.

Can you flatten a rippled bottom?

We go back to the seashore, a shore where the sea moves up and down the beach under influence of the tide. We arrive some time after high water. The higher parts of the beach are already drying but at several places seawater pools are left on the beach. From these pools, which consist of oblong troughs (runnels) between shore-parallel banks (ridges), the water is flowing seaward through small gullies, with a width of typically one or two metres (Fig. 2.9); the current speed is 20 to 50 cm/s. Looking at the bed of these gullies we see that the sand bed is not flat; it is completely covered with small ripples. The ripple wavelength is about 10 cm and the ripple height is in the order of 1 cm. Behind each ripple, flow vortices are visible which retain sediment in the lee of the ripple. In this way the ripple pattern moves downstream with a velocity in the order of 1 cm/s. Where do these ripples come from? Let us see what happens if we remove them! So we take off our shoes and flatten the stream bed with our feet. At first the water gets very turbid, but after a short time the bottom is visible again and locally the ripples have almost disappeared. Almost, because we see that new ripples start growing again. Within a few minutes the original situation is restored. The ripples are back! Apparently a very powerful mechanism exists through which ripples are generated starting from a flat bottom. What is this mechanism? We will see this in Chapter 3. For

Fig. 2.9. Ridges and runnels on a meso-tidal beach at low water (Goeree, Holland). Small meandering gullies drain the runnels to the sea.

the moment we just note that apparently the channel bed morphology is subject to strong symmetry-breaking processes.

Competition between stream channels

Fortunately we have taken a shovel with us. The gully that discharges the pool into the sea makes a big meander. What happens if we shortcut this meander? We dig a shortcut channel of a size similar to the meandering gully. As soon as we have finished the water starts flowing through the new channel. At the same time the flow through the meandering gully decreases. At the bifurcation a sill starts growing, decreasing still further the flow through the meandering channel. After some time the meandering channel is almost completely abandoned by the flow. We watch the evolution of the shortcut channel for a while. Some gradual changes in its morphology become apparent. A slight curvature develops, first at the entrance of the shortcut channel and later it also develops further downstream. This experiment (see Fig. 2.10) shows at least two things. First, a system of two competing channels is not necessarily stable, even if the two channels have a similar size. The channel which captures most of the flow will continue to grow at the expense of the other channel. In other words, if we start with a system of two symmetrical channels and deepen one of the

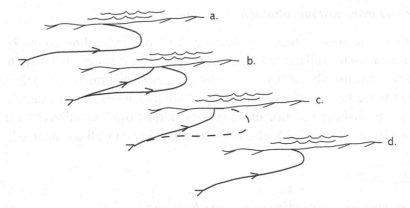

Fig. 2.10. A meandering draining gully on the beach some time after HW, before (a) and after (b) digging a shortcut. The initial meander is abandoned (c), but after a while the shortcut gully starts developing a similar meander (d).

channels at the expense of the other, then the system moves away from its original symmetrical state towards a one-channel system. Second, we see that an initially straight channel tends to develop into a meandering channel. In our experiment the incoming flow was not perfectly aligned along the axis of the shortcut channel. We see that meandering starts developing here. It shows that the system increasingly deviates from along-channel symmetry after a slight perturbation somewhere along the channel.

Can we steer morphodynamic feedback?

Our beach experiments show that processes of symmetry breaking and pattern formation are inherent to the natural interaction between flow dynamics and bed sediment dynamics. In our beach experiments we changed the initial morphological state, but we did not steer the evolution of the system. The system responded to change by enhancing initial deviations from a symmetrical state. The experiments showed that the natural dynamics of the system may generate a positive feedback to deviations from symmetry. Of course, we would like to know what exactly the nature of this positive feedback is. Does such feedback always exist, or only under certain conditions? Can we predict the evolution of a system subject to morphodynamic feedback? For coastal management purposes we would like to know whether it is possible to adjust coastal development to morphodynamic feedback, or whether, how and in which cases it is possible to steer this feedback.

Learning from past morphologic evolution

A first clue to answer these questions can be found by scaling up our beach experiments to the full coastal system. However, this requires intervention at a scale which is usually far beyond available experimental means. We may look instead to the history of past large-scale coastal changes. As an example we will discuss in the last section the long-term morphologic evolution of the delta of the River Rhine in The Netherlands, which has been well documented.

2.3.2. Stability

Space-symmetry versus time-symmetry breaking

The process we call space-symmetry breaking is, in a physical sense, of a different nature than the process of time-symmetry breaking. The two notions are similar in the mathematical sense, because both time and space-symmetry breaking refer to nonlinearity; nonlinearity of morphodynamics with respect to temporal and spatial evolution. However, breaking of time symmetry and breaking of space symmetry have different causes. Temporal symmetry breaking is a consequence of topography, while spatial symmetry breaking may develop from perfectly uniform spatial conditions. Both symmetry breaking processes evolve over time through morphodynamic interaction.

Response to spatial perturbation

Space-symmetry breaking refers to the growth of morphologic patterns which differ from the spatial structures existing originally in the morphology, or in the external conditions imposed on the system [325]. As an example we consider the morphodynamic system consisting of water flowing uniformly over a flat sloping seabed consisting of loose sediment grains (down-slope sheet run-off). The grains are entrained by the flow down the seabed slope and, if the slope is sufficiently long, this downslope transport is a steady, spatially uniform process. This situation does not conflict with any physical law. However, we know from experience that downslope transport of water and sediment can also take place in other ways. The flow may be concentrated in a channel or in a number of channels. The seabed may also exhibit cross-flow ripples, which move downstream with the flow. These other downslope transport modes obey the same physical laws as the uniform mode. It seems obvious to relate these different transport modes to different initial morphologic conditions. But is that the whole story? The question as to how these different initial conditions come about remains unanswered. Moreover, the beach experiments discussed earlier show that changing the initial morphology does not necessarily change the transport mode; the bottom ripples in a gully reappear very soon after they have been eliminated! But is it conceivable that a morphodynamic system switches autonomously between different transport modes?

The marble on a hemisphere analogy

The answer to this question can be illustrated by an analogy. This analogy consists of a simple experiment with a perfectly round marble which we place on top of a larger perfectly round hemisphere, in such a way that it stays in equilibrium, see Fig. 2.11. Is that possible? Theoretically it is, because at the top of the hemisphere there is a location which is exactly horizontal. At this location the marble will stay in equilibrium. There is a practical problem, though. The location of exact horizontality is infinitely small! So, in practice, there is no chance that we can realise this equilibrium. Equilibrium at the top of the hemisphere does not conflict with any physical law and yet the system will spontaneously depart from this situation. The marble on a hemisphere is an example of an unstable equilibrium. Any perturbation of this theoretical equilibrium, even if infinitely small, will cause the marble to roll down. So we can predict with certainty that the marble will not stay on the top of the hemisphere. But can we also predict the path followed by the marble when it rolls down? And can we predict the position where the marble will finally arrive? The issue of predictability will be discussed later.

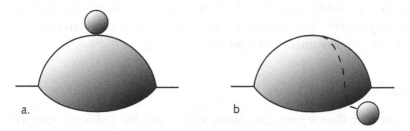

Fig. 2.11. A perfectly round marble on top of a perfect hemisphere. Any perturbation, even infinitely small, will cause the marble to roll down. If the direction of the perturbation is unknown the path and the final location of the marble cannot be predicted.

Instability cannot persist

Under certain conditions, uniform flow over a sloping, loose-grained sediment seabed is unstable and comparable with the situation of a marble on a hemisphere. It is an unstable mode if, for instance, the flow strength is slightly larger than the threshold for sediment motion [211, 233]. In that case it is a transport mode possible in theory, but not in practice, because a range of infinitesimal perturbations of the flat seabed will tend to grow. Growth of such perturbations will finally lead to a morphology in which the flow is concentrated in one or more meandering channels with cross-flow ripples migrating over the seabed. The mechanism by which such perturbations grow out to a ripple morphology and by which channel meanders are generated is discussed in Chapter 3.

Static versus dynamic equilibrium

The marble on a hemisphere analogy mimics the instability of uniform flow over a flat sediment bed only to a certain degree. An essential difference exists between the two systems, which is related to the energy dissipated in the system and to the feedback produced by the instability. If the hemisphere is fixed, no energy will be transferred from the marble to the hemisphere; the marble on globe then represents a static unstable equilibrium. This contrasts with the fluid flow over a flat seabed, which dissipates energy by exerting traction and lift forces on the sediment grains. A small flow perturbation caused by an incidental non-uniform redistribution of sediment grains may grow by extracting energy from the basic downslope flow, while increasing the perturbation of the sediment distribution on the seabed. In this way the seabed disturbance will grow towards a finite amplitude, generating a flow-seabed pattern which depends on the intrinsic dynamics of flow-seabed interaction. This finite-amplitude perturbation may evolve to a non-steady equilibrium bedform, which migrates over the seabed with eventually a fluctuating amplitude.

Forced instabilities

Any change in the external conditions will modify the sediment transport pattern. If the seabed topography was initially in morphodynamic equilibrium it will be unstable afterwards and evolve to a new equilibrium compatible with the new external conditions. A shell on the subtidal beach causes a local perturbation of the beach morphology and generates a characteristic scour pattern in the wake of the shell. On a larger scale, a jetty extending from the beach into the sea causes a pattern of erosion and sedimentation extending in longshore direction over a distance of 5 to 10 times the jetty length. Incoming long waves which are reflected by a steep beach produce a standing wave pattern which causes an evolution towards a correspondingly undulating bar-trough pattern on the seabed. These are examples of forced instabilities of an initially unperturbed seabed morphology. If we capture the physical laws of morphodynamic evolution in a mathematical model with sufficient accuracy, then the system's response to a forced instability can be well predicted.

Free instabilities

The situation is different for a free instability of the morphodynamic system (Fig. 2.12). In this case the system evolves spontaneously from one unstable equilibrium mode to another (stable or unstable) equilibrium mode without being forced by any intervention in the external conditions. Instability of the initial equilibrium implies that any infinitesimal fluctuation from the equilibrium (within a certain range of possible fluctuations) will tend to grow exponentially, with a growth rate depending on the characteristics of the fluctuation (the wavelength, for instance). The development of free instabilities in an initially uniform system causes spontaneous symmetry breaking. In practice, spatial and temporal fluctuations around an equilibrium always exist. The smaller the

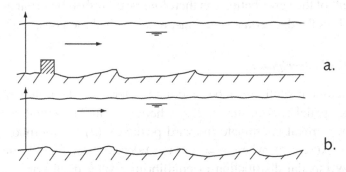

Fig. 2.12. Forced (a) and unforced or free (b) instability of the seabed. In the first case the seabed structures develop in response to flow perturbation produced by the obstacle; the length scale of the seabed pattern is related to the scale of this forced flow perturbation. In the second case the seabed pattern responds to self-produced flow perturbation.

initial fluctuation and the smaller the growth rate, the longer it will take before a given departure from the unstable equilibrium is reached. This is due to the exponential nature of initial growth after perturbation. However, a finite departure will always be reached if sufficient time is available. Predicting morphologic evolution in the case of free instabilities will be discussed later; we will see that there are intrinsic limits to predictability, contrary to the case of forced instabilities.

2.4. Linear Stability Analysis

Morphodynamic feedback can only be modelled analytically for a highly idealised topography and for a limited range of spatial and temporal scales. The general morphodynamic problem cannot be solved by analytical methods, due to the strong nonlinearities and to the broad range of space and time scales involved. Figure 1.2 gives an impression of the spatial and temporal scales at which coastal morphodynamic processes take place.

Initial response

The principles of morphodynamic feedback can be illustrated for the initial response of the system to an infinitesimal perturbation of the basic equilibrium state. In the initial phase of development the perturbation is so small that the flow pattern is hardly changed and the amplitude of the flow response is almost linearly proportional to the amplitude of the bed perturbation. The initial development of the perturbation can therefore be described by linear equations, which can be solved, in some situations, by analytical methods.

Mathematical description

Morphodynamic evolution can be described in terms of the following quantities: (1) the spatial coordinates x, y, z, where x, y are horizontal coordinates and z is the vertical coordinate (upward positive), (2) the morphology, $z = -h(x, y) + z_b(x, y, t)$, where $z = -h(x, y)$ describes an equilibrium seabed morphology (spatial distribution of equilibrium water depth) and $z_b(x, y, t)$ the departure from equilibrium; (3) the sediment flux $\vec{q} = \vec{q}_0(t) + \vec{q}'(x, y, t)$, averaged over a period which is long compared to the wave period or the tidal period; $\vec{q}_0(t)$ relates to the equilibrium state and $\vec{q}'(x, y, t)$ is the response to the perturbation. The x- and y-components $q^{(x)}$, $q^{(y)}$ are expressed as sediment volume per unit time and unit width (m^2/s).

Sediment balance equation

The equilibrium sediment flux $\vec{q}_0$ is spatially uniform (independent of x, y); the departure from the equilibrium sediment flux, $\vec{q}'(x, y, t)$, is zero if z_b is zero. The sediment flux gradient, $\vec{\nabla}.\vec{q}' = \partial q'^{(x)}/\partial x + \partial q'^{(y)}/\partial y$, equals the amount of erosion ($\vec{\nabla}.\vec{q}' > 0$) or sedimentation ($\vec{\nabla}.\vec{q}' < 0$) per unit time and unit seabed area, if we assume that the averaging time is sufficiently long to allow us to neglect changes in suspended sediment load with respect to erosion or sedimentation. We refer to the porosity of the deposited material as p. The change in morphology per unit time follows from the sedimentation and erosion and is thus related to the sediment flux $\vec{q}'$ by:

$$(1 - p)\partial z_b(x, y, t)/\partial t + \vec{\nabla}.\vec{q}' = 0. \tag{2.1}$$

This sediment balance equation describes the morphologic evolution $z_b(x, y, t)$ of the seabed. It is a morphodynamic equation if we can express $\vec{\nabla}.\vec{q}'$ as an explicit function of $z_b(x, y, t)$. Such an analytical expression can be established only in a few, highly idealised, cases, and numerical techniques normally have to be used.

A one-dimensional idealisation

Here we will analyse the solution of Eq. (2.1) for such a highly idealised situation. We first assume that the morphology and the sediment flux are uniform in one horizontal direction, y, and we call $q^{(x)} = q/(1 - p)$. The sediment balance equation then reads:

$$\partial z_b(x, t)/\partial t + \partial q'(x, t)/\partial x = 0. \tag{2.2}$$

The porosity p is assumed to be constant. We restrict the analysis to the initial departure from equilibrium, during which z_b is very small compared to the equilibrium depth h. The depth h is approximated by a constant; this approximation is equivalent to considering only perturbations with a characteristic scale (wavelength) which is much smaller than the spatial scale of variation of the equilibrium morphology. We further assume that the sediment flux depends on the local, instantaneous flow velocity $u(x, t)$; this excludes fine sediments for which lag-effects in the suspended sediment distribution are important. Finally we simplify the problem by considering steady boundary conditions.

Initial morphodynamic response

Now we investigate the morphodynamic response to an initial perturbation which consists of an infinitesimal undulation of the seabed, $z_b(x, 0) = \epsilon h \cos kx$, where k is the wavenumber of the undulation and ϵ a number much smaller than 1. This seabed perturbation produces a perturbation q' in the equilibrium sediment flux q_0, which can be written as a power series of the small parameter ϵ:

$$q'(x, t) = \epsilon q_1(x, t) + \epsilon^2 q_2(x, t) + \cdots, \tag{2.3}$$

where $q_1, q_2, \ldots$ have the same order of magnitude as the undisturbed flux q_0. The first term q_1 is linearly related to the perturbation z_b; the spatial pattern of q_1 therefore contains only the wavelength of the perturbation. This implies that the *initial* response to the seabed perturbation $z_b(x, 0)$ generates no other wavenumbers than the initial wavenumber k. Therefore we may write

$$z_b(x, t) = \epsilon h f(t) \cos(kx + \phi(t)) \tag{2.4}$$

and

$$q_1(x, t) = \beta q_0 f(t) \cos(kx + \phi(t) + \delta), \tag{2.5}$$

where $f(t)$ is a function describing the time evolution of the perturbation amplitude ($f(0) = 1$) and $\phi(t)$ a function describing the migration of the perturbation ($\phi(0) = 0$). The magnitude of the sediment flux perturbation is proportional to the magnitude of the seabed perturbation; however, the phase may be different. The coefficient β and the phase difference δ depend on the sediment transport mechanism and on the perturbation $u'(x, z, t)$ of the flow field. The flow field perturbation can be obtained by solving the flow equations to first order in ϵ; in Chapter 3 the parameters β and δ will be determined in this way for different types of seabed perturbation z_b. Here we will stay with a qualitative interpretation. The phase difference δ is crucial for the stability of the seabed, as will soon become clear.

Conditions for instability

Substitution of (2.3), (2.4) and (2.5) in the sediment balance Eq. (2.2) yields at order ϵ:

$$f_t = \sigma_i f, \quad \phi_t = -\sigma_r, \quad \text{with } \sigma_i = \sigma_r \tan \delta = -k\beta q_0 \sin \delta / h, \qquad (2.6)$$

where the subscript $_t$ indicates differentiation with respect to time t. The first equation yields an exponential growth (and therefore instability) of the perturbation if $\sin \delta < 0$ or $-\pi < \delta < 0$. In this case, according to (2.5), the x-derivative of the sediment flux is negative at the crest of the perturbation ($kx + \phi = 0$). Sediment transport therefore converges at the crests of the seabed perturbation, which means that the perturbation amplitude will grow (Fig. 2.13). If $0 < \delta < \pi$, the sediment flux diverges at the crests and the seabed perturbation decays. The second equation of (2.6) implies that the phase of the seabed perturbation increases or decreases linearly with time. According to (2.4), this linear phase increase/decrease corresponds to a migration of the bed perturbation with constant velocity. The seabed perturbation migrates downstream if the flow perturbation is positive at the crest (increase of flow velocity) and otherwise upstream. Most seabed perturbations migrate downstream; however, upstream migration is theoretically possible (in the case of antidunes, for instance).

The role of inertia

The perturbation increases if the sediment flux has its maximum not too far upstream of the perturbation crest ($-\pi/2 < \delta < 0$) or its minimum not too

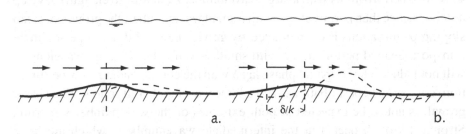

a.   b.

Fig. 2.13. Seabed perturbation at time $t = 0$ (solid line) and some time later (dashed line). The arrows indicate direction and magnitude of the sediment flux. (a) If the location of the greatest sediment flux coincides with the crest, then the seabed perturbation will migrate but not grow. (b) If the sediment-flux maximum is located upstream of the crest, then the perturbation will grow and migrate.

far downstream ($-\pi < \delta < -\pi/2$). But why should maximum or minimum of the sediment flux be shifted relative to the crests and troughs of the seabed perturbation? There are several reasons for such a phase shift, which will be discussed more in detail in Chapter 3. The influence of inertia (momentum conservation) in relation to frictional momentum dissipation is essential. Inertia delays the flow response to a seabed perturbation in remote regions (far from the seabed) relative to nearby regions (close to the seabed).

Flow over a bottom ripple

The effect of inertia can be illustrated by the flow over a bottom ripple. The water depth is smaller at the ripple crest than upstream or downstream; the flow therefore has to accelerate at the upstream ripple slope. The surface flow has not yet reached its maximum strength at the ripple crest, but it is still accelerating. However, averaged over the water column the flow cannot be accelerating at the ripple crest, because the total water flux would then be greater downstream than upstream of the crest. This implies that the flow near the bottom decelerates at the ripple crest. The near-bottom flow maximum is shifted upstream, and so is sediment bedload transport. We have $\delta < 0$, which is the condition required for initial growth of a perturbation. This instability mechanism is further analysed in Sec. 3.3.

Wavelength with the strongest initial growth

The growth rate σ_i and the migration rate σ_r of the perturbation depend on the wavenumber k because the coefficient β and the phase δ in (2.6) both depend on k. Seabed perturbations with a large wavenumber k (small wavelength) develop a steeper slope than perturbations with a small wavenumber; the growth of steep sloping perturbations is counteracted by gravity-induced downslope sediment transport. Seabed perturbations with small wavenumber k (large wavelength) will not induce a large spatial phase lag δ with the corresponding flow perturbation, compared to perturbations with a large wavenumber. It follows that strong growth is not to be expected at both extremes of the wavenumber spectrum. Strongest growth occurs at the intermediate wavenumbers, which are large enough to produce a significant phase lag δ and small enough to avoid a strong gravity-induced down-slope transport. This dynamic balance determines the range of perturbation wavelengths which will be most likely observed in natural situations. The resulting seabed pattern thus develops according to feedback effects inherent to the physical dynamics of the system; it is not imposed by the external conditions. The wavelengths present in the seabed pattern may be entirely absent in the spatial pattern of the external forcing or in the spatial pattern of the initial morphology.

Pattern formation and self-organisation

In the previous section a qualitative description is given of the process leading to spontaneous growth of ripples or sand dunes on a uniform seabed topography. This process is triggered by a spatial shift between the initial seabed perturbation and the ensuing flow response. Such a shift may come about in several ways, depending on the length scale of the perturbation, the water depth, the presence of lateral boundaries, the average seabed slope and the character of the basic flow (steady or oscillating). Different types of flow response may generate different types of seabed instability and may lead to different types of morphologic patterns (different length scale, different orientation). Most morphologic patterns observed in the coastal environment can be related to instability produced by the flow response to seabed perturbation: Sand dunes and ripples, beach cusps, rip cells, breaker bars, meandering and braiding channels, tidal flats, sandbanks, the equilibrium topography of tidal rivers and tidal basins, etc. These morphologic patterns have a characteristic wavelength which is not present in the external conditions; they also tend to return spontaneously after being wiped out by natural events or artificial interventions. The ability of a system to generate patterns which are independent of the external conditions is often referred to as self-organisation capacity. Pattern formation is the term most commonly used in the literature on coastal morphodynamics; however, this term does not distinguish between free and forced pattern formation processes. Self-organisation or pattern formation originating from spontaneous symmetry breaking plays an important role in shaping the coastal environment.

2.4.1. *Perturbation Growth Towards a Finite Amplitude*

What happens after initial growth?

Exponential growth cannot go on for ever, so after some time the growth rate has to decrease. The linear approximations (2.4, 2.5) do not hold for large perturbations and the exponential growth law (2.6) therefore ceases to be valid.

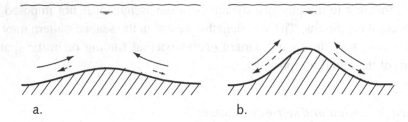

a.   b.

Fig. 2.14. (a) In the first stage of development, the flow response to the seabed perturbation results in a contribution to sediment-transport convergence (solid arrows) which is greater than the contribution of gravity-induced downslope transport to sediment-transport divergence (dashed arrows). (b) In the final stage both contributions become comparable.

This is a mathematical way to express the growing importance of physical processes which can be ignored only when the perturbation is very small, because they are related to the perturbation in a nonlinear way (quadratic or higher order). An example of such a process is the turbulence generated by flow separation in the wake behind the crest of the seabed perturbation, which increases the steepness of the lee-side slope [91]. In oscillating flow, both slopes of the perturbation will become steeper, especially near the crest. At a certain stage of evolution, sedimentation at the crest due to convergence of near-bottom flow will be exceeded by erosion due to gravity-induced downslope transport [156] (Fig. 2.14). Wave-induced seabed stirring in the zone around the crest may further stimulate downslope transport. Observations indicate that sand dunes in the North Sea, for instance, are less developed in regions of high wave intensity [212, 307].

Growth-limiting processes

Perturbation growth is also affected by a shift in sediment transport mode. An increase of the shear stress at the crest will favour suspended-load transport over bedload transport. In Chapter 3 it is shown that some seabed instabilities, such as ripples and sand dunes, depend on bedload transport for their growth. But when the dune height increases the flow velocity at the crest may become high enough to bring sand in suspension. Suspended load transport may then overtake bedload transport and interrupt the growth mechanism.

There are often several processes that may limit the growth rate of seabed patterns at an advanced stage of development. How the different growth-limiting mechanisms compare to each other is not well known in general, because of the great complexity of the underlying processes. It should also be mentioned that growth limitation does not apply to all types of seabed instability. Bars in estuaries, for instance, may continue growing till they reach above sea level.

Nonlinear perturbation growth

In the initial phase of emergence the growth rate of a perturbation does not depend on time. This is illustrated in the simple one-dimensional example (2.6). The linear stability analysis does not include processes that decrease the growth rate in a later stage of development. For inclusion of these processes higher order terms of the ϵ-power expansion (2.3) must be considered. Theoretically this is possible, as long as the expansion series converges sufficiently. Inclusion of higher order terms provides information about the evolution of the perturbation at finite values of the amplitude. Nonlinear feedback in response to the initial perturbation in the equation of motion produce secondary perturbations with new wavelengths; the perturbation then loses its initial sinusoidal shape. The most important nonlinear feedback terms will often have a quadratic character, producing secondary perturbations of one-half wavelength. If these secondary perturbations are also unstable and start growing, then the initial sinusoidal perturbation turns into a complex, asymmetric seabed pattern with finite amplitude. Field observations show that in nature the morphology of most bedforms is indeed much more complex than a single sinus shape [91]. The analysis of perturbation growth to a finite amplitude can be carried out using numerical techniques.

Weakly nonlinear analysis

Sometimes a semi-analytic approach is possible, using a mathematical technique known as weakly nonlinear approximation. This technique consists of seeking a fictive system, related to the real physical system, which is stable except for perturbations in a narrow wavelength band. This fictive system differs from the real system by changing the value of one of the model parameters. This stability parameter could be the friction coefficient, for instance. The initial exponential growth of the unstable perturbation is a valid approximation even for finite amplitude in the fictive system, because the contribution of perturbations with wavelengths outside the narrow unstable band remains small.

If the real system is not very different from the fictive system, then its finite amplitude behaviour can be derived by solving the linear stability problem for a linear expansion of the stability parameter around the fictive value. This method has been applied successfully for obtaining a semi-analytical description of the finite amplitude behaviour of perturbations in several idealised physical systems, for instance, the growth of bars in a straight river with horizontal bottom [392].

A simple nonlinear evolution equation

The growth rate resulting from a weakly nonlinear analysis is not constant, but depends on the perturbation amplitude. The simplest form it can take is an expression of the form $\sigma_i - f^2$, where f is a factor representing the time dependence of the perturbation amplitude. The amplitude evolution equation then reads:

$$f_t = f(\sigma_i - f^2). \tag{2.7}$$

Initially the perturbation behaves exponentially, since we may approximate $f_t \approx \sigma_i f$ for $f \ll \sigma_i$. For negative values of σ_i, $f = 0$ is a stable solution. For positive values of σ_i the original state $f = 0$ is not stable; stable solutions are given by $f = \pm\sqrt{\sigma_i}$. In the differential equation $\sigma_i = 0$ is a bifurcation point. Moving σ_i beyond the bifurcation point induces spontaneous pattern formation. We already know that the wavelength of the pattern is a characteristic inherent to the physical dynamics of the system. Now it appears that the amplitude of the pattern is also related to inherent system characteristics, which are not imposed solely by the external conditions and which are independent from the initial state of the system.

Cyclic morphologic behaviour

In the simple one-dimensional example the system moves to a morphology characterised by a rhythmic seabed pattern with a constant amplitude; this pattern will normally migrate. Migration implies that the system does not evolve to a fixed state; in the final state the evolution of the seabed has a cyclic character. Cyclicity of morphologic patterns is a general characteristic of many coastal systems. This cyclicity is not only related to pattern migration, but also to nonlinear feedback in the system, which prevent any stable equilibrium developing.

This will be illustrated in the next section, which deals with the instability of a two-channel system. This example also shows that the direction in which a system initially moves, starting from an unstable equilibrium, strongly depends on the initial perturbation and the external conditions, as can be expected from the marble on a hemisphere analogy.

Predictability

If we accept that many features of coastal morphology originate from spontaneous symmetry breaking, then one may wonder if coastal morphology is predictable. In the literature much consideration has been given to this question, see for instance [208]. Whether a given morphology is stable or unstable depends on the external conditions, for instance, the strength of current velocities, the water depth, the wave height, the wave period and the angle of wave incidence. These conditions usually fluctuate and conditions of stability and instability may therefore alternate. The period of alternation can be short, in the order of hours to days or weeks, for fluctuations caused by tides or weather conditions. The period of alternation can also be very long, in the order of decades to centuries, corresponding to fluctuations caused by large-scale morphologic evolution or by sea-level rise. If unstable conditions last for a sufficient time, morphologic patterns will develop. The development of small-scale patterns may take place over short periods of hours to days, but the development of large scale patterns may take periods of decades to centuries. In the foregoing it has been shown that the initial growth of morphologic patterns is strongest for wavelengths in a certain range of scales, which can be determined theoretically. However, the initial morphology also matters; if the wavelengths present in the initial morphology are close enough to the wavelength of maximum instability, then these former wavelengths will dominate the latter for a substantial period. Bottom cores of coastal sandbanks show that they often consist of recent deposits with a much older kernel [212]; this suggests that in these cases positive morphodynamic feedback to remnant bedforms dominates over the development of a new bedform pattern corresponding to the wavelength of fastest growth. If external conditions fluctuate over periods comparable to the time scale of bedform growth, a strong competition exists between forced instability (related to the initial topography) and free instability (related to the wavelength of strongest growth). A further complicating factor is the interaction of forced and free instabilities, because of the mutual dependency of the flow responses to perturbations of similar scale. Predicting coastal morphology is therefore an extremely complex problem, requiring very accurate morphologic modelling. The periods of alternating stability and instability often can be predicted only in statistical sense, but not in a deterministic sense. It is hence to be expected that accurate predictive modelling is not possible at all. In that case one may conclude that predicting morphology is subject to limitations which cannot be overcome.

2.5. Instability of a Two-Channel System

To illustrate the inherent instability of coastal morphology we will consider the instability mechanism of a two-channel system as an example. This mechanism is based on competition between mutually interacting flow paths in tidal basins. Tidal flow is seldom confined to a single well-defined channel. In many tidal basins the flow can follow different pathways, often corresponding to flood-dominated or ebb-dominated channels. At some locations flood and ebb flow follow the same channel, but at other locations they concentrate in distinct parallel channels. These channels compete for conveying the tide and this competition determines the respective cross-sectional areas of these channels. The relative importance of the channels is not always the same, and shifts may occur between the channels over the course of time. In Fig. 2.15 this is illustrated for the Western Scheldt estuary. The same phenomenon is observed in river deltas; the instability mechanism is in both situations similar, though not identical. In the following we concentrate on the latter, somewhat simpler situation. We will take the Rhine delta as an example and start by describing the history of the two major branches of the delta (see Figs. 2.16–2.18). This history sheds some light on the large-scale morphologic evolution of the Rhine delta and gives some clues for answering the question: Why did it develop in this way?

2.5.1. *History of the Rhine delta*

It takes more than a few decades to change the Rhine delta

Upon entering The Netherlands, the River Rhine flows through a large coastal plain built up with fluvial deposits (sand, clay) in the upstream reaches and with marine deposits and peat in the downstream reaches. Just after leaving Germany, about 100 km from the sea, the Rhine divides into two branches, the Nederrijn river and the Waal river. The land between the two branches is called Betuwe, land of the Batavians. The Batavians settled there shortly before the

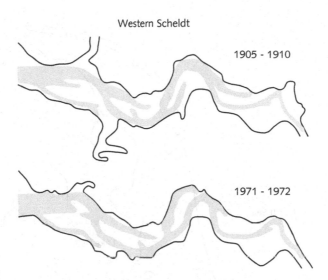

Fig. 2.15. Evolution of the Western Scheldt channel system. The main tidal channel has a secondary channel throughout the basin. The relative importance of the two channels may change over the course of time; such a change has occurred in the central part of the basin during the past century.

Roman era, more than 2000 years ago. The river branch Waal was the southern border of the Batavian territory, forming a natural barrier to the Roman conquerors. After some unsuccessful attempts to cross the river, the Roman general Drusus decided to build a dam on the Waal, near the bifurcation point with the Nederrijn. This intervention made it possible to wade across the Waal by decreasing the discharge through it, in favour of the Nederrijn (AD 0–50). After their defeat the Batavians were educated in Roman culture and engineering. A few decades later they rebelled against the Roman domination under their Batavian leader Julius Civilis. Using their new skills they managed to restore the discharge through the Waal river and to recover the southern defence line of their territory. This would not have been possible if an irreversible change had occurred in river morphology during the Roman episode; we may conclude that the time scale for substantial morphological change of the Rhine delta is greater than a few decades.

A storm surge modifies the Rhine outlet

Dike building in The Netherlands started under the Romans, but it took more than 1000 years before the river plains in the Netherlands were protected against

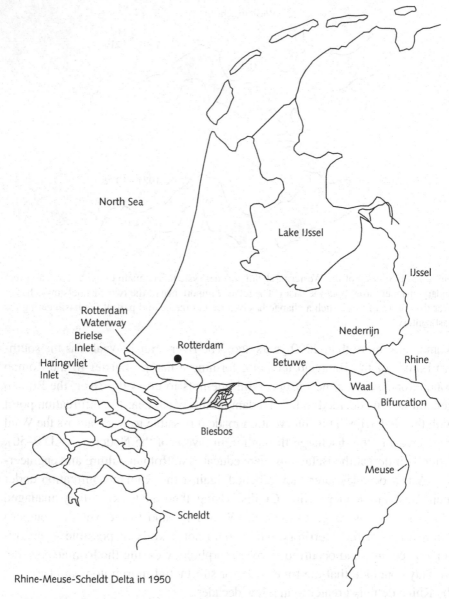

Rhine-Meuse-Scheldt Delta in 1950

Fig. 2.16. The Rhine delta in The Netherlands around 1950, indicating the bifurcation into the Waal and Nederrijn branches near the German border. The Haringvliet-Biesbos is the tidal delta created by the storm surge in 1421 and has since served as the main connection of the Waal to the North Sea. The Nederrijn got a new connection to the North Sea when the Rotterdam Waterway was constructed in 1870.

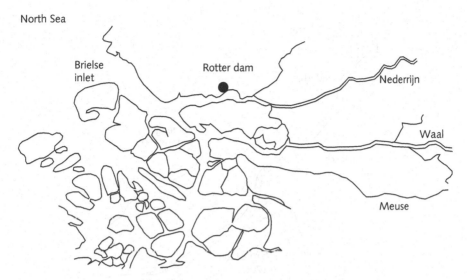

Fig. 2.17. The lower reach of the Rhine delta in the Middle Ages, before the storm surge of 1421. The Brielse inlet was the main river outlet to the North Sea. South of the Brielse inlet the delta was formed by a mosaic of islands, polders, tidal flats and marshes, all embedded in a complex network of tidal channels and creeks.

flooding by dikes. Water boards, elected by landowners and financed through taxes, were in charge of maintaining this infrastructure, with the support of monasteries. In the Middle Ages, land reclamation extended to the lower reaches of the delta, where the tide penetrated through an extensive system of tidal channels and creeks. Drainage of reclaimed land was accompanied by peat extraction, soil oxidation and subsidence, increasing the vulnerability of the reclaimed land to flooding. Around the 14th–15th century (AD) The Netherlands was afflicted by an intermittent civil war during which water infrastructure was neglected. In 1421 the Dutch coast was struck by a huge storm surge (St. Elisabeth Flood) that flooded many polders situated at the lower reach of the Waal river. The political situation prevented rapid dike restoration; this allowed tidal scouring of deep channels that penetrated far into the Rhine delta. The Waal river now became connected to the sea by a broad estuary, the Haringvliet. Finally the Haringvliet was closed in 1970 with discharge sluices for high river runoff, almost five and a half centuries after its formation.

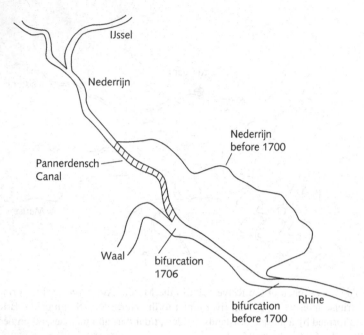

Fig. 2.18. The bifurcation of the Rhine into the Waal and Nederrijn rivers. In the 16th and 17th century the upstream part of the Nederrijn had become so shallow that it only carried a very small share of the Rhine discharge. A new bifurcation point was created by the construction of the Pannerdens canal, shunting the upstream part of the Nederrijn.

Upstream and downstream impact of the Haringvliet outlet

In the centuries following the creation of the Haringvliet estuary, the Waal river started carrying an increasing share of the Rhine discharge, at the expense of the other branch, the Nederrijn. During high river floods the Waal bifurcation became enlarged, while the Nederrijn bifurcation continued to shoal; in the 17th century the Nederrijn carried hardly any river discharge. At the end of the 17th century a new connection between the Rhine and the Nederrijn was created by the construction of the Pannerdens Canal. In the meantime there were many changes in the lower delta. In the Middle Ages the Rhine delta had several outlets, which were interconnected by an intricate system of tidal channels. In the Roman era, the largest outlet, Helenium, was located south of The Hague. In the Middle Ages it had shifted more than ten kilometres southward and was called Brielse outlet. After the St. Elisabeth flood, the Brielse outlet had to compete with the Haringvliet outlet. The Brielse outlet then started shoaling and the navigation channel to Rotterdam became obstructed. Early in the 19th century a canal was constructed connecting Rotterdam to the Haringvliet. However, due to increasing draught of the ships, the canal soon became inadequate.

Creation of a new artificial Rhine outlet

In the middle of the 19th century a young engineer, named Caland, proposed the construction of a canal through the dunes, north of the Brielse inlet, connecting Rotterdam directly to the sea. A small initial short-cut channel would suffice; Caland counted on the scouring power of tidal currents and river discharge to further deepen and enlarge the canal. Today we would call this approach 'building with nature'. In fact, what Caland proposed was similar to the experiment we did on the beach at small scale. The canal, called Rotterdam Waterway, was finished in 1870. Unfortunately, Caland's concept did not work. In 1877 the entrance of Rotterdam Waterway was obstructed by large shoals; the strength of the currents was insufficient to clear the inlet. In the same period similar problems had occurred at dredged navigation channels in the Clyde and the Tyne estuaries (UK). Based on experience gained with dredging in the UK, the decision was taken to widen the Rotterdam Waterway (from 50 to 100 m) and to deepen the channel to 10 m. In addition, the harbour moles were extended further seaward and training walls were built to concentrate the current. Later the Brielse inlet was also closed. After implementation of these improvements, the result anticipated by Caland was achieved; around 1950 the new Rhine delta was close to morphologic equilibrium. At that time the river beds in the lower delta were constrained almost everywhere within groynes and embankments and most river meanders were cut off; dredging was only necessary to maintain the depth in the harbours.

Conclusions from the Rhine delta history

What can we learn from this brief history of the Rhine delta? In Sec. 2.3 we described an experiment with a shortcut of a meandering beach gully. This experiment is similar to the competition between the Rhine delta branches after the creation of a new outlet. However, the spatial and temporal scales are completely different: In the order of half an hour for the beach gullies and in the order of a few hundred years for the Rhine delta. The large morphologic time scale for the Rhine delta is related to the low sediment discharge of the Rhine river; this discharge amounts to some 1–2 million m³ a year. The surface area of the Rhine branches is approximately 200 km². Hence, deposition of the total sediment load would yield an average 1–2 m rise of the river bed in 200 years; this amount of sedimentation is needed to create substantial morphologic feedback.

The morphodynamic history of the Rhine delta shows that:

* In the absence of geographical constraints, the Rhine river could build a delta of several bifurcating and meandering channels;
* Shifts occurred between different channel configurations; such shifts were triggered, in particular, by erosion/sedimentation processes in the coastal zone (influence of tides, storms and waves);
* A perturbation of sufficient strength and duration was required to trigger these shifts;
* Large-scale evolution of the river delta was intimately related to the social context.

The Rhine delta is not a unique example of competing outlet branches; similar delta processes occur in other low coastal plains, for instance, the deltas of the Danube, Nile, Niger, Ganges, MacKenzie, etc. [500].

2.5.2. *Principles of Channel Competition*

How can we understand the behaviour of delta river systems? What kind of mechanisms determine competition between channel branches? Do the same principles hold for river deltas and for tidal deltas? We will try to answer these questions by considering an idealised situation, which can be investigated with a simple analytical model. The model describes steady flow through a split channel system; the system consists of two straight parallel channels of finite length. We perform the following physical-mathematical experiment. Initially the interconnected parallel channels are identical and have the same depth, width and length. Then we change the depth of one of the channels slightly and compute the response of the system to this perturbation using a simple sediment transport model and assuming spatially uniform depth in each of the channels. This experiment yields a stability criterion for the two-channel system. The model is an oversimplification of reality and is meant only for demonstrating the basic principle of instability related to channel competition. First we will address some underlying assumptions in a qualitative discussion.

The physical instability mechanism

What happens when the flow velocity is decreased in one of the parallel channels and increased in the other? When the flow enters the channel with decreased velocity the sediment load will quickly decrease; sediment will be deposited at the channel entrance and a shoal will be formed. The location of velocity decrease and related deposition will then shift to the downstream side of the bar; this morphodynamic feedback implies that the bar will grow in the downstream direction. Meanwhile a scour hole will develop at the entrance of the other channel; this scour hole also starts growing in downstream direction. Now we may ask if this sedimentation/erosion process will lead to a new equilibrium, when the bar in the first channel and the hole in the second channel finally extend along the full length of the parallel channels. The answer depends on the flow response to the depth decrease in the first channel and the depth increase in the second channel. If the flow velocity increases in the shoaling channel and decreases in the scouring channel then the flow velocity in both channels will eventually become equal and the sedimentation/erosion process will stop. But in reality we expect the opposite response: The flow velocity will further decrease in the shoaling channel and increase in the scouring channel. The reason is that frictional momentum dissipation increases with decreasing depth. The shoaling of the first channel decreases the flow velocity in this channel compared to the other channel; the initial flow perturbation is thus enhanced and finally one of the channels will close (Fig. 2.19). This occurs when the current velocity in the shoaling channel falls below the critical threshold for sediment transport; the bar at the entrance will then grow until the channel is completely shut. This is what almost happened to the Nederrijn river branch. We can conclude that a two-channel system is unstable and tends to develop into a single-channel system.

Lateral channel accretion

In the previous section we have made the implicit assumption that channel infill in response to a decrease of flow velocity takes place through shoaling, i.e., through vertical channel accretion. In theory, channel infill may take place also

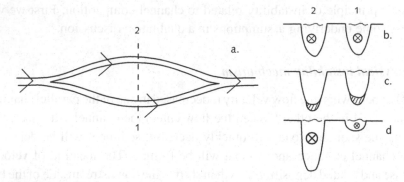

Fig. 2.19. Perturbation of an initially symmetric split-channel system. The depth is slightly decreased in channel 1 and slightly increased in channel 2. Instability means that the asymmetry is amplified and that finally channel 1 is eventually completely filled.

through sedimentation along the channel banks, i.e., through lateral channel accretion. In the hypothetical case that channel infill is entirely stored as lateral bank accretion no substantial increase of frictional dissipation will occur (as long as the channel width remains much larger than the channel depth). This mode of channel infill therefore leaves the flow velocity unaffected; in the absence of frictional feedback the sedimentation process will come to an end. However, this case is only of theoretical interest, because in natural streams both depth and width adapt to river discharge, roughly according to a square-root law [264, 248, 390]. An initial velocity decrease therefore leads to a decrease of both channel width and channel depth. So even if we allow for changes in channel width, a feedback process does exist as described in the previous section, leading to instability of the two-channel system.

River deltas

If multichannel systems are unstable, then why do they exist in nature? For instance, many large rivers develop at the mouth a multibranch delta system. In the absence of a multibranch system all fluvial sediment will be deposited around the single outlet, where a seaward protrusion of the outflow channel develops. After some time high river floods will seek a shorter path to the sea and create one or several secondary channels. The flow diversion to the new channel decreases the flow velocity in the original branch, which may silt up and finally be closed. According to the previously described feedback process this will happen if the flow velocity is stronger in the new branch than in the original branch. River deltas with a very high sediment load, such as the Mississippi and the Yellow river, typically have a single outlet branch, which is clogged shortly after the creation of a new outlet branch. Now it is the new outlet branch that starts protruding, while the old delta protrusion erodes. After some time this may lead to rehabilitation of the original branch. If rehabilitation of the original branch occurs before its complete closure, the river delta will be formed by a two-branch or by a multibranch system with long-term periodic shifts in the discharge distribution over the different branches. This is the most common situation for rivers with moderate sediment load; an example is the Danube Delta [500].

Braiding rivers and tidal channels

In wide tidal basins, multichannel systems are an ubiquitous feature; the same is true for rivers with a very large width-to-depth ratio. In Chapter 3, it will be shown that multichannel systems may come about as an inherent instability of the channel bed (Sec. 3.5). Their initial formation is triggered by a lateral channel bed perturbation and by positive feedback from the ensuing perturbation of the lateral flow distribution. Further growth of the instability is stimulated by flow meandering through positive feedback between the sediment transport pattern in meandering flow and the formation of channel bed meanders. In tidal basins the formation of a multichannel system is also enhanced by the alternation of flow direction, because the ebb flow and flood flow tend to follow different pathways through a meandering channel. It is clear that these features are not captured by the earlier one-dimensional description of channel flow. There are nevertheless many indications that channel competition takes place in wide braiding rivers and tidal basins. This competition is illustrated in Fig. 2.15, showing the shifts in relative channel importance which have taken place during the past century in the Western Scheldt basin. In some estuaries, the main and secondary tidal channels have been separated by the construction of training walls; subsequent rapid shoaling of the secondary channels has been observed, for instance, in the Seine, Mersey, Ribble and Lune estuaries [308, 18]. These observations suggest that a multichannel system does not survive if the flow loses its two-dimensional character.

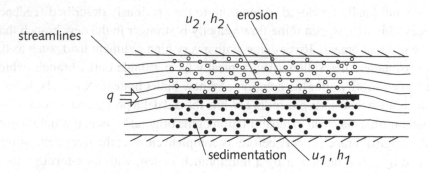

Fig. 2.20. Tidal channel split over a length l in two parallel channels of initially equal width and depth. At $t = 0$ the morphology is perturbed by slightly decreasing the depth of channel 1 and increasing the depth of channel 2: $h_2 > h_1$. The discharge in channel 2 becomes larger than in channel 1 for two reasons: The cross-section is larger and the friction is smaller. Therefore the streamlines sketched in the figure bend slightly towards channel 2. The flow velocity in channel 2 is higher than in the upstream channel, while the flow velocity in channel 1 is lower. A scour hole will develop at the entrance of channel 2, and a bar at the entrance of channel 1. In the course of time, these features will extend along the length of the split channel section, due to morphodynamic feedback. Scouring of channel 2 causes an increase of the flow velocity in that channel and, by continuity, a decrease of flow velocity in channel 1. Accretion of channel 1 will not restore equilibrium, because channel 2 responds with further scouring.

A simple model

In this section the previous considerations will be illustrated by a mathematical model referring to an idealised system of two parallel channels with the same initial length l, width b, depth h and cross-sectionally averaged flow velocity u (Fig. 2.20). Upstream and downstream of the split channel system we have a single channel with width $2b$ and depth h. The total discharge through both channels, $Q = 2bhu$, is kept constant. For the sake of simplicity we will consider all widths as fixed quantities and we set $b = 1$. This is not entirely realistic; it would be more realistic to assume that the width responds to a variation in the discharge in each of the channels in roughly the same way as the channel depth. However, introducing such an assumption in the model does not change the result of the stability analysis. Other simplifying assumptions:

- uniformity of depth, neglect of inertial acceleration terms, neglect of channel curvature and cross-sectional flow distribution, sediment transport formulation, do influence the stability criterion, but cannot be assessed by the simple model.

Now we assume a small depth perturbation in the two split channels; we will distinguish these channels with indices 1 and 2. Then we have in channel 1: $h_1 = h + \epsilon h_1'$, $u_1 = u + \epsilon u_1'$, and similar definitions for channel 2. Here, ϵ is an infinitesimal small quantity. The continuity equation reads

$$Q = 2hu = h_1 u_1 + h_2 u_2.$$

We assume that the channel length is much larger than the depth ($l/h \gg c_D^{-1}$, where $c_D \approx 0.003$ is the friction coefficient, see Appendix B.2). We also assume that the flow velocity is almost uniform in this cross-section. In that case the momentum balance implies that the surface slope over the length of the channel is proportional to the frictional momentum dissipation; this dissipation is assumed to be proportional to the square of the velocity and inversely proportional to depth (see also Appendix B.4, Eq. (B.33)). The surface slope in both channels is the same, because the water levels at both channel ends are the same. The momentum balance therefore is reduced to

$$u_1^2/h_1 = u_2^2/h_2. \tag{2.8}$$

Eliminating h_1, h_2 from both equations yields

$$u_1 = \frac{Q h_1^{1/2}}{h_1^{3/2} + h_2^{3/2}}, \quad u_2 = \frac{Q h_2^{1/2}}{h_1^{3/2} + h_2^{3/2}}. \tag{2.9}$$

From these formulas it follows that the depth perturbations h_1', h_2' produce velocity perturbations u_1', u_2' which to first order in ϵ are given by

$$u_1' = -u(h_1' + 3h_2')/4h, \quad u_2' = -u(h_2' + 3h_1')/4h. \tag{2.10}$$

Sediment transport formulation

We will examine now the sediment balance equations for both channels. We make the following assumptions:

- The sediment transport per unit width, q, is related to the flow velocity u by

$$q = \alpha u^{n+1},$$

where n is a number which depends on the sediment transport mode. For bed-load transport the exponent $n \approx 2$; for suspended load transport experimental evidence indicates $n \geq 3$, see Sec. 3.2.4. More accurate formulations of the sediment flux include a threshold velocity for incipient sediment motion; here we use the simpler formula.

- At the upstream channel junction, each channel takes a share of the upstream sediment transport proportional to the discharge entering each channel (resp. $h_1 u_1 / Q$ and $h_2 u_2 / Q$), see Fig. 2.20. Other formulations for the sediment transport distribution at channel junctions, proposed in the literature [481, 46], are not basically different. However, it should be mentioned that the theoretical stability criterion of a multichannel system is strongly dependent on the formulation of the transport distribution at the channel junction.

- Sedimentation or erosion due to convergence or divergence of the sediment flux in each channel are assumed to extend over the entire length of the split channel; the channel depth will be assumed to be uniform, except for local fluctuations corresponding to dunes or bars.

Stability criterion

With these assumptions the sediment balance equations read:

$$lh_{1t} = \alpha \left(u_1^{n+1} - \frac{h_1 u_1}{hu} u^{n+1} \right), \quad lh_{2t} = \alpha \left(u_2^{n+1} - \frac{h_2 u_2}{hu} u^{n+1} \right). \quad (2.11)$$

To first order in ϵ we find, after substitution of (2.10):

$$\frac{l}{\alpha} u^{-n-1} h'_{1t} = - \left(1 + \frac{n}{4} \right) h'_1 - \frac{3n}{4} h'_2,$$

$$\frac{l}{\alpha} u^{-n-1} h'_{2t} = - \frac{3n}{4} h'_1 - \left(1 + \frac{n}{4} \right) h'_2. \quad (2.12)$$

The solution of these coupled equations is a sum of two exponential functions, $\exp \sigma_{i1} t$, $\exp \sigma_{i2} t$. The arguments σ_{i1}, σ_{i2} of these exponential functions are given by the eigenvalues of the matrix formed by the right-hand side of the above set of equations. If one of these arguments is positive the solution will be unstable. If the determinant, $((1 + n/4)^2 - (3n/4)^2)$, is negative then the eigenvalues have opposite signs: This is the case if $n > 2$. It appears that the stability of the split channel system depends on the prevailing mode of sediment transport. For bedload transport ($n = 2$) the split channel system remains stable.

Rivers are more stable than tidal channels

This result is a direct consequence of the assumption that each channel branch receives a share of the upstream sediment flux proportional to its discharge. The velocity in each channel branch is proportional to the 1/2-power of the depth and the discharge is proportional to the 3/2-power (see 2.9); therefore, if $n = 2$, the outgoing sediment flux changes with depth at the same rate as the incoming sediment flux and no sedimentation or erosion will occur. For suspended load transport ($n > 2$) the determinant is negative and an initial perturbation of the split channel system will grow. The share of bedload in the total sediment transport in rivers is generally greater than in tidal basins. For that reason one should expect greater stability of multichannel systems in rivers than in tidal basins. In the simple model, instability leads to complete sediment infill of one of the channels. Inclusion of a minimum critical transport velocity in the sediment transport formulation would lead to closure of just the entrance. The formation of multichannel morphology is further discussed in Sec. 3.5.4.

2.5.3. The Ameland-Reef Cycle

In this section the principle of channel competition is illustrated for a small tidal basin under natural conditions. In the mid-1990s a large ebb-delta shoal ($2\,km^2$) joined the Wadden Sea island Ameland. Soon after assimilation a high beach berm developed around the shoal, preventing direct tidal submersion, except during the highest tides. The term 'reef' stems from the presence of this berm. Tidal intrusion into the low-lying central part of the shoal occurred through an inlet at the downdrift side of the shoal; the reef formed a small tidal basin with a flood delta at the centre of the shoal. The length of the inlet channel grew due to downdrift inlet displacement. A large meander also developed; it cut into the dunefoot and a restaurant at that location had to be removed, see Fig. 2.21. After ten months a second meander had developed; the old meander still existed, but was abandoned in favour of a secondary channel that had cut through the tidal flat at the inner meander bend, see Fig. 2.22 (top). One and a half years later the beach berm was broken and a new inlet was created, much closer to the flood delta. The old inlet channel lost the competition with the new channel and closed a year later, see Fig. 2.22 (bottom). The new inlet channel has now started shifting in downdrift direction. The whole morphologic cycle described above took place without human interference; it can be considered as

Fig. 2.21. Top: The Ameland reef in January 1999, at low water. The shoal is surrounded by a beach berm, with a single tidal inlet situated at the downdrift side. The dominant wave incidence is from the northwest, at the upper right side of the picture. The centre of the shoal, which is below low tidal level, is filled and emptied by a meandering inlet channel. The outer channel bend is eroding the dune foot. A former inlet situated more closely to the centre of the shoal was closed up. Bottom: Closer view of the channel meander, which in January 1999 was cutting into the dune foot. Photographs on this page and next page are from an aerial monitoring campaign conducted by the Survey Department of Rijkswaterstaat Noord-Nederland.

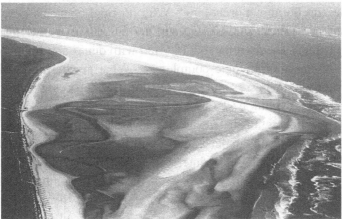

Fig. 2.22. Top: The Ameland reef at the end of October 1999. The inlet has shifted further downdrift and the length of the inlet channel has increased. Closer to the inlet a second meander has developed, which interferes with the first meander. A channel through the shoal at the inner bend is shunting the first meander; this meander is being abandoned. Bottom: The Ameland reef in May 2001. The beach berm has been breached and a new inlet is created much closer to the centre of the shoal. The former inlet channel has been abandoned and the former inlet is closed. The new inlet has started shifting in downdrift direction. The tidal prism has decreased due to sand infill of the central shoal basin.

a large-scale natural field experiment. It illustrates several feedback processes responsible for morphodynamic instability, the channel-competition instability in particular.

2.6. How Does the Sea Shape the Land?

In the previous sections we have illustrated by a few examples that sea and land play in the coastal zone a game with a sometimes surprising outcome. We have also introduced the most important principles of this game: Space-symmetry breaking and time-symmetry breaking. For one particular phenomenon, the competition between parallel channels, we illustrated how these principles work. But in this chapter most questions related to 'how does the sea shape the land?' have been set aside. Taking the Ameland-reef development as an example, such questions include: What is the origin of migrating shoals in the coastal zone, are they a common phenomenon and how do they survive in the highly energetic coastal environment? What processes are responsible for the formation of tidal inlets, can tidal inlets become stable and if so, then what is the equilibrium morphology? Why do tidal channels develop, why are they not straight but meandering and are meanders always unstable and why? Why is the seabed not flat, but covered with dunes and ripples and what is the role of this morphologic complexity? What are the timescales of all these morphologic processes and how do short term processes influence long term morphologic evolution? To answer such questions we need a basic under-standing of the interaction between sea and land. In the following chapters some contributions to this understanding will be introduced. These contribu-tions refer to current-topography interaction, tide-topography interaction and wave-topography interaction. They will be discussed separately, because the interplay between these elements is very complex and therefore beyond the scope of this book. We will mainly concentrate on idealised situations, limiting the range of morphodynamic interactions. This may sometimes give a false impression of simplicity. The Ameland-reef cycle can serve as an example to remind us of the complexity of the real-life dynamics of the coastline.

Chapter 3

Current-Topography Interaction

3.1. Abstract

Why is the seabed not flat?

Should we not expect the energy permanently transmitted by currents to the seabed to erode and flatten sedimentary bottom structures? Observations of intra-tidal morphology and maps of marine topography contradict this expectation. In some cases, the morphology of a sedimentary seabed is related to pre-existing structures, for instance, when scour holes are formed near obstacles. For most seabed structures there are no such straightforward relationships; their existence cannot be explained by topographic or hydrographic constraints. Laboratory experiments show that sedimentary structures, such as ripples or dunes develop on the sediment bed even under conditions of almost perfect initial uniformity. Scientists have always been challenged by this phenomenon, but were long unable to give a plausible explanation.

Seabed instability

It is only in the second half of the past century that the major underlying principles of seabed pattern formation were first explained; these principles relate to the nonlinearity of flow-topography interaction. The same principles apply to a broad range of phenomena of pattern formation in nature, including, for instance, the development of sea waves. At present many topographic features of the seabed are thought to originate from nonlinear flow-topography interaction and related seabed instability. However, the dynamics of this interaction are quite complicated. Therefore, mathematical descriptions are often greatly simplified. It is difficult to ensure that all essential physical processes are adequately included and hence the results should be interpreted with great care. These models should therefore not be considered as substitutes for physical reality, but as tools for improving our ability to interpret phenomena observed in the field. The importance of correctly interpreting information about seabed topography is considerable. Although the accuracy of sounding techniques has greatly improved, the cost of frequent seabed surveys can be generally prohibitive. Interpreting seabed maps helps to identify areas of rapid change with related temporal and spatial scales, which is highly relevant to navigation, mining, pipeline installation and other offshore activities.

From sand grain to sandbank

The chapter starts with current-topography interaction at the smallest scale: The scale of individual sand grains. If the shear stress over a sediment bed is strong enough sediment grains are dislodged from their position in the bed matrix and are entrained by the flow. Sediment transport is the process through which currents and seabed topography interact. At the scale of individual grains turbulent fluid motions play an essential role in the sediment transport process. This process also depends on the properties of the sediment grains and their interaction in the bed matrix. There are no theories based on first principles capable of reliable transport predictions in field situations. For the description of sediment transport in practical situations several formulas have been derived from laboratory experiments and field observations. In the first part of the chapter an introduction is given to some formulas commonly used in practice and to their range of applicability. The main body of the chapter is devoted to the analysis of feedback processes responsible for the formation of morphologic bed patterns, such as ripples, dunes and sandbanks. In this analysis we use the empirical sediment transport formulas, assuming that for the large-scale bed features, the dynamics of sediment transport at the grain scale is not strongly influenced by the large-scale interaction between seabed topography and flow pattern.

Dune and sandbank families

The process of sediment transport is related to the fine-scale structure of the seabed; ripples on the seabed are both a cause and a result of sediment transport. Seabed ripples are the smallest of a large family of rhythmic bedform patterns, known as the 'dune family', and which includes structures which are sometimes designated as 'megaripples' or 'sandwaves'. The members of the dune family share the characteristic that the formation process is strongly related to the vertical structure of the flow field (Fig. 3.1(a)). Rhythmic bedform patterns also exist at much larger scales. These structures are known as the 'sandbank family', with members such as 'bars' and 'ridges'; they are more specifically related to the horizontal structure of the flow field (Fig. 3.1(b)). Differences among the sandbank family members relate to large-scale topographic features, such as average seabed slope and lateral flow constraints (channel boundaries). Several (but not all) of these structures occur both in steady and oscillating currents. Often different families and family members coexist, in which case they are superimposed [267], as in Fig. 3.2. They impose different kinds of modifications

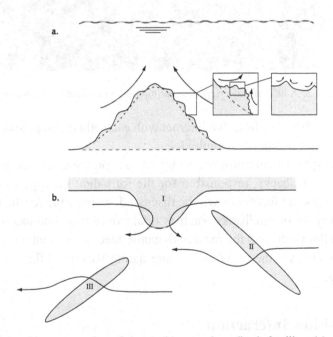

Fig. 3.1. Schematic representation of the sanddune and sandbank families. (a) A large dune on the seabed with superimposed smaller megaripples, on which smaller current ripples are superimposed. The vertical scale is exaggerated with respect to the horizontal scale by one order of magnitude. (b) Plan view of a coast with protruding headland (I), a linear ridge on the sloping inner shelf, bending towards the shoreface (II) and a tidal ridge on the outer shelf (III), inclined to the left (Northern hemisphere) relative to the main tidal flow axis. The arrows indicate average flow patterns induced by the different bedforms.

Fig. 3.2. A field of wave ripples superimposed on an field of beach bars.

on the unperturbed flow field that interact with each other; compound structures therefore exhibit additional complexity.

The principles of current-topography interaction are described first for the smaller scale feedbacks, responsible for the formation of ripples and dunes; then the larger scale feedbacks will be discussed, responsible for the formation of different types of sandbanks, such as tidal ridges and shoreface-connected ridges. Finally we discuss the morphodynamic feedbacks in laterally bounded flow, responsible for the formation of alternating bars, tidal flats and channel meandering.

3.2. Bed-Flow Interaction

3.2.1. *Flow Layers*

Turbulence

Water motion is governed by the principles of the conservation of mass and momentum. The mass and momentum balances are nonlinear, nonlinearity being inherent to the principle of momentum conservation and to the free motion of the water surface. Due to this nonlinearity the flow response to a perturbation can receive a positive feedback from the flow pattern it generates and grow by extracting energy from the unperturbed flow. This is the source of high instability, in particular at spatial and temporal scales which are smaller than those of the constraining boundary conditions (i.e., topographic scales and scales of external forcing due to tides, waves, etc.). This unstable water motion is known as turbulence and turbulent motion makes flow phenomena extremely complex. In practice, it is almost impossible to resolve the full turbulent flow structure from the governing equations. The question is whether this is necessary, if we are interested only in flow properties at larger scales, i.e., flow properties at the scale of topography or the scale of external forcing. In this case, instead of the detailed flow structure we only need to model the average impact of turbulence on mass and momentum conservation. This requires making assumptions about the statistical properties of turbulence; the validity of these assumptions has to be checked always against observations.

Focus on conditions of strong turbulence

Many models have been developed to derive statistical properties of turbulence in relation to large scale mass and momentum conservation. These properties are not the same everywhere, and it therefore makes sense to distinguish different flow regions. It appears, for instance, that the characteristics of turbulence are different in different zones of the water column; these different zones are known as flow layers, see Fig. 3.3. For turbulent flow over rough surfaces three layers are often distinguished. In the case of density stratification, due to salinity or suspended sediment, it may be necessary to introduce additional layers. Water flow over a flat bottom is considered hydraulically rough if the grain Reynolds number is much larger than 1,

$$Re_g = u_* d / v \gg 1. \tag{3.1}$$

In this equation v is the kinematic viscosity ($v = 1 - 1.5 \times 10^{-6} \, \text{m}^2/\text{s}$), d the median grain diameter and u_* the friction velocity, which is related to the bottom shear stress τ_b by

$$u_* = \sqrt{\tau_b / \rho}, \tag{3.2}$$

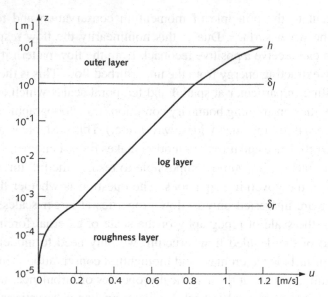

Fig. 3.3. For steady flow three boundary layers with different turbulence characteristics can be distinguished: Roughness layer $\delta_r = 2.5\, d$, logarithmic layer $\delta_l = 0.1\, h$ and outer layer ($h = 10$ m in this example). A characteristic velocity profile ($\bar{u} = 1$ m/s) is shown for flow over a flat rough bed of medium sand (grain diameter $d = 250\,\mu$m); z is the distance from the bottom.

where ρ is the fluid density. In coastal waters the condition (3.1) is often satisfied. This is true in particular for situations where the seabed is strongly stirred by highly energetic currents and waves, i.e., in situations where large amounts of sediment are carried by the flow. Much of the morphodynamic activity of coastal systems takes place under such conditions. In this book we therefore will consider mainly conditions of hydraulically rough near-bottom flow.

Roughness layer

The bottom flow layer, also known as 'roughness layer', extends right from the sediment bed up to a thickness δ_r, which is on the order of the average maximum height of bottom irregularities: A few times the median grain diameter d for a flat bed [410] or the average ripple height $2z_{bmax}$ for a rippled bed [415]. The roughness layer is characterised by turbulent flow structures generated by small-scale bottom irregularities such as sediment grains and bottom ripples.

These turbulent structures correspond to a very irregular process of alternate growth and separation of eddies in the wake of the bottom irregularities. The characteristic vertical scale of the turbulent eddies is similar to the height of the bottom irregularities. The thickness of the roughness layer δ_r is often designated in the literature as 'equivalent roughness' or 'Nikuradse roughness' k_s. For a rippled bed an adequate criterion for hydraulic roughness is considered to be [326]

$$Re = u_* \delta_r / v > 70. \tag{3.3}$$

The shear stress τ in the roughness layer is approximately constant and equal to the bottom shear stress,

$$\tau \approx \tau_b = \rho u_*^2. \tag{3.4}$$

Shear stress is mainly due to vertical exchange of horizontal flow momentum over the roughness layer. We assume that this vertical exchange, which is produced by turbulence, can be parameterised as a vertical diffusion process,

$$\tau = \rho N u_?. \tag{3.5}$$

The generalised viscosity coefficient N in the roughness layer is considered independent of z, to a first approximation. Combining (3.4) and (3.5) we find

$$u(z) = C u_* z / \delta_r, \tag{3.6}$$

where $C = \delta_r u_* / N$ is a constant depending on the geometry of the roughness elements and on bed sediment motion. In the roughness layer, the flow velocity decreases to zero at the bottom (assuming that the seabed is at rest); the average velocity is approximately a linear function of the distance z to the bottom. For uniform steady flow, the velocity gradient depends on the layer thickness δ_r and the friction velocity u_*.

For a flat bed with fixed spherical roughness elements (sand grains with diameter d), $\delta_r = d$ and $C = 8.5$ [316]. For a rippled seabed (roughness height δ_r of 1 to 2 centimetres [370]) with moving sand, field observations indicate values of C between 5 and 8 [415, 328]. It is worth noting that the velocity profile (3.6) is similar to the profile for a flat smooth bottom, where vertical momentum exchange occurs through viscosity. In that case we have a viscous sublayer of thickness $\delta_r = C v / u_*$ and $C = 11$ [316].

The thickness of the roughness layer is not constant; strong perturbations are caused by coherent turbulent motion (vortices) generated in the higher turbulent flow layers. These vortices produce an inrush of high-velocity fluid into the roughness layer ('sweeps') or ejections of low-velocity fluid out of this layer ('bursts'). Sweeping and bursting events strongly enhance sediment motion and suspension of bed material [429, 234] and are probably the major mechanism for suspending sand in tidal inlets [417].

Logarithmic layer

Above the roughness layer we have a flow layer where turbulent eddies are constrained by the distance to the bottom; as a first approximation the average eddy size increases linearly as a function of this distance. We will assume that the shear stress in the logarithmic layer can also be parameterised as a diffusion type process, see Eq. (3.5). This is a reasonable approximation under conditions of smooth spatial and temporal gradients, see also Sec. 3.2.2. It should be noticed that turbulent fluctuations in fluid density are ignored in Eq. (3.5); this so-called Boussinesque approximation is valid only for non-stratified conditions. The effect of salinity stratification is discussed in Sec. 4.7.2. The diffusion coefficient N in the logarithmic layer is known as eddy viscosity; observations show that it can be approximated by a linear function of the distance to the bottom,

$$N = \kappa u_* z, \tag{3.7}$$

where $\kappa = 0.4$ is the von Karman constant. As the shear stress τ is approximately constant in a zone close to the bed, it is related to the friction velocity u_* by (3.4). It follows from (3.5, 3.7) that the velocity profile takes a logarithmic form,

$$u(z) = \frac{u_*}{\kappa} \ln \frac{z}{z_0}. \tag{3.8}$$

The parameter z_0 is the bed roughness length. Observations in the sea show that the logarithmic shape applies to approximately the lower 10 percent of the water column [418]. Some studies report a greater thickness of the logarithmic layer, up to about half the water depth [295]. At the transition between the roughness layer and the logarithmic layer, $z = \delta_r$, the velocities (3.6) and (3.8) should match. This results in the following equation for the bed roughness length z_0,

$$\delta_r = z_0 \exp(\kappa C). \tag{3.9}$$

For a flat bed ($C = 8.5$) we find $\delta_r/z_0 = 30$. If we take the thickness of the roughness layer δ_r as 2.5 times the median grain diameter d [139] we find $d/z_0 \approx 10$, which is consistent with results from field studies [297]. For a rippled bed with moving grains ($C = 5$–8) Eq. (3.9) yields for δ_r/z_0 values between 7 and 25, with $\delta_r \approx 15\,\text{mm}$.

The logarithmic law (3.8) yields a good representation of the velocity profile for steady flow. In the coastal zone, currents are often modulated by an oscillating wave component. It has been shown by Sleath [411] that the logarithmic law remains a fair representation of the steady velocity profile in such situations.

Outer layer

Higher up, more than 10 percent of the total water depth above the bottom, the eddy viscosity is constrained not only by the distance to the bottom, but also by the distance to the water surface. This upper part of the water column is known as the outer boundary layer. In shallow coastal waters it often extends up to the water surface. In deeper waters the thickness is limited by the effects of earth's rotation and by the tidal timescale (in the case of tidal flow). The thickness δ_t of the tidal boundary layer is then given by [418]

$$\delta_t = 0.0038 \frac{\omega U_{max} - f U_{min}}{\omega^2 - f^2}, \tag{3.10}$$

where U_{max}, U_{min} are the maximum and minimum values of the depth-averaged tidal velocities through a tidal cycle, f is the Coriolis parameter and ω is the tidal frequency in radians. Observations show that for steady uniform flow the velocity profile in the outer layer can be represented by a power law distribution [418],

$$u(z) = u_s \left(\frac{z}{h}\right)^{1/7} \approx 1.14\bar{u} \left(\frac{z}{h}\right)^{1/7}, \tag{3.11}$$

where $\bar{u}$ is the depth-averaged velocity and h the water depth. Matching the velocities at the transition $z = \delta_l$ of the outer layer and the logarithmic layer yields

$$\bar{u} = 0.88(u_*/\kappa)(\delta_l/h)^{-1/7}\ln(\delta_l/z_0). \tag{3.12}$$

The friction factor c_D ('drag coefficient') relates the bottom shear stress τ_b to the depth-averaged velocity $\bar{u}$ by:

$$\tau_b = \rho c_D \bar{u}^2. \tag{3.13}$$

According to (3.2) we also have

$$c_D = (u_*/\bar{u})^2. \tag{3.14}$$

Friction factor c_D

From (3.12) we find a relationship between the friction factor c_D and the roughness length z_0. For large values of δ_l/z_0 the dependency on δ_l can be ignored:

$$c_D \approx 0.03(z_0/h)^{2/7}. \tag{3.15}$$

From observations of steady flow over flat mobile and immobile beds a best fit yields [418] $c_D \approx 0.02(d/h)^{2/7}$, where d is the median grain diameter. This is equivalent to (3.15) if we assume $z_0/d \approx 0.2$. The friction coefficient c_D usually ranges from 0.001 to 0.01 [439], depending on several variables: Sediment type, grain diameter, bottom irregularities (ripples), sediment transport rate and flow strength. In shelf seas and estuaries the friction coefficient is found typically on the order of 0.002–0.003 [225, 407, 439].

For instance, at Georges Bank (USA Atlantic coast, some 100 km offshore Cape Cod) the sandy sea floor (medium grain diameter 250–1000 μm) at 80 m depth is usually covered by ripples formed after high wave events, with 1–2 cm height and 15–20 cm wavelength. The maximum strength of the tidal currents $u_{1.2}$ at 1.2 m above the bottom is on the order of 1 m/s. From velocity profile measurements of the semidiurnal tidal currents the friction coefficient defined by $(u_*/u_{1.2})^2$, was found to be $3 \pm 0.1 \times 10^{-3}$, without important seasonal changes [486]. The corresponding roughness length z_0 ranges between 0.05–0.09 cm, which is consistent with a roughness layer thickness of 1–2 cm and $C \approx 8$ (see Eq. (3.9)).

The relationship (3.15) indicates that the friction factor is also a function of the water depth. Since the roughness length z_0 hardly depends on the depth

[370], Eq. (3.15) can be used to estimate the depth dependency. Often the empirical formula of Manning is used instead, which reads

$$c_D = gn^2/h^{1/3}, \tag{3.16}$$

where g is the gravitational acceleration and n is Manning's bed roughness coefficient. The friction factor increases with decreasing depth.

High friction coefficients in the range 0.02–0.06 have been measured in the uprush and downrush flow of broken waves on the beach [368]. Not only the small water depth (typically less than 0.5 m) is responsible for these high values, but also turbulence produced by wave-breaking. However, under highly energetic conditions the friction coefficients may also be reduced instead of increased, as indicated by field observations in the North Sea [225]. This phenomenon is attributed to near-bed stratification, due to sediment resuspension under high waves [172]. Drag reduction will also occur when a high-concentration layer of fine cohesive sediment is formed near the bed, see Sec. 3.2.3.

Eddy viscosity N

An estimate of the eddy viscosity N in the outer boundary layer can be derived from the momentum balance, which states that the shear stress in steady uniform (non-stratified) flow is a linear function of depth,

$$\tau(z) = \rho N u_z = \rho u_*^2(1 - z/h). \tag{3.17}$$

By introducing the velocity profile (3.11) in this equation it appears that N in the outer layer is approximately a parabolic function of z with a maximum at mid-depth. Its depth-averaged value can be estimated from

$$\bar{N} \approx h u_*^2/\bar{u} = \sqrt{c_D} h u_* = c_D h \bar{u}. \tag{3.18}$$

In uniform steady flow the eddy viscosity is closely related to the friction factor.

Density differences strongly reduce vertical mixing by inhibiting turbulent diffusion. Density differences due to fresh-water inflow are a common feature in estuaries and in near-coastal waters, while further offshore density stratification may occur due to temperature gradients over the water column. The eddy viscosity has completely different magnitude and depth dependency in the presence of (even weak) density stratification, compared to homogeneous conditions, see Sec. 4.7.2.

Skin friction and form drag

Sediment transport is related to the bottom shear stress. One should distinguish the shear stress contribution related to the grain structure of the bed and the contribution related to the ripple structure [157]. The former is known as 'skin friction' or 'effective shear' and corresponds to the tractive stress exerted by the flow in the viscous sublayer or in the roughness layer on bed grains. The latter is known as 'form drag' and corresponds to energy losses of the flow in the wake of bedforms (ripples, dunes). The skin friction includes viscous drag and pressure drag that arise due to flow around individual particles on the bed. The skin friction is particularly important for incipient sediment motion through bedload transport, which requires dislodging sediment grains from their equilibrium position in the bed matrix. The form drag is unimportant for transporting sediment particles as bedload, because the length scale that characterises variations in the associated pressure field is much larger than the scale of individual particles [309]. Form drag makes a major contribution to frictional dissipation of momentum and is mainly responsible for sediment transport through suspended-load sediment transport. Due to its influence on the total turbulent intensity, the form drag also contributes indirectly to skin friction. For a rippled bed, form drag is typically several times larger than skin friction [418, 486].

Non-uniform or non-steady flow

In the previous section the vertical flow profile was discussed for steady uniform flow. The terms 'steady' and 'uniform' have to be understood in a statistical sense, i.e., averaged over the temporal and spatial scales of turbulence. Water flow in the marine environment is continuously accelerating or decelerating, either because of time varying forcing (waves, tides, etc.) or because of topography. The theoretical shapes (3.6, 3.8, 3.11) of the velocity profile seldom occur in practice. This is because turbulent motion does not adapt instantaneously to variation in forcing; it takes some time before a new equilibrium is established between flow profile and turbulent stresses. This time delay is longer for large turbulent eddies than for small eddies and is therefore strongly depth dependent. For instance, observations in the Irish Sea of the tidal variation of turbulent energy dissipation as a function of depth, show an adaptation phase delay of around 2 hours between bottom (-70 m) and surface for mixed winter conditions and a phase delay of more than 4 hours for stratified summer conditions [407].

The flow profile near the seabed will thus adapt faster to fluctuations in the surface pressure gradient than the flow profile higher in the water column; far away from the bed the fluid keeps it original momentum longer than close to the bed. When the flow is accelerated (or decelerated), the fluid near the bed is accelerated (or decelerated) first, followed later by the fluid higher in the water column. Compared to steady uniform flow, the difference between near-bottom and near-surface velocities is smaller during acceleration and larger during deceleration. Accordingly, turbulent stresses in the fluid are smaller in the acceleration phase than in the deceleration phase. The frictional time delay of flow adaptation plays a crucial role for the emergence of seabed structures, as will be shown later.

Wave boundary layer

In the case of high frequency wave motion (wind waves) the layer structure is quite different from steady flow. An introduction to wave dynamics is given in Chapter 5 on Wave-Topography interaction and in Appendix D. We could postpone the discussion of sediment transport under waves to that chapter. We feel, however, that the differences between current-induced sediment transport and wave-induced sediment transport can be better understood by presenting the two cases together.

A turbulent boundary layer will not develop throughout the water column under high-frequency wave motion, because of the shortness of the oscillation period, see also Chapter 5. This implies that wave-induced flow is almost frictionless (potential flow), except in a thin layer at the bottom, see Fig. 3.4. For a smooth bottom the flow is viscous and the thickness of the layer is approximately given by

$$\delta_w \approx \sqrt{2\nu/\omega}, \qquad (3.19)$$

where ν is the viscosity and ω is the wave radial frequency; this corresponds to a layer thickness of at most a few millimetres. In the coastal environment,

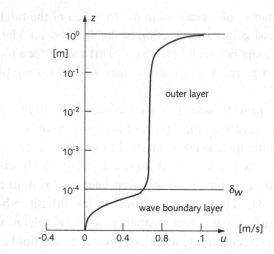

Fig. 3.4. For wave-orbital flow two layers can be distinguished: The wave boundary layer $\delta_w = \sqrt{2\nu/\omega}$ and the outer layer ($h = 10$ m in this example). In the outer layer a characteristic wave-orbital velocity profile $u = a\omega \cosh(kz)/\sinh(kh)$ is shown; z is the distance from the bottom.

however, the magnitudes of bed roughness and wave-orbital velocity are such that the wave boundary layer will more often be turbulent. In the case of a rippled bottom a fair empirical estimate of the thickness of the turbulent bottom boundary layer is given by [434, 179, 277]

$$\delta_w \approx 100\, z_{bmax}^2/\lambda, \tag{3.20}$$

where $2z_{bmax}$ is the average ripple height and λ the average ripple wavelength. For small ripples with a typical height of 1.5 cm and wavelength of 10 cm Eq. (3.20) yields a bottom layer thickness d_w in the order of 5 cm; a similar value is predicted for large wave ripples with a typical height of 5 cm and a wavelength of 1 m.

Momentum dissipation in the wave boundary layer

In the boundary layer wave-orbital momentum is dissipated by bottom shear; the wave-induced bed shear stress can be related to the wave-orbital velocity by a quadratic relationship, similar to that for steady flow,

$$\tau_w = \frac{1}{2}\rho f_w U_b^2, \tag{3.21}$$

where U_b is the amplitude of the wave-orbital velocity at the bottom and f_w is the wave friction factor. Several empirical relations have been established for estimating the friction factor [418]; f_w depends mainly on the ratio r of wave-orbital amplitude and bed roughness length and varies approximately as $f_w \approx 0.2r^{-0.5}$. Typical values range between 0.01 and 0.1.

Streaming

Longuet-Higgins [287] first showed that progressive waves produce a forward drift velocity u_s (called 'streaming') in the wave boundary layer above a smooth seabed. The maximum forward drift is reached on top of the viscous wave boundary layer and amounts to

$$u_s \approx (3/4)U_1^2/c, \tag{3.22}$$

where U_1 is the maximum wave-orbital velocity at the top of the viscous layer and c the wave propagation speed. This near-bottom fluid drift contributes to net sediment transport in the wave propagation direction. Under high waves the top layer of the sediment bed may become fluidised and form a mobile sediment sheet. The streaming velocity extends through this fluidised sediment layer and moves this layer in the onshore direction [153].

Streaming is caused by dissipation of wave-orbital momentum in the wave boundary layer. Due to frictional momentum dissipation the horizontal flow in the wave boundary layer responds faster to fluctuations in the hydrostatic pressure gradient than the wave-orbital flow higher in the water column. The phase difference between the horizontal and vertical wave-orbital velocities u and w, which is 90° far above the wave boundary layer, therefore increases down the water column. This results in a net wave-induced stress $\langle uw \rangle < 0$, corresponding to a net downward transport of forward horizontal momentum. This net downward momentum transport reaches a maximum near the top of the wave boundary layer. The downward increase of $|\langle uw \rangle|$ induces a net forward momentum deficit and a corresponding fluid acceleration opposite to the wave propagation direction. In the wave boundary layer the net downward momentum transport decreases to zero at the bottom ($w = 0$), causing an excess of net forward momentum and a corresponding net forward fluid acceleration. These net fluid accelerations (in the propagation direction near the bottom and against the propagation direction just above the wave boundary layer) produce the fluid drift called streaming. This fluid drift is zero at the bottom, maximum forward near the top of the wave boundary layer and rapidly decreasing further upward. The net excess and deficit of forward momentum are exactly balanced throughout the water column by the frictional shear stress produced by this net fluid drift. The vertical streaming profile can be derived mathematically from the balance of frictional and wave-induced stresses in the case of a viscous wave boundary layer [157].

Laboratory measurements show that for rough beds the forward near-bottom drift is strongly reduced [37]; for rippled beds the near-bottom drift is even reversed and opposite to the direction of wave propagation. This effect can be explained by considering the influence of wave asymmetry on the eddy viscosity [445]; wave asymmetry reduces the phase-lead of the near-bottom velocity. For a rippled seabed, the process of vortex shedding at the ripple crests also opposes the development of shoreward streaming in the wave boundary layer [95, 300].

3.2.2. Sediment Characteristics

Sedimentary particles

Seabed material in coastal areas generally consists of very different kinds of particles. Many of these particles are produced by soil erosion in inland catchment basins and are called clastic sediments. They are made of minerals, for instance, kaolinite (clay mineral), feldspar (silt mineral) or quartz (sand mineral). Other particles have a biotic origin, for instance shell and coral debris, peat, detritus and plankton. An important characteristic by which sediment particles can be distinguished is their diameter. This diameter differs widely, from clay and silt (grain diameter approximately 10^{-6}–10^{-5} m) to sand (diameter approximately 10^{-4}–10^{-3} m) and pebbles, gravel and cobbles (diameter approximately 10^{-2}–10^{-1} m). The density of sediment particles ρ_{sed} is on average 2.5 to 3 times the density of water; however, heavier minerals with a much higher density may also be present.

Settling velocity

Without turbulence or upwelling water motion, sediment particles will move downward, because their density is higher than the density of water. Several

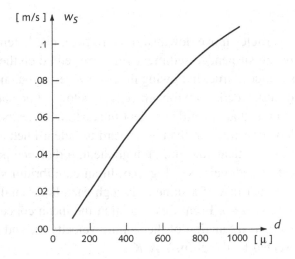

Fig. 3.5. Settling velocity at $10°\,C$ as a function of grain diameter for sand particles.

empirical formulas have been established for the settling velocity w_s of sand particles, for instance the expression [471]

$$w_s = \frac{10\nu}{d}\left[\sqrt{1 + 0.01d_*^3} - 1\right], \quad d_* = \left(\frac{g\Delta\rho}{\rho\nu^2}\right)^{1/3} d, \qquad (3.23)$$

shown in Fig. 3.5. In this expression d and d_* are the grain diameter and the dimensionless grain diameter, ν is the kinematic viscosity of water ($\approx 1.35\times 10^{-6}\,\mathrm{m^2 s^{-1}}$ at $10°$ and $\approx 10^{-6}\,\mathrm{m^2 s^{-1}}$ at $20°$), $\Delta\rho/\rho = \rho_{sed}/\rho - 1 \approx 1.5$ is the average relative density difference of sediment particles and seawater and $g = 9.8\,\mathrm{ms^{-2}}$ is the gravitational acceleration.

For particles with a small dimensionless grain diameter (sand grains with a diameter smaller than $200\,\mu$m or larger particles with a lower density) the expression (3.23) can be simplified to

$$w_s \approx 0.05g(\Delta\rho/\rho)d^2/\nu. \qquad (3.24)$$

For these particles the fall velocity is proportional to the ratio of the submerged mass and the grain diameter of the particle.

Turbulent uplift

Since sediment particles move downward on average, one is tempted to conclude that in time all suspended sediment will have settled on the bottom. In a
fluid at rest this is indeed true. In flowing fluids however, turbulence causes the
sediment to stay suspended, even though vertical velocity fluctuations average
to zero. The reason is that upward turbulent fluctuations transport fluid with a
higher suspended concentration than downward turbulent fluctuations; hence,
in the presence of a vertical concentration gradient, turbulence produces a net
upward transport of sediment, see Fig. 3.6. In an equilibrium situation with
steady current, the net flux of sediment through each horizontal plane in the
water column will be zero. From this condition the mean concentration profile can be derived, as a function of the settling velocity w_s and the turbulent
intensity (represented by the diffusivity K).

Suspended sediment profile

The vertical velocity of a sediment particle w_s consists of a mean downward
motion $\langle w \rangle = -w_s$ and a fluctuating component w' due to turbulence, with
zero mean, $\langle w' \rangle = 0$. The sediment concentration $c(z)$ can also be represented

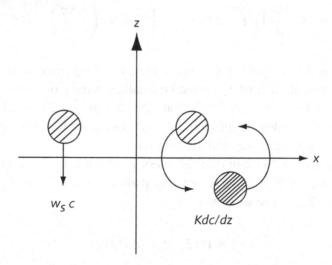

Fig. 3.6. Schematic representation of the balance between vertical sediment flux components.
Upward transport by turbulent diffusion compensates downward transport due to particle
settling.

by a mean concentration $\langle c \rangle$ and a fluctuating component c'. The mean vertical flux through any given horizontal plane equals $\langle cw \rangle = -\langle c \rangle w_s + \langle c'w' \rangle$. The last term is the flux caused by turbulent diffusion; it is generally assumed that in the absence of stratification, the turbulent motion of water particles can be characterised by a random walk. In this case the flux is approximately proportional to the vertical gradient of the mean concentration,

$$\langle c'w' \rangle \approx -K d\langle c \rangle / dz, \tag{3.25}$$

where z is the distance from the bottom and K is the turbulent diffusivity. The condition that the mean vertical sediment flux equals zero implies

$$w_s c = -K dc/dz. \tag{3.26}$$

For simplicity we have left out the brackets denoting turbulence time averaging. In the logarithmic boundary layer the turbulent diffusivity K can be represented by an expression similar to that for the turbulent viscosity N, $K = \kappa u_* z$, where $\kappa = 0.4$ is the von Karman constant and u_* is the friction velocity. If the settling velocity w_s is assumed to be constant, the solution of (3.26) yields a power-law distribution for the suspended sediment equilibrium profile in a stationary current, see Fig. 3.7,

$$c(z) = c_a (z/z_a)^{-w_s/\kappa u_*}. \tag{3.27}$$

Sediment tends to be almost uniformly distributed over the water column if the settling velocity is low and the turbulent diffusivity is high; in the opposite

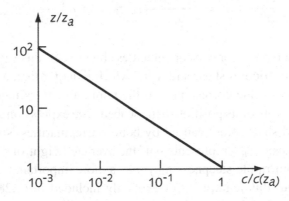

Fig. 3.7. Suspended sediment equilibrium profile in a double-logarithmic plot, according to (3.27), for typical values $w_s = 0.02\,\text{m/s}$ and $u_* = 0.03\,\text{m/s}$.

case sediment concentration has a strong vertical gradient. In general most of the suspended load is confined to a near-bed layer with a thickness of 10 to 20 times the ripple height [75].

Near-bed concentration

Realistic mathematical descriptions of the sediment uptake process need to include explicitly the turbulent fluid and particle motions in the near-bed zone. Models based on large-eddy simulation techniques have been developed which are capable of reproducing many features which are also observed in the laboratory. These models confirm the important role of coherent turbulent structures and bottom ripples; sediment is carried up into the flow at the upslope side of a ripple and advected downstream by the current and upward by the vertical velocity component of streamwise vortices [507]. These models also show that significant differences exist between the sediment uptake processes for straight-crested two-dimensional ripples and for three-dimensional ripples.

However, these models do not provide explicit expressions for the near-bed sediment concentration. Expressions for the near-bed concentration to be used in practice have been derived from laboratory experiments. In the case of suspension by currents alone one may use, for instance, the formula of Van Rijn for the near-bed reference volumetric concentration c_a at the reference level z_a [471]

$$c_a = 0.015 \frac{d\tau_*^{3/2}}{z_a d_*^{0.3}}, \quad \tau_* = \frac{\tau_s - \tau_{crit}}{\tau_{crit}}, \quad z_a = min\left[\frac{h}{100}, z_{bmax}\right]. \quad (3.28)$$

This expression is valid for current velocities above a certain critical threshold, producing a skin friction shear stress, τ_s, which is larger than a critical value τ_{crit}. This critical value corresponds to the minimum stress required to lift a sediment particle out of its position in the seabed. The expression also takes into account the bed shear stress caused by bottom irregularities, such as bottom ripples and dunes; z_{bmax} represents half the average height of bottom ripples or dunes. Bottom-ripple steepness may play a more important role than ripple height [202], but this feature is not explicitly included in (3.28). For strong currents the near-bottom reference concentration varies approximately with the third power of velocity, according to (3.28).

Wave-induced suspension

Wave action may also contribute to suspension of sediment. As mentioned before, wave-generated turbulence and diffusivity are basically different from current-induced turbulence and diffusivity; this is due to the shortness of the wave period relative to the time scales of turbulent motion [246, 326, 411]. The physics of sediment suspension by waves or by the combination of waves and a steady current is quite complex. In the frictional bottom boundary layer, waves and currents interact nonlinearly and produce together a higher skin friction than the sum of the individual contributions. The strong accelerations associated with wave-orbital motion also influence seabed erodibility.

Many models, most with a strong empirical character, have been proposed for describing sediment suspension and sediment transport by waves and currents [471, 94, 418]. Numerical simulations with a simplified mathematical model show that particles are picked up from the bed and carried into suspension when the low-speed turbulent streaks start to break up and generate small incoherent vortices that increase mixing. Low-speed streaks are generated at the end of the acceleration phase when the flow is maximum, and break down during the deceleration phase of the wave cycle [479].

In the near-coastal zone, wave-orbital velocities are generally stronger than steady current velocities. In that case it is often assumed that wave action is mainly responsible for sediment suspension (expressed by a wave-stirring function) and transport is mainly caused by the steady current. However, such a simplification should be used cautiously. For instance, residual sediment transport may also be produced by waves alone as a result of wave asymmetry and wave-induced currents [314].

Graded sediment

The expression (3.28) is valid for sand with a well-defined median grain diameter d. In practice the sediment bed contains a mixture of different grain sizes ('graded sediment'). Although the sediment fraction with the finest grains will be suspended more easily than the coarsest fraction, initial motion of the fine fraction may be inhibited by sheltering effects; the coarse grains decrease the skin shear stress experienced by the fine grains [111, 333, 408]. The critical shear stress of the coarse fraction may, in contrast, slightly decrease. However, even a small addiction of fine cohesive sediment (on the order of 5 to 10 %) may drastically increase the critical shear stress for erosion, also for the coarse fraction. In the next section a short overview will be given of the salient properties influencing the transport of fine cohesive sediments.

3.2.3. Cohesive Sediment

Flocculation of fine sediment depends on many parameters

For small non-cohesive sediment particles the mean settling velocity follows approximately Stokes' law

$$w_s = \frac{\nu}{18d}d_*^{\,2}, \tag{3.29}$$

where the dimensionless settling velocity d_* is given by (3.23). This expression is valid for Reynolds numbers of order 1, i.e., for very fine unaggregated sediment particles (10^{-6} m). The settling velocity according to (3.29) is so small that in practice fine sediments always remain in suspension. The fraction of fines that always remain in suspension is called washload. Fine sediment particles have a tendency to flocculate, however, and flocs with a much larger diameter are formed (Fig. 3.8). Due to their large diameter, flocs have a higher settling velocity than their component small particles, even though the floc density is

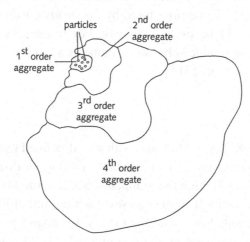

Fig. 3.8. Flocs are made up of aggregates of various orders. After [262].

much smaller than the density of the composing small particles [315]. Flocculation is affected by a number of factors [125]:

- ## Concentration

Flocculation and floc size both increase when particles collide more often [262]. Observations show that flocs formed at higher particle densities have a larger settling velocity than flocs formed at lower concentration; the settling velocity increases approximately linearly or quadratically with increasing concentration. At higher density (in particular, close to the seabed) flocs impede each other's downward motion and the effective settling velocity decreases with increasing concentration [340]. This is called 'hindered settling', see Fig. 3.9.

- ## Turbulence

Turbulence enhances the collision frequency of sediment particles and thus enhances floc growth. However, in cases of very strong turbulence and shear stresses (e.g. close to the seabed) flocs break up; the settling velocity then decreases and sediment particles will go in suspension more easily [311, 125]. For a given degree of turbulence a balance is established between floc coagulation and breakup.

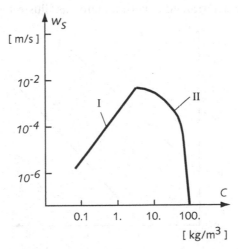

Fig. 3.9. Flocculation as a function of the concentration, after [312]. High sediment concentrations produce larger flocs and higher settling velocities (upgoing branch I, fit to experimental data: $w_s = 5.10^{-4} c^{1.3}$ m/s). Floc settling is hindered at high suspension density (downgoing branch II, fit to experimental data $w_s = 2.610^{-3}(1 - 0.008\,c)^{4.65}$ m/s).

• *Salinity*

Clay particles have electric surface charges through which particles may stick together. This process is influenced by the presence of Cl^- ions [131] and metallic or organic coatings [223]. However, in field situations the influence of salinity on flocculation cannot be isolated from other influences.

• *Presence of organic matter, bacteria and other organisms*

Organic molecules (polysaccharides in particular) excreted by organisms, such as zooplankton, bind to the surface of particles which clumb together into large aggregates, capturing smaller particles and detritus. Bacterial colonisation may also stimulate floc growth [1, 282].

Need for experimental determination

Flocs are easily disturbed; their size and composition are strongly variable. The diameter may be up to 1000 times the diameter of the composing particles. However, flocs this large are very fragile. The settling rate is not only affected by the diameter, but also by the density (water content of a floc) and drag (hydrodynamic form). In practice, settling velocities of flocculated sediment particles have to be determined experimentally, as illustrated in Fig. 3.10.

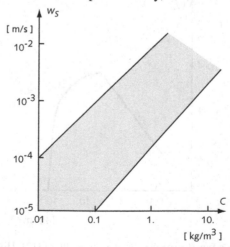

Fig. 3.10. The domain of measured median settling velocities of mud flocs in different estuaries as a function of concentration, after [469]. At each concentration the measured values span a range exceeding one order of magnitude.

Fluid mud

Due to particle settling, a suspension may develop close to the seabed with significantly higher density than the water above (assuming there is sufficient turbulent mixing in the bottom-layer). This occurs, in particular, during periods of slack tide and near the seawater intrusion limit in estuaries. The density gradient suppresses turbulent exchange between the turbid lower layer and the upper layer, stimulating the formation of a lutocline (sharp interface between turbid layer and overlaying water, see Fig. 3.11). The drag experienced by the overlaying fluid is substantially decreased, even at low concentrations of a few hundred mg/l [183, 279]. During neap tides, accelerated settling due to flocculation in the upper layer and hindered settling in the lower layer may both contribute to transforming the turbid layer into a layer of fluid mud with increased effective viscosity due to floc interaction [127]. The maximum in the settling velocity curve (Fig. 3.9) implies that mass settling to the fluid mud layer converges to a concentration close to 10 kg/m^3 [128]. Fluid mud layers may be entrained by near bottom currents over considerable distances without being dispersed over the water column [254].

Deposition

Settling of flocculated particles in the water column does not always lead to seabed deposition. The reason is that the shear stress experienced by flocs close to the bottom may be too high for keeping the flocs intact; if flocs break up the settling velocity of the constituent particles becomes too low and the

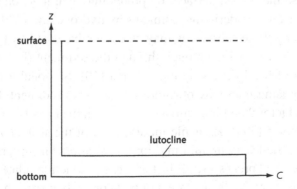

Fig. 3.11. Schematic suspended sediment profile with lutocline.

settling process is interrupted. Deposition is only possible if the bottom shear stress τ_b is below a certain critical value, τ_{De}. Below the critical shear stress for deposition one may introduce a settling probability, given by $1 - \tau_b/\tau_{De}$ [130]. Experimentally found values for τ_{De} are in the range $0.05 - 0.6\,\text{N/m}^2$. The deposition rate De can then be written

$$De = cw_s(1 - \tau_b/\tau_{De}), \hspace{3cm} (3.30)$$

where c is the sediment concentration just above the bed and w_s is the settling velocity.

Bed consolidation

The rate at which deposited material is brought in suspension by the current depends on several factors. One of the most important factors is consolidation and packing of deposited sediment. The degree of consolidation increases as the deposit gets older; due to its own weight the sediment is compressed and dewatered. In a loosely packed water-saturated seabed, pore pressure fluctuations disrupt contacts among grains, facilitating entrainment of particles by bottom shear stress [315]. Bed erosion is reduced if the pores between larger sediment particles are filled with fine sediment particles, as the bottom becomes isolated from the water column and wave-induced pressure fluctuations hardly penetrate the seabed. Sediment suspension is also influenced by the presence of adhesive exopolymeric substances, produced by micro-organisms, for instance, diatoms and cyanobacteria. These substances stabilise the seabed surface, by producing algal mats on tidal flats [7, 455] and stabilise the underlying sediments by hydrogelation [292]. Stabilisation of the bed by these substances may be counterbalanced by the activity of organisms like worms, which plough through the sediment (bioturbation) and the production of fecal pellets by macrofauna [10]. Seasonal variation in the density of stabilising microphytobenthos can alter net sediment deposition on a mudflat by a factor 2 and interannual variation in the density of bioturbating clams by a factor 5 [495]. Biogenic modification of the seabed also occurs at greater depth in shelf seas, prohibiting ripple formation during spring and summer [486]. Chemical processes, for instance gas production related to bacterial activity, also play a role. The net effect of biota on seabed erosion may be positive or negative, but no model provides a reliable simulation of these processes.

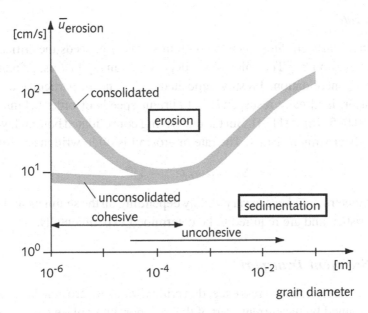

Fig. 3.12. Erosion and deposition regimes as a function of depth-averaged current strength and average grain size. Consolidated cohesive sediment beds can resist erosion until high current velocities. Redrawn after [356].

Erosion

As discussed already, bed erosion is primarily caused by turbulent shear stresses over the bed, with an important role of large turbulent eddies ('sweeps' and 'bursts'). Below a certain critical shear stress, which depends on bed cohesion, bed density and sediment diameter, no sediment is eroded from the bed. Freshly deposited material is easily resuspended, but for old deposits the critical erosion shear stress can be quite high depending on the degree of burial and compaction. Figure 3.12 shows an empirical curve of the minimum (critical) current strength needed to initiate bed erosion, as a function of grain diameter for different types of sediment particles [356]. This figure does not take into account the influence of waves on bed erosion, which may be particularly important, for instance on tidal flats. When waves are present, erosion takes place at much weaker currents than in the absence of waves [82, 105, 388]. This is attributed, inter alia, to the wave-induced cyclic loading and pore-pressure amplification discussed above. Cyclic loading may lead to the disruption of the bed structure and to liquefaction of the upper bed layer [8, 313].

Erosion rate

The seabed starts eroding when the bottom stress τ_b exceeds the critical shear stress for erosion τ_{cr}. The value of τ_{cr} depends on many factors, primarily the degree of consolidation. Freshly deposited sediment, in particular the upper 'fluff' layer, is already resuspended at current speeds of order 0.1 m/s (shear stress $\approx 0.025\,\mathrm{Nm^{-2}}$) [134]. But the underlying consolidated bottom layers can be strongly erosion resistant. The rate of erosion is often written as [344, 312]

$$Er = \mu(\tau_b/\tau_{cr} - 1)^m. \tag{3.31}$$

The parameters μ and m can vary widely depending on the sediment and seabed characteristics and are required to be determined experimentally.

3.2.4. Sediment Transport

As soon as the skin shear stress exceeds a critical value for erosion, bottom material is entrained by the current. Part of the transport takes place over the bottom (rolling or saltating) and the other part is suspended by turbulence. In the former case we refer to bedload transport and in the latter case to suspended-load transport. The ratio between bedload and suspended-load transport depends on the strength of the flow velocity and on the sediment diameter, see Fig. 3.13.

Transport by steady current

No theoretical models are capable of providing accurate and generally applicable estimates of the amount of material set in motion by fluid flow over a sediment bed [129]. The sediment transport formula currently used are therefore based on empirical relationships derived from laboratory experiments and validated with field observations. Many formulations have been proposed, but no formulation is better than the others for all situations. For non-sloping or slightly sloping bottoms and sufficiently far above the threshold of motion, the total instantaneous sediment transport $\vec{q}$ can be cast in the form

$$\vec{q} \propto C\vec{u}, \tag{3.32}$$

where C is the concentration integrated over depth. This total sediment load is often expressed in $\mathrm{m^{-1}}$ (sediment volume per square meter). For the current velocity $\vec{u}$ a representative value near the seabed is chosen, such that the right

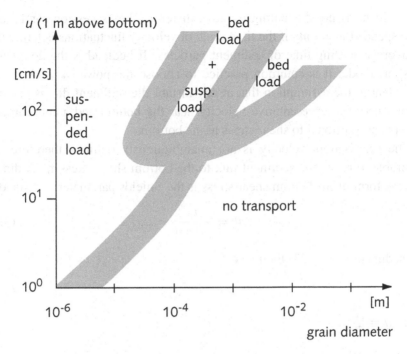

Fig. 3.13. Relationship between flow velocity (1 m above the bed), grain size and transport mode for uniform noncohesive sediment.

hand side of (3.32) represents a reasonable approximation of depth-integrated horizontal sediment transport. If the sediment is mainly transported as bedload, the representative transport velocity u corresponds to the velocity at which grains are rolling and saltating over the seabed. For sand transport, the sediment load C is found to vary as a power $n > 1$ of the depth-averaged current velocity,

$$C \propto |\vec{u}|^{n-1}, \quad \vec{q} = \alpha |\vec{u}|^{n-1} \vec{u}. \tag{3.33}$$

The coefficient α is mainly dependent on seabed and sediment properties; for practical applications the value has to be determined experimentally. Typical values of the exponent n and the proportionality constant α are in the range (for medium sand)

$$n = 3 - 6, \quad \alpha = (0.5 - 5).10^{-4} \, [\text{m}(\text{m/s})^{1-n}]. \tag{3.34}$$

Equation (3.33) relates sediment transport to the instantaneous velocity, averaged over the turbulence time scale. This only holds for situations where the flow velocity is not fluctuating too fast; therefore the expression can be used for suspended load only if the time scale of velocity fluctuations is larger than the average settling time of sediment particles. If bedload is the dominating transport mode, it is common practice to choose the power $n = 3$. This is equivalent to the assumption that at any instant the sediment flux is given by the product of a representative velocity near the bottom and a concentration which is proportional to shear stress at the bottom.

The near-bottom velocity is not unambiguously defined; therefore it is preferable to relate the sediment flux to the bottom shear stress τ_b. A dimensionless form of the bottom shear stress is the Shields parameter ϑ defined by

$$\vartheta = \frac{\tau_b}{gd\Delta\rho}.\tag{3.35}$$

The sediment flux (3.33) then reads

$$q \propto \vartheta^{n/2},\tag{3.36}$$

where $q \equiv |\vec{q}|$.

Bedload transport

A distinction is often made between bedload transport q_b and suspended-load transport q_s. Both types of transport may occur simultaneously; in that case the total sediment flux is generally taken as the sum of both contributions:

$$q = q_b + q_s.$$

Bedload transport is dominant if the flow velocities are relatively low or if sediment is medium to coarse grained. At low current speed, bedload transport is mainly due to the downdrift motion of small-scale seabed structures [456], which are formed by interaction of the bed and the near-bed flow (see Sec. 3.3). If the flow velocity is just above the threshold for motion, the critical velocity u_{cr} for erosion needs to be incorporated in the transport formula (3.33). Typical values for the near-bottom critical velocity range between 0.2 m/s and 0.4 m/s. For medium to coarse sand the critical velocity u_{cr} can be derived from the empirical threshold Shields parameter $\vartheta_{cr} \approx 0.055$ [123], or equivalently $c_D u_{cr}^2 \approx 0.055 gd(\Delta\rho/\rho)$. The bed slope may also play a role; downhill transport is enhanced relative to uphill transport because of gravity. This effect can be incorporated in the sediment transport formula by including a gravity term proportional to the seabed slope $\vec{\nabla}h$

$$\vec{q}_b = \alpha(|\vec{u}| - u_{cr})^{n-1}(\vec{u} + \gamma|u_\perp|\vec{\nabla}h), \quad u_\perp = \vec{u}.\vec{\nabla}h/|\vec{\nabla}h|, \qquad (3.37)$$

where $\vec{u}$ is the near-bottom flow velocity. In this expression gravity effects are assumed to be proportional to the magnitude of the cross-slope velocity component. Direct measurements of the proportionality parameter γ are not available; comparisons of predicted and measured transport rates indicate that the value should be close to 1 (somewhere between 0.3 and 3). Sometimes it is assumed that gravity effects can also be triggered by an along-slope flow velocity; in that case one may use the sediment transport formula

$$\vec{q}_b = \alpha(|\vec{u}| - u_{cr})^{n-1}(\vec{u} + \gamma|\vec{u}|\vec{\nabla}h). \qquad (3.38)$$

Suspended-load transport

The amount of sediment in suspension does not adjust immediately to a change in the flow strength. When the velocity is suddenly reduced, it takes some time before sediment particles have settled and a new equilibrium suspension has established; this time is known as settling lag T_s. When the velocity is suddenly increased, more sediment is eroded from the seabed. Again it takes some time before an equilibrium is reached; this is the resuspension time lag T_e. For suspension transport, and in particular for transport of fine sediment, a formulation should be used that takes into account the settling and resuspension time lags. For fine noncohesive sediment one may use a model based on the concept of equilibrium concentration; an example of such a model is

$$C_t + \vec{\nabla}.\vec{q}_s \approx \frac{1}{T_s}[C_{eq}(u) - C] \quad \text{for } C > C_{eq}, \qquad (3.39)$$

$$C_t + \vec{\nabla}.\vec{q}_s \approx \frac{1}{T_e}[C_{eq}(u) - C] \quad \text{for } C < C_{eq}. \qquad (3.40)$$

These differential equations describe sedimentation and erosion as an exponential adjustment processes of the instantaneous sediment load $C = C(t)$ to an equilibrium that depends on the instantaneous current velocity. For practical applications the settling time lag T_s and the resuspension time lag T_e have to be determined experimentally. The equilibrium load C_{eq} may be derived from the suspended sediment profile (3.27, 3.28); this yields

$$C_{eq} = \frac{0.015 d \tau_*^{3/2}}{d_*^{0.3}} \frac{(h/z_a)^{(1-w_s/\kappa u_*)} - 1}{1 - w_s/\kappa u_*}. \tag{3.41}$$

The suspended sediment flux can be estimated by inserting the concentrations derived from (3.39, 3.40) into the expression (3.32). For steady flow we have

$$\vec{q}_s = C_{eq}\vec{u}. \tag{3.42}$$

For cohesive sediment the concept of equilibrium concentration is often not applicable, because of the temporal and spatial variability of bed erosion properties. In that case the sediment concentration should be determined by introducing the deposition and erosion expressions (3.30) and (3.31) as sink and source terms in the sediment balance equation,

$$C_t + \vec{\nabla}.\vec{q}_s = Er - De. \tag{3.43}$$

Total-load transport

In practice it is often impossible to make distinction between bedload and suspended-load transport. In this case one may apply a formula including both types of transport. A total-load formula for q (sediment volume per metre and second, m^2/s) has been derived empirically by Engelund and Hansen [139]:

$$q = \frac{0.04 c_D^{3/2}}{g^2 (\Delta\rho/\rho)^2 d} u^5, \tag{3.44}$$

where c_D is the friction factor (drag coefficient) and $\Delta\rho$ the density difference between sediment and water. The different expressions for bedload (3.38), suspended-load (3.42) and total-load transport (3.44) are compared in Fig. 3.14 for 250 μm and, as a function of the depth-averaged velocity u. The sediment flux increases greatly at velocities higher than 1 m/s, mainly due to a strong increase of the suspended load. According to these expressions, bedload and suspended-load transport are of comparable magnitude only at velocities between 0.4 m/s (critical velocity for incipient sediment motion) and 0.8 m/s.

Conditions for significant sediment transport

The strength of currents in natural systems fluctuates with time due to storms, tides, waves and river floods. The great sensitivity of sediment transport to

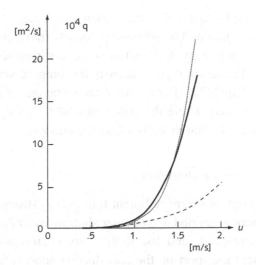

Fig. 3.14. Comparison between expressions for bedload transport (3.38, dashed line), suspended-load transport (3.42, dotted line) and total-load transport (3.44 solid line) for 250 μm sand, in the absence of a bottom-slope, as functions of the depth-averaged velocity u. The following values were used: Bedload exponent $n = 3$, bedload factor $\alpha = 10^{-4}$ m^{-1}s^2, reference level $z_a = 1$ cm, depth $h = 10$ m, friction factor $c_D = 0.003$, settling velocity $w_s = 0.03$ m/s, critical velocity for incipient sediment motion $u_{cr} = 0.4$ m/s. For $u > 0.8$ m/s suspended-load transport strongly increases with velocity and dominates bedload transport.

the flow velocity implies that conditions with strong currents are far more significant for morphologic evolution than periods with weak currents. For instance, the periods of maximum flood and ebb currents are more significant than the periods of slack water and spring tide conditions are more significant than neap tide conditions. Periods of storms and river floods in many cases contribute more to long-term morphologic change than periods of average conditions, even if the frequency of storms and river floods is low. Short-term morphologic change under extreme conditions often deviates from the long-term morphologic trend; the trend will in general be partly or fully restored under average significant conditions.

Sediment transport under waves

It has already been noted that the presence of waves may strongly enhance sediment transport rates, due to wave-induced stirring of the seabed. In the absence of a mean current most of the suspended sediment remains confined in a thin layer above the bottom, but in the presence of a mean current it is lifted higher in the water column. The suspended load is transported by the mean current, even if the current by itself is too weak to stir up sediment from the bed. The details of this transport process are quite complex, see for instance the monograph of Van Rijn [472]. For practical computations often a formulation of the type (3.32) is used, where the sediment load C mainly depends on the amplitude of the wave-orbital velocity along the bottom.

Sediment transport on the shoreface

The shoreface is a region where sediment transport is strongly influenced by wave action; offshore generated waves enter shallow water (water depth much smaller than the wavelength) and disturb the seabed before and after breaking. Downslope sediment transport on the shoreface is approximately in equilibrium with wave-induced onshore sediment transport (see Chapter 5). Sediment is transported both as bed load and suspended load; for this situation semi-empirical expressions have been derived. The formula for bedload transport is related to the work required to lift a sand grain out of a sloping seabed matrix; the formula for suspended-load transport is related to the phenomenon of auto-suspension (particles settle less easily if the streamlines diverge from the bottom). These expressions, originally proposed by Bagnold [19], were developed into a form similar to (3.38) by Bailard [20]

$$\vec{q}_b = \alpha_b \left(\langle |\vec{u}|^2 \vec{u} \rangle + \gamma_b \langle |\vec{u}|^3 \rangle \vec{\nabla} h \right), \tag{3.45}$$

$$\vec{q}_s = \alpha_s \left(\langle |\vec{u}|^3 \vec{u} \rangle + \gamma_s \langle |\vec{u}|^5 \rangle \vec{\nabla} h \right), \tag{3.46}$$

with the coefficients

$$\alpha_b = \frac{\epsilon_b c_D}{g \tan \varphi_r \Delta \rho / \rho}, \quad \gamma_b = \frac{1}{\tan \varphi_r}, \quad \alpha_s = \frac{\epsilon_s c_D}{g w_s \Delta \rho / \rho}, \quad \gamma_s = \frac{\epsilon_s}{w_s}.$$

In these expressions u is the cross-shore near-bottom velocity, which contains a fluctuating component (the wave-orbital velocity) and a steady component, which is generally smaller than the fluctuating component. The brackets stand for time averaging over the wave period. The coefficients ϵ_b, ϵ_s are efficiency parameters with values on the order of 0.1 and 0.02 respectively. In the bed-load formula φ_r is the angle of repose; this is the angle at which spontaneous avalanching may occur in the absence of bed shear stress ($\varphi_r \approx 30°$). Application of the formula requires that the bed slope $\beta = |\vec{\nabla} h|$ is much smaller than $\tan \varphi_r$. In the suspension transport formula the seabed slope particularly affects the fine sediment fraction; this fraction settles less rapidly and is kept more easily in suspension in periods of seaward wave motion than landward motion. The formula for suspended load only makes sense if the bed slope β is much smaller than the ratio for autosuspension $w_s / |\vec{u}|$.

Suspension and settling lag

Both sediment transport expressions (3.45, 3.46) are very rough models of wave-induced sediment transport; for instance, the instantaneous and local adaptation of sediment concentration to orbital wave velocities is a questionable oversimplification. In reality, wave periods are so short that instantaneous adaptation cannot occur. It has been observed, for example, that the magnitude and direction of wave-related sediment transport may change depending on the geometry of ripples; under asymmetrical waves, transport is directed shoreward for flat ripples and seaward for steep ripples. This can be explained by the adaptation delay of the sediment concentration [477, 75, 249]. This delay is short for flat ripples, where no flow separation occurs; in that case a net shoreward motion under asymmetrical waves results from the stronger shoreward than seaward orbital wave motion. For steep ripples, flow separation at the ripple crest during shoreward orbital motion and vortex shedding at the reversal of shoreward orbital motion to seaward orbital motion produce a high suspended load which is afterwards carried seaward by the offshore orbital motion. This may result in net seaward sediment transport of the fine sediment fraction. The suspension-settling lag effect can be incorporated to only some degree in the parameters ε_b, γ_b, ε_s and γ_s, which are tuned to match observed sediment transport rates.

3.2.5. Bed Evolution

If the spatial distribution of the sediment flux is known it is possible to predict the evolution of the sediment bed. The bed elevation relative to a reference level is given by $z_b(x, y, t)$. The change of elevation per unit time follows from the sediment balance of the water column above the bed, i.e., from the difference between the sediment fluxes entering and leaving a reference volume. This leads

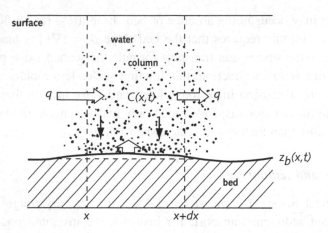

Fig. 3.15. Schematic representation of bed evolution as a sink/source term in the sediment balance of the water column.

to the following sediment balance equation, see Fig. 3.15:

$$(1 - p)z_{bt} + C_t + \vec{\nabla}.\vec{q} = 0. \tag{3.47}$$

In this equation $\vec{q}$ is the total volumetric sediment flux (bedload and suspended load), C is the total sediment load and $1 - p$ is a factor representing the porosity of the seabed. The sediment volume related to change of bed elevation is generally much greater than the sediment volume related to change of the sediment load C. The change in the total load, C_t, can therefore be ignored in the sediment balance if we consider time averages over periods of significant seabed erosion or accretion:

$$(1 - p)z_{bt} + \vec{\nabla}.\vec{q} = 0. \tag{3.48}$$

This equation defines the relationship between bed changes and flow hydrodynamics and can be solved by introducing for q an appropriate sediment transport formula (for instance, 3.32, 3.37, 3.38, etc.). Equation (5.64) forms the starting point for morphodynamic modelling.

Bed evolution due to suspended load alone

In the case of dominant suspended-load transport the bed evolution equation can be related to erosion-deposition by averaging (3.39, 3.40) over a time interval (indicated by angle brackets $\langle \rangle$) which is sufficiently long for ignoring temporal changes in suspended load; this yields

$$(1 - p)z_{bt} = \left\langle \frac{C - C_{eq}(u)}{T_{s,e}} \right\rangle, \qquad (3.49)$$

where $T_{s,e}$ is the adaptation time scale for settling or erosion, depending on which of these two processes prevails at a given time or location.

3.3. Dunes and Ripples

3.3.1. *Qualitative Description*

Dependence on flow strength

The seabed under flowing water is seldom flat. The 'regular' seabed is characterised by bottom structures with a great variety of shapes and horizontal and vertical scales, depending on sediment type, sediment diameter and flow velocity (see Table 3.1). Ripples and dunes are the most common structures. Large dunes are also known as sandwaves and small dunes are sometimes referred to as megaripples. The seabed becomes rippled when the flow is relatively weak (15–50 cm/s); the lower value holds for fine sand. Ripples have a typical wavelength of 10–50 cm and a typical height of a centimetre (generally less than 2 cm). Dunes are formed when the flow is somewhat stronger (typically in the order of 50 cm/s to 1 m/s). They appear in water depths of about 1 m or more; in steady flow their size is positively correlated with water depth, but it is typically independent of flow strength [152]. Ripples and dunes may exist simultaneously; in that case the ripple pattern is superimposed on the much

Table 3.1. Dune-type bedforms at different flow regimes, after [47, 91, 370].

	Ripples	Megaripples	Low-energy dunes	High-energy dunes
spacing [m]	0.1–0.5	0.5–10	6–30	>10
height [cm]	1–2	3–50	10–100	>50
geometry	2D–3D, highly variable	sinuous to 3D, scour pits in troughs	straight to sinuous, uniform scour in troughs	straight to sinuous
flow velocity [m/s]	0.15–0.5	0.7–1.5	0.3–1	0.7–1.5

larger dune pattern. The flow velocity required for the formation of ripples and dunes depends primarily on sediment characteristics. Higher flow velocities are required for coarse grained seabeds and for seabeds containing fine cohesive materials. In these cases there will be little ripple formation. Ripple and dune formation is also influenced by waves; strong waves (high wave-orbital velocities near the bottom) inhibit ripple and dune formation [306]. At the incipient stage of formation, ripples and dunes can be characterised as two-dimensional structures in a vertical plane parallel to the flow; the crest line is almost straight and perpendicular to the flow direction. Ripples and dunes get a more complex structure when they further develop and when the flow strength increases; the crest line becomes sinuous with numerous interruptions and the trough becomes highly irregular, with scour pits and spurs.

Dunes on the North Sea bottom

Dunes and ripples are common features of flow over a sandy bed. They are observed in rivers, estuaries, tidal inlets and in coastal seas. Dunes occur in regions with moderate tidal flow, where maximum flow velocities exceed 0.5 m/s [218]. Dunes occur, for instance, in almost the entire Southern Bight

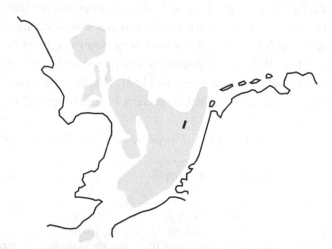

Fig. 3.16. Occurrence of dunes in the Southern Bight of the North Sea (shaded area), from [225]. The region of occurrence coincides roughly with the region where maximum tidal bottom stresses are between 0.5 and 2 N/m^2, see Fig. 3.17. The dash indicates the location of the seabed map of Fig. 3.18.

Fig. 3.17. Distribution of maximum bottom stress (N/m²) due to semi-diurnal tidal currents in the Southern Bight of the North Sea, from [350].

of the North Sea (see Fig. 3.16), in particular in the region where the average maximum tidal bottom stress is between 0.5 and 2 N/m² [306] (see Fig. 3.17). The seabed of the Southern Bight is sandy with medium grain diameter of 250–500 μm in the southern part and medium grain diameter of 125–250 μm in the northern part. Seabed sediments north of the Southern Bight are very fine, with a medium grain diameter below 125 μm. Figure 3.18 shows a detailed map of a 1 by 5 km strip of the North Sea bottom, obtained with a multibeam sonar. The seabed is covered by dunes, which are part of a large dune field. A field of megaripples is superimposed on the dunes. The orientation and wavelength of both dunes and megaripples are well defined in a statistical sense, but the detailed structure of both dunes and megaripples has a three-dimensional character.

Scale relationships for dunes

How are bedforms related to flow and sediment characteristics? Many investigations have been devoted to this question. However, measuring bedforms is not easy, because they rarely appear as isolated two dimensional structures. We often encounter compound structures, with ripples, megaripples and smaller dunes superimposed on larger dunes. To make it more difficult, flow conditions

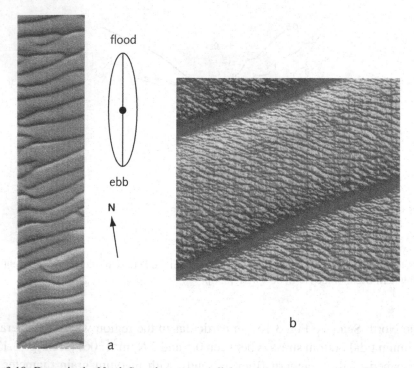

Fig. 3.18. Dunes in the North Sea; the crests are light and the troughs are dark. The left image shows a north-south oriented bottom strip of 1 × 5 km at 25–30 m depth (the arrow indicates the north), located 60 km offshore the northern Holland coast (see Fig. 3.16). The tidal ellipse is shown for mean springtide; tidal currents are almost rectilinear. In this part of the North Sea the maximum flood flow is slightly stronger than the maximum ebb flow and the mean flow (a few cm/s) is also directed northward. The average dune spacing is about 250 m; the dune height is about 5 m. The dunes are not rectilinear and bifurcate frequently. The crests run east-west; the dune-normal makes a small cyclonic angle to the tidal current axis. A more detailed picture is shown in the right image; it reveals megaripples running across the dune. The megaripple-normal makes a small anticyclonic angle to the tidal current axis. At the dune trough the megaripples make an angle of more then 45° to the trough line. It can also be seen that the dunes are asymmetric: The trough (dark) lies close to the crest (light) at the flood-side. From [467].

are typically variable, so what flow conditions should be considered representative? The general conclusion of several extensive phenomenological studies is that the dune wavelength is related mainly to the flow depth, while the ripple-wavelength is not. The following relationship has been proposed between the wavelength λ of dunes and the flow depth h [501, 6]

$$\lambda \approx 6h. \tag{3.50}$$

However, the scatter is high, $\lambda \approx (1-16)\,h$, and there are doubts about the general validity of this relationship. For example, dunes superimposed on tidal sandbanks in the Southern Bight of the North Sea have similar wavelengths near the sandbank crest (at small depth) and near the swales (at large depth) [331]. Moreover, the typical wavelength of these dunes is much larger than indicated by the relationship (3.50); observed wavelengths are on the order of $\lambda \approx (10 - 50)h$ [468].

The commonly observed height H of dunes is on the order of

$$H \approx h/6. \tag{3.51}$$

The scatter in the relationship for the dune height is rather less than for the dune wavelength [152].

The migration direction and migration rate of dunes depend on tidal flow asymmetry and on the strength of the dominating flow direction; observed migration rates range from a few metres per year to several hundred metres per year or sometimes even more. In unidirectional or in asymmetric tidal flow the dune profile is also strongly asymmetric, with gentle stoss-side slopes (on the order of a few degrees) and steep lee-side slopes (on the order of ten up to a few tenths of degrees) [91].

Complexity of ripple and dune patterns

Why are dunes and ripples not straight-crested and why is the crest line often at an oblique angle to the flow direction? There are several answers to these questions, but so far they remain rather speculative. Flume observations clearly show that the migration rate of ripples is not constant along the ripple crest; this is probably due to the three-dimensional structure of the vortices generated at the lee of the ripples. However, dune crests may exhibit the same type of sinuosity, see Fig. 3.19. It has been suggested that differences in migration speed along the crest are the major cause. If it is assumed that the migration rate depends on the crest angle then it is possible that a small initial variation in the crest angle is amplified and finally leads to break-up of the crest line [89]. Non-uniformity of ripple height also induces differences in migration speed [370]. Another suggestion is that differences in migration speed are produced by smaller bedforms superimposed on the dune [91].

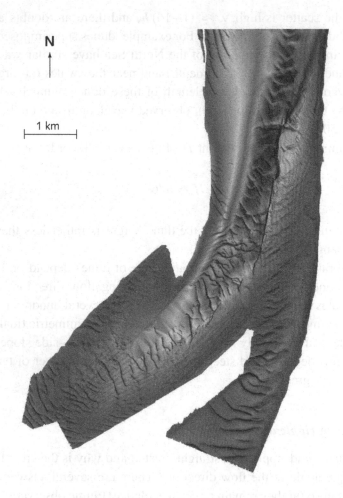

Fig. 3.19. Seabed topography of the southern Schulpengat channel in the ebb-tidal delta of Texel Inlet based on detailed bottom soundings. The location of the channel in the ebb-tidal delta can be found in the topographic map of Fig. 3.49, between km 106–110 on the x-axis and km 544–550 on the y-axis; the North-Holland coast borders the channel on the right. The main channel (depth 25 m) is ebb-dominated; it turns to the southwest and terminates in a shoal (depth 10 m), where flow divergence is visible in the dune pattern. The channel bed is covered by dunes; the crests are sinuous, bifurcating and even linguoid in the northern part of the channel. In this deeper northern part the average dune wavelength is 150 m and in the southern shoaling area less than 100 m. A channel bifurcates to the south along the coast; the dune wavelength here is also on the order of 150 m although the depth (10 m) is less than in the main channel. Almost everywhere in both channels the southern dune slope is steeper than the northern slope, indicating ebb-dominant tidal flow. The seabed is covered with medium sand, but erosion resistant clay and peat layers are present below the surface at some places.

Several hypotheses have been formulated for the obliqueness of the average crest angle. Smaller bedforms often have a different orientation than the larger bedforms on which they are superimposed [266, 147], see for an example Fig. 3.18. If the large bedforms are not perpendicular to the flow then the near-bed flow will be deflected and this deflection may produce a different orientation of smaller superimposed bedforms [298, 199]. It has been observed, for instance, that in the swales between large linear bedforms, the flow tends to be aligned with the swale direction. Another suggestion is that obliqueness of large bedforms relative to the main flow direction may result from non-alignment of flood and ebb flow [384]. It is not clear, however, if these hypotheses apply to all situations where sinuous and oblique crest lines are observed.

3.3.2. Feedback Mechanism

Seabed instability

In a tidal environment the ripple and dune structure of the seabed is visible at low water. The obvious and widespread appearance of these structures has long been a puzzling phenomenon which has challenged many researchers. It is only a few decades ago that an explanation based on first physical principles was given. Although the full complexity of ripple and dune mechanics has not yet been explained, at present general agreement exists that bedform generation results from the instability of bed-flow interaction. This process relates to the initial formation of ripples and dunes, which may start from a plane seabed, horizontally uniform flow and a flow strength just exceeding the minimum required for sediment transport.

Vertical flow structure

It is generally assumed that the development of ripples and dunes is related primarily to the vertical flow structure rather than the horizontal flow structure; the latter is primarily associated with the development of sandbanks. Therefore we will first investigate the influence of a local seabed structure on the vertical flow distribution. Because of the similarity of dune genesis mechanisms for tidal flow and steady unidirectional flow, we will concentrate first on the latter, simpler case. Before introducing equations, we will discuss a qualitative conceptual model. We will consider a dune of very small height as a two-dimensional perturbation (uniform in the direction perpendicular to the flow) of an otherwise perfectly plane seabed. We will show that this perturbation induces a modification of the flow field such that sediment transport converges at the crest of the perturbation. In other words, from the interaction between the seabed and the flow field results a positive feedback promoting the growth of the bottom perturbation.

Frictional delay in the vertical plane

We consider the initial phase of dune growth, where the near-bottom flow follows the seabed and where no flow separation occurs at the lee of the dune. We also assume that the flow over the dune is subcritical; the flow velocity is much smaller than the propagation speed of surface disturbances (small Froude number). This is the normal situation in coastal environments; supercritical flow will be discussed later. In the case of subcritical flow the water depth at the dune crest is lower than at the trough. Therefore the flow must accelerate to maintain a constant water flux when crossing the dune. Flow acceleration has to overcome inertia and therefore an additional pressure gradient is required. This pressure gradient is provided by a local inclination of the water surface, which slightly dips at the dune crest. The surface pressure gradient increases the velocity gradient near the bottom (primarily in the roughness layer and in the logarithmic layer), where the velocity decreases to zero. This increase of the near-bottom velocity gradient produces additional momentum dissipation (momentum transfer to the bottom), which opposes near-bottom flow acceleration. Therefore the flow along the bottom will decelerate before reaching the dune crest. It takes some distance, however, before the velocity gradient and frictional momentum dissipation are substantially increased in the outer layer; the near-surface velocity is therefore still accelerating at the dune crest.

The water level dip creates an opposite pressure gradient downstream of the dune crest, which enhances the deceleration of the near-bottom flow. The flow acceleration near the surface is turned to flow deceleration; this occurs only some distance downstream of the dune crest, where the velocity gradient and frictional momentum dissipation have sufficiently developed. The crucial point for seabed stability is that the near-surface velocity has a maximum downstream of the dune crest, while the near-bottom velocity has a maximum upstream of the dune crest, as shown in Fig. 3.20.

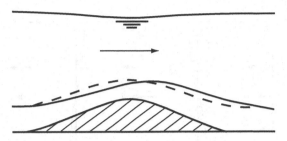

Fig. 3.20. Flow over a dune; the figure depicts schematically two near-bottom streamlines. The dashed streamline corresponds to a hypothetical situation in which frictional lag does not exist; in that case the momentum balance is symmetrical upstream and downstream of the dune crest, and so is the streamline. The solid streamline displays the effect of frictional lag. The flow does not respond instantaneously to the influence of increased friction; horizontal and vertical acceleration are both delayed when passing the dune. Inertia concentrates the flow in the near-bottom layer at the upstream side of the dune and the opposite happens at the downstream side of the dune. The near-bottom flow will thus be stronger at the upstream slope of the dune than at the downstream slope. The near-bottom flow velocity therefore reaches a maximum upstream of the dune crest, and at the dune crest the near-bottom flow is already decelerating. The sediment transport capacity is decreasing at the crest and sediment will be deposited. The dune crest therefore will tend to grow.

Momentum excess and momentum deficit in the vertical plane

A second, equivalent, explanation for flow asymmetry around the dune crest is shown in Fig. 3.21. Here we assume a fully grown dune with great length (small dune slope and no flow separation). Upon acceleration over the dune crest, the flow will adjust its vertical profile such that frictional momentum dissipation remains approximately uniform. The flow acceleration from dune trough to dune crest is then greater near the surface than near the bottom. The picture is exactly reversed in the region between dune crest and trough. However, this picture does not yet satisfy the momentum balance. More momentum is leaving the trough-crest region than entering it, especially near the surface. The momentum deficit averaged over the water column is balanced by a negative surface pressure gradient, caused by a small water level dip at the dune crest. However, a momentum deficit remains in the upper part of the vertical and a momentum excess appears in the lower part of the vertical. The momentum excess in the lower part of the vertical drives a secondary circulation towards the dune crest which has its maximum upstream of the crest. The near-bottom flow at the crest is therefore beyond its maximum and decelerating. After the

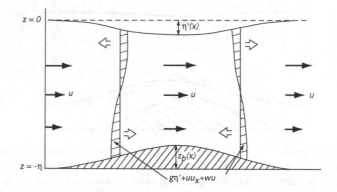

Fig. 3.21. Flow over a dune. Solid arrows indicate flow velocities in a section across the dune. The open arrows indicate the pressure gradient distribution resulting from inertia opposing flow acceleration over the crest, partially compensated by a local water level depression.

crest the secondary near-bottom flow is directed away from the crest; flow deceleration is enhanced downstream of the dune crest.

Sediment flux convergence at the dune crest

If we assume that sediment transport is governed by near bottom flow then sediment transport will already be converging upstream of the crest. Sediment is deposited at the dune crest and the crest height will grow, see Fig. 3.22. Hence, an initial seabed perturbation in the form of a linear dune perpendicular to the flow induces a modification of the flow structure which contributes to growth of the dune. The crucial assumption is that the flow adaptation needed to reach an equilibrium flow profile takes more time (i.e., a greater distance) at the surface than near the bottom. This will be the case if the near-surface flow and near-bottom flow interact by turbulent diffusion of momentum, such as in a non-stratified outer flow layer.

Small-scale bedforms

The foregoing description of seabed instability pertains to the response of the outer flow layer to an emerging seabed perturbation. In this description the term 'near-bottom flow' has to be understood as the flow at the lower boundary of the outer layer. The roughness layer and logarithmic layer also respond to a seabed perturbation, but not necessarily in the same way, because the characteristics

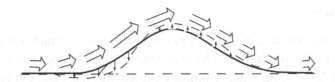

Fig. 3.22. Transport convergence at the dune crest and resulting dune growth.

of turbulence are different in these layers. However, we may expect that in each of these layers a phase delay in the flow response between the upper and lower layer boundaries will occur as well. In that case bedforms may develop which are related to the flow perturbation in each of these layers. It will be shown that these bedforms have smaller wavelengths than dunes; there is some evidence that they can be associated to emerging ripples or megaripples, see Sec. 3.3.4.

3.3.3. *The Dune Genesis Model of Engelund and Fredsøe*

Flow response to a small seabed undulation

For a more quantitative analysis of the assumptions behind the model we will now investigate the flow equations, and in particular the streamline pattern, following a model formulated by Engelund and Fredsøe [138, 154]. We introduce the streamfunction $\psi(x, z)$, defined by the condition that the flux between two streamlines is constant:

$$\int_{z_1}^{z_2} u(x, z)dz = \psi(x, z_2) - \psi(x, z_1), \quad \text{or} \quad \psi_z = u. \qquad (3.52)$$

We will continue to use our crude two-dimensional dune representation and ignore the lateral dimension. We consider a flat seabed at $z = -h$, perturbed by a dune of low amplitude and wavelength λ, represented by

$$\Re z_b(x) = \epsilon h \Re e^{i(kx - \sigma t)} = \epsilon h e^{\sigma_i t} \cos(kx - \sigma_r t), \qquad (3.53)$$

with $\epsilon \ll 1$. In this expression $k = 2\pi/\lambda$ is the wavenumber, $\sigma_r/k \equiv \Re\sigma/k$ is the migration velocity and $\sigma_i \equiv \Im\sigma$ is the growth rate, which can be either positive or negative.

Flow equations

We assume that the total water flux per unit width is constant and the same everywhere along the channel. The wavelength of sand dunes is generally on the order of 2π times the water depth, and seldom much greater. For this reason vertical velocities and longitudinal gradients in the turbulent stresses cannot be ignored in the equations of motion. These equations are (D.8)

$$u_x + w_z = 0, \tag{3.54}$$

$$uu_x + wu_z + \frac{1}{\rho}p_x = N(u_{zz} + u_{xx}), \tag{3.55}$$

$$uw_x + ww_z + \frac{1}{\rho}p_z + g = N(w_{zz} + w_{xx}), \tag{3.56}$$

where u, v, w are the velocity components in x, y, z directions and p is the pressure. The first equation describes the mass balance or continuity of flow; inflow equals outflow in each elementary cell. The two other equations describe conservation of momentum; an excess of momentum inflow or outflow in each elementary cell is compensated by turbulent diffusion of momentum. The eddy viscosity N is taken as a constant. This is a reasonable assumption for the outer flow layer but in the logarithmic layer near the bottom the flow field is not well described by this assumption. Therefore the near-bottom flow will not be modelled explicitly; as discussed later, it will be dealt with by imposing a condition on momentum dissipation at the bottom. The continuity equation (3.54) imposes on the streamfunction the following condition

$$\psi_x = -w.$$

The pressure is eliminated from the equations of motion by considering the vorticity

$$\zeta = u_z - w_x = \psi_{xx} + \psi_{zz}.$$

By taking the z-derivative of Eq. (3.55) and substracting the x-derivative of Eq. (3.56) we obtain

$$u\zeta_x + w\zeta_z = N(\zeta_{zz} + \zeta_{xx}). \tag{3.57}$$

This equation is a vorticity balance; an excess of inflow or outflow of vorticity in each elementary cell is balanced by turbulent diffusion of vorticity. In terms of the streamfunction this equation reads

$$\psi_z(\psi_{xx} + \psi_{zz})_x - \psi_x(\psi_{xx} + \psi_{zz})_z = N\left(\psi_{zzzz} + 2\psi_{zzxx} + \psi_{xxxx}\right). \quad (3.58)$$

Before stating the boundary conditions we first scale the variables with depth h and/or with u_{0b}, the velocity near the bottom in the unperturbed flow: $x \rightarrow hx, z \rightarrow hz, z_b \rightarrow \epsilon hz_b, \eta \rightarrow h\eta, u \rightarrow u_{0b}u, w \rightarrow u_{0b}w, \psi \rightarrow hu_{0b}\psi, N \rightarrow hu_{0b}N, k \rightarrow k/h$. This change of variables leaves Eq. (3.58) unchanged.

Boundary conditions for the outer flow layer

At the surface and bottom the flow has to satisfy the following boundary conditions.

- the water surface is a streamline,

$$2F^2\eta + u^2 + w^2 = 0 \quad \text{(Bernouilly)} \quad \text{and} \quad w = u\eta_x \quad \text{at} \quad z = \eta, \quad (3.59)$$

where $F = \sqrt{gh}/u_{0b}^2$ is the Froude number.
- no momentum is transfered through the water surface,

$$Nu_z = 0 \quad \text{at} \quad z = \eta, \quad (3.60)$$

- the bottom surface is a streamline,

$$w = uz_{bx} \quad \text{at} \quad z = -1 + \epsilon z_b, \quad (3.61)$$

- transfer of momentum to the seabed $N\zeta$ is proportional to the square of the bottom velocity u_b^2, with the same friction coefficient $c_b = N\zeta_0$ as for the unperturbed flow; this yields

$$\zeta = \zeta_0 u_b^2 \quad \text{at} \quad z = -1 + \epsilon z_b. \quad (3.62)$$

Slipping velocity at the bottom of the outer layer

The last condition is used instead of $u = 0$ at $z = -1$, because $z = -1$ does not correspond to the physical seabed. As the large velocity gradients in the logarithmic layer near the bottom are not resolved in the model it is assumed that $z = -1$ corresponds to the lower boundary of the outer layer; the bottom velocity u_b represents the flow velocity at this lower boundary. Accordingly, the perturbation z_b does not, in reality, represent the perturbation of the seabed, but the perturbation of the lower boundary of the outer flow region. The bottom boundary conditions are essential, as they relate the perturbation of the flow directly to the perturbation of the seabed. It can be verified that the set of boundary conditions (3.59–3.62) automatically ensures continuity of the depth-integrated water flux.

Infinitesimal perturbation and linearisation

Now we will determine to first order in ϵ the perturbation of the streamline with wavenumber k which matches the dune perturbation z_b; therefore we choose ψ of the form

$$\psi(x, z) = \psi_0(z) + \epsilon\psi_1(x, z) + \epsilon^2\psi_2(x, z) + \cdots, \quad \psi_1 = z_b(x)\chi(z). \quad (3.63)$$

Similarly we define $u_1 = \psi_{1z}$, $w_1 = -\psi_{1x}$. This means that ψ_0 represents the basic flow, in the absence of bed perturbation; ψ_1 represents the first order perturbation of this basic flow caused by the presence of the perturbation $z_b(x)$. Substitution in (3.58) yields the first-order linear differential equation

$$\psi_{0z}(\psi_{1xx} + \psi_{1zz})_x - \psi_{1x}\psi_{0zzz} = N(\psi_{1zzzz} + 2\psi_{1zzxx} + \psi_{1xxxx}). \quad (3.64)$$

After substituting $\psi_{0z} = u_0$, $\psi_1 = z_b(x)\chi(z)$ we get

$$\chi_{zzzz} - \left(2k^2 + \frac{ik}{N^*}\right)\chi_{zz} + \left(k^4 + \frac{ik}{N^*}(k^2 - \beta^2)\right)\chi = 0. \quad (3.65)$$

The parameters β and ν are defined as

$$\beta^2 = -u_{0zz}/u_0, \quad N^* = N/u_0.$$

These parameters depend on z, and therefore no analytical solution of (3.65) generally exists. However, we will assume that this z-dependence is only weak. This is equivalent to the assumption that the flow velocity in the outer flow water layer is almost constant or slowly varying with depth (β of order 1 or smaller). In the following, β and N^* will be taken constant.

Solving the linear flow equation

If we consider (3.65) as a linear differential equation with constant coefficients then the solution has the form:

$$\chi = c_1 e^{\kappa_1 z} + d_1 e^{-\kappa_1 z} + c_2 e^{\kappa_2 z} + d_2 e^{-\kappa_2 z}. \tag{3.66}$$

Substitution yields

$$\kappa_1^2 = k^2 - \beta^2, \quad \kappa_2^2 = \frac{ik}{N^*} + k^2 + \beta^2.$$

As $N^* \ll 1$ we have $|\kappa_2| \gg 1$, except for perturbations z_b of very small wavelength (much smaller than water depth).

The constants $c_1 - d_2$ can be found from the boundary conditions at first order in ϵ. We first eliminate η from the Bernouilly condition, to evaluate the surface boundary conditions; this yields $\eta_1 \approx -F^2 u_1$. The boundary condition (3.59), imposing the water surface as a streamline, then becomes

$$\psi_1 = F^2 \psi_{1z} \quad \text{at} \quad z = 0.$$

By substitution of (3.66) we get

$$c_1 + d_1 + c_2 + d_2 = F^2 [\kappa_1 (c_1 - d_1) + \kappa_2 (c_2 - d_2)]. \tag{3.67}$$

The boundary condition of zero momentum transfer through the water surface (3.60) becomes

$$\psi_{1zz} = 0 \quad \text{at} \quad z = 0,$$

i.e.,

$$\kappa_1^2 (c_1 + d_1) + \kappa_2^2 (c_2 + d_2) = 0. \tag{3.68}$$

Imposing the seabed as a streamline, the boundary condition (3.61) yields

$$\psi_1 = -z_b(x) \quad \text{at} \quad z = -1,$$

i.e.,

$$c_1 e^{-\kappa_1} + d_1 e^{\kappa_1} + c_2 e^{-\kappa_2} + d_2 e^{\kappa_2} = -1. \tag{3.69}$$

Origin of flow convergence at the dune crest

The boundary condition specifying the momentum dissipation at the bottom (3.62) requires an evaluation of the basic flow above the unperturbed bed level, at $z_{bottom} = -1 + \epsilon z_b$. We determine u_0 and u_{0z} above the unperturbed bed to first order in ϵ by means of a Taylor expansion of the flow velocity near the bottom. We may expect rapid convergence of this Taylor expansion because the velocity variation at the lower boundary of the outer flow layer is not very strong. We thus may write

$$u_0(z_{bottom}) \approx 1 + \epsilon u_{0z}(-1)z_b, \quad u_{0z}(z_{bottom}) \approx u_{0z}(-1) + \epsilon u_{0zz}(-1)z_b.$$

The boundary condition (3.62) now becomes

$$\zeta \approx u_{0z} + \epsilon \left(z_b u_{0zz} + \psi_{1zz} + \psi_{1xx}\right) = \zeta_0 u_b^2$$

$$\approx u_{0z}\left(1 + \epsilon(z_b u_{0z} + \psi_{1z})\right)^2 \quad \text{at} \quad z = -1. \tag{3.70}$$

It appears that, due to this boundary condition, the perturbed flow depends crucially on the vertical variation of the basic flow in the near-bottom zone (u_{0z}, u_{0zz}). The vertical variation of the basic flow implies an increase in flow velocity $u_0(z_{bottom})$ along the upstream flank of the perturbation and a related increase of vorticity dissipation ($\propto u_0^2(z_{bottom})$). The boundary condition (3.62) states that a balance must exist between transfer of vorticity to the bottom and vertical vorticity diffusion $N\zeta$ in the flow region above the upstream flank. This is realised by a flow adjustment u_1 which opposes the increase of flow velocity $u_0(z_{bottom})$ along the upstream flank and which enhances the velocity shear ζ. The velocity shear ζ and the related downward momentum transfer $N\zeta$ are more enhanced in the lower part of the outer layer than higher in the water column; the flow adjustment u, is driven by the ensuing momentum deficit in the lower part of the outer layer. The flow perturbation u_1 therefore produces in the region around the crest a downstream velocity decrease and a corresponding convergence of sediment transport.

Retaining only first order terms in ϵ in (3.70) and after substitution of (3.66) we find

$$(\kappa_1{}^2 - k^2)\left(c_1 e^{-\kappa_1} + d_1 e^{\kappa_1}\right) + (\kappa_2{}^2 - k^2)\left(c_2 e^{-\kappa_2} + d_2 e^{-\kappa_2}\right)$$

$$- 2u_{0z}\left[\kappa_1\left(c_1 e^{-\kappa_1} + d_1 e^{\kappa_1}\right) + \kappa_2\left(c_2 e^{-\kappa_2} + d_2 e^{-\kappa_2}\right)\right]$$

$$= 2u_{0z}{}^2 - u_{0zz} \quad \text{at} \quad z = -1. \tag{3.71}$$

Flow perturbation of the outer layer

The coefficients c_1, d_1, c_2, d_2 are easily determined from the equations (3.67–3.71), especially by noting that c_2, d_2 are much smaller than c_1, d_1, because $|\kappa_2| \gg 1$. Near the bottom we ignore $c_2 \exp(-\kappa_2)$ compared to $d_2 \exp(\kappa_2)$. The result reads, in the original dimensional variables

$$\begin{pmatrix} c_1 \\ d_1 \end{pmatrix} = \frac{1}{2} \frac{\pm 1 + \kappa_1 h F^2}{\sinh \kappa_1 h - \kappa_1 h F^2 \cosh \kappa_1 h}$$

$$d_2 = -2 \frac{iN}{kh^2 u_0} e^{-\kappa_2 h} \left(r_2 + r_1 \kappa_1 h \frac{\cosh \kappa_1 h - \kappa_1 h F^2 \sinh \kappa_1 h}{\sinh \kappa_1 h - \kappa_1 h F^2 \cosh \kappa_1 h} \right), \quad (3.72)$$

with $r_1 = h \frac{u_{0z}}{u_0}$, $r_2 = h^2 \left(\left(\frac{u_{0z}}{u_0} \right)^2 - \frac{1}{2} \left(\frac{u_{0zz}}{u_0} + \beta^2 \right) \right)$ at $z = -h$.

The perturbation of the near-bottom velocity to first order in ϵ, is given by

$$u(z_{bottom}) = u_{0b} + z_b u_{0z}(-h) + u_1(-h),$$
$$u_1(-h) = \psi_{1z}(-h) \approx u_{0b} z_b \left(\kappa_1 (c_1 e^{-\kappa_1 h} - d_1 e^{\kappa_1 h}) - \kappa_2 d_2 e^{\kappa_2 h} \right). \quad (3.73)$$

We first consider the case of small Froude numbers ($\kappa_1 h F^2 \ll 1$), which is most common in the tidal environment. We then find for the near-bottom velocity, see Fig. 3.23

$$u(z_{bottom}) = u_{0b} + u_{nolag} + u_{lag},$$
$$u_{nolag} = z_b \left(u_{0z}(-h) + u_{0b} \kappa_1 \coth \kappa_1 h \right),$$

$$u_{lag} = -2 \frac{z_b u_{0b}}{h^2} \sqrt{\frac{N}{k u_0}} e^{-i\pi/4} \left(r_2 + r_1 \kappa_1 h \coth \kappa_1 h \right). \quad (3.74)$$

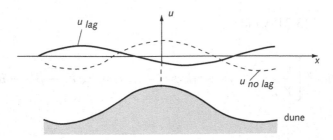

Fig. 3.23. The different contributions to the near-bed flow velocity over a dune. The contribution of u_{lag} is responsible for dune growth.

The term u_{nolag} indicates that the velocity is slightly increased at the crest of the dune perturbation. This term does not contribute to a phase shift between flow pattern and dune pattern and therefore it will not affect growth or dissipation of the perturbation. The term u_{lag} corresponds to a decrease of the velocity, which is greatest one eighth wavelength downstream of the dune crest. The model thus predicts a slowing down of the current velocity around the dune crest. Assuming that the sediment flux responds almost instantaneously to this flow deceleration then sediment will be deposited at the wave crest and the dune perturbation will grow. This corresponds to the situation shown in Fig. 3.21 and 3.20. The expression (3.74) also shows that u_{lag} scales with $u_{0b}z_b/h$ if it is assumed that the eddy viscosity N scales with hu_{0b} and the wavenumber k^{-1} with h.

Perturbation growth rate

Using expression (3.74) it is possible to derive an estimate of the wavelength of the fastest growing bottom perturbations. Therefore we evaluate bedload transport over the perturbation, using the transport formula (3.37)

$$q_b = \alpha u^n (1 - \gamma z_{bx}) \approx \alpha \gamma u_{0b}{}^n \left(\frac{n(u_{lag} + u_{nolag})_x}{\gamma u_{0b}} - z_{bxx} \right). \qquad (3.75)$$

The transport formula includes a term for down-slope transport due to gravity effects, which opposes dune growth. The growth rate is given by (3.127),

$$\Im\sigma = \Re[z_{bt}/z_b] = -\Re[q_x/z_b]. \qquad (3.76)$$

Substitution of (3.74) yields

$$\Im\sigma = \alpha \gamma u_{0b}{}^n \left[\frac{n}{\gamma} \sqrt{\frac{2Nk}{u_0}} (r_2 + r_1 h \sqrt{k^2 - \beta^2} \coth h \sqrt{k^2 - \beta^2}) - h^2 k^2 \right]. \qquad (3.77)$$

To evaluate this expression we introduce values for the constants N^*, β, r_1, r_2 and γ derived from the expressions for uniform steady flow in the outer flow layer (see Sec. 3.2.1). We take the lower boundary of the outer layer at δ_l as unperturbed bottom and for u_0 we use the expression (3.11), where z has to be replaced by $z - h$. This yields

$$\beta^2 = -\overline{(u_{0zz}/u_0)} = 6/49(h\delta_l)^{-1}, \quad N^* = \overline{(N/hu_0)} \approx c_D$$

$$u_{0z}/u_0 = (7\delta_l)^{-1}, \quad u_{0zz}/u_0 = -6(7\delta_l)^{-2} \quad \text{at } z = -h + \delta_l.$$

Substitution in (3.77) and neglecting terms of higher order in δ_l/h gives the expression

$$\Im\sigma = \alpha\gamma u_{0b}{}^n \frac{h^2}{\delta_l^2} \left(\sqrt{\frac{0.013n^2 c_D h k'}{\delta_l \gamma^2}} - k'^2 \right), \quad \text{with} \quad k' = \delta_l k. \tag{3.78}$$

The qualitative behaviour of the growth rate $\Im\sigma$ as a function of wavenumber k' is shown in Fig. 3.24.

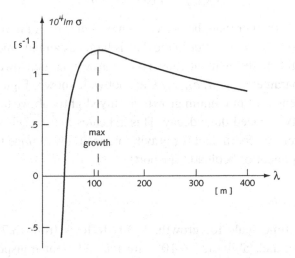

Fig. 3.24. Qualitative behaviour of the growth rate $\Im\sigma$ (3.78) as a function of the wavelength $\lambda = 2\pi\delta_l/k'$, using the values $h = 10$ m, $\delta_l = 1$m, $c_D = 0.0025$, $\gamma = 1$ and $n = 3$. For very large wavelengths the growth rate decreases to zero, because the phase shift between flow and bed perturbation becomes very small; for small wavelengths the growth rate is negative because of down-slope gravity transport. The wavelength corresponding to the largest growth rate is the most likely wavelength of naturally occurring dune patterns.

Dune wavelength

The maximum growth rate occurs for perturbations of wavelength

$$\lambda = \frac{2\pi\delta_l}{k'} \approx 67 \left(\frac{\gamma^2}{n^2 c_D} \right)^{1/3} \left(\frac{\delta_l}{h} \right)^{1/3} \delta_l. \qquad (3.79)$$

The wavelength of the fastest growing perturbations is approximately proportional to the thickness δ_l of the logarithmic layer. This reflects the close relationship between dune development and the turbulence-induced vertical profile of the flow velocity. It is a mutual dependency, because of the form drag exerted on the flow by ripples and dunes. In steady flow the thickness of the logarithmic layer is proportional to the water depth h; this is the only length scale for steady flow in the outer layer, if it is assumed that the eddy viscosity N scales with $h u_0$. If we choose the down-slope transport parameter $\gamma = 1$, the friction coefficient $c_D \approx 0.0025$, the sediment transport exponent $n = 3$ and the thickness of the logarithmic layer $\delta_l \approx 0.1\,h$, then we find the model estimate

$$\lambda_{mod} \approx 110\delta_l \approx 11\,h. \qquad (3.80)$$

This is somewhat larger than the wavelength of dunes found in rivers and tidal flows, $\lambda_{obs} \approx 6\,h$, but the scatter in the data is large, as already mentioned. The uncertainty in the model estimate (3.80) is probably not much lower; the precise values of the parameters γ, n, c_D, δ_l/h are not well known. Figure 3.24 shows that the wavelength of maximum growth is only slightly above the wavelength range of gravity-induced dune decay. This indicates that λ_{mod} is very sensitive to the parameters γ (coefficient for gravity-induced down-slope transport) and n (velocity exponent of bedload transport).

Growth time scale

The e-folding time scale for growth, $1/\Re\sigma$, follows from (3.78). Choosing $u_{0b} = 0.5\,\text{m/s}$ and taking $\alpha = 10^{-4}$ for the sediment transport parameter (3.34), we find a time scale of a few hours; growth from an almost flat bottom will take at least several times longer. It should be remembered, however, that the model only refers to the initial growth of dunes. There is no *a priori* reason that the initial wavelength will be identical to the wavelength of fully developed dunes. The structure of fully developed dunes also appears to be much more complex than the simple 2D-structure assumed in the model. An analysis of dune growth for finite dune amplitudes indicates that the results of the initial, linear stability analysis remain qualitatively valid. The growth rates predicted in the weakly nonlinear case are slightly smaller than for the linear case and the wavelength of maximum growth is slightly larger, while the sensitivity to gravity-induced down-slope transport is lower [241].

Migration rate

The migration rate $\Re\sigma/k$ follows from

$$\Re\sigma/k = -\Im(z_{bt}/kz_b) = \Im(q_x/z_b). \qquad (3.81)$$

The contribution of u_{nolag} to migration is larger than the contribution of u_{lag}. Substitution of (3.74) yields

$$\Re\sigma/k = \alpha n u_{0b}^{n-1} u_{nolag}/z_b. \qquad (3.82)$$

If we consider a depth of 10 m and a near bottom velocity of 0.5 m/s and using the same parameter values as before, we find a downstream dune migration rate on the order of 1 m a day.

Dune genesis in tidal flow

First we consider the hypothetical case that tidal acceleration and deceleration do not strongly affect the vertical flow profile. Then ebb flow and flood flow can be considered as successive periods of stationary flow, with reversed flow direction. Dune growth occurs in both periods, similar as for steady flow without flow reversal, and the wavelength of strongest growth is given by the same expressions (3.79) and (3.80). An essential difference between unidirectional flow and tidal flow is the migration speed of dunes; in the former case migration is much faster than in the latter case. For a symmetric tidal wave and in the absence of a steady current the tidally averaged structure of the flow perturbation u_1 will be symmetric around the dune crest. If we use (3.74) for the average near-bottom flow perturbation during flood ($u_0 > 0$) and ebb ($u_0 < 0$), we find, after taking the tidal average:

$$\langle u(z_{bottom}) \rangle = -u_{circ} \sin kx, \qquad (3.83)$$

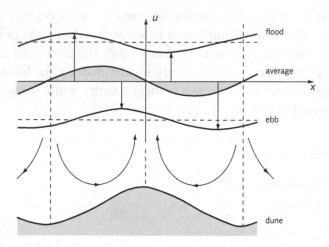

Fig. 3.25. Near-bed velocity over a dune in a tidal channel. Upper part: Rectification of the flood and ebb velocities by the dune-delayed flow contribution u_{lag} and tide-averaged flow rectification. Lower part: Corresponding vertical tide-averaged circulation pattern.

with

$$u_{circ} = \epsilon u_{0b} \sqrt{\frac{2N}{h^2 k u_0}} \, (r_2 + r_1 \kappa_1 h \coth \kappa_1 h) \, .$$

The expression (3.83) shows that the average flow pattern in the tidal case corresponds to a series of circulation cells with near-bottom flow directed towards the dune crests (see Fig. 3.25).

The above description is too simplistic, however. Tidal acceleration and deceleration has a significant influence on the vertical velocity profile. The time scale T_u for adaptation of the flow profile depends on the eddy viscosity N and the depth h and can be approximated by $T_u \approx h^2/N$. Using for N the estimate (3.18) we find $T_u \approx h/c_D \bar{u}$. The ratio β of the frictional adaptation time scale T_u and the tidal time scale $\omega^{-1} = T/2\pi$ is thus given by $\beta \equiv T_u \omega = h\omega/c_D u$. For $c_D = 0.0025$, $\bar{u} = 1 \, \text{m/s}$ and $h = 20 \, \text{m}$ the value of β is close to 1. Therefore, during most of the tidal period the vertical distribution of shear stress will be different from the equivalent steady flow. The tidal variation of the shear stress interferes with the spatial variation related to flow acceleration and flow deceleration over the bottom perturbation. On the continental shelf (great depth, $\beta \geq 1$) this effect will be more significant than in tidal basins (small to moderate depth, $\beta \ll 1$).

An Engelund–Fredsøe type model can also be formulated for analysing seabed instability in the tidal case, but the mathematical treatment is more complex. The results of numerical and analytical investigations of seabed instability under tidal conditions show many similarities with the steady flow case [217, 169]. The tidally averaged flow distribution over the bed waves exhibits the same structure of residual eddies, with near-bottom flow directed towards the dune crests (Fig. 3.25). The fastest growing wavelength depends linearly on the thickness of the logarithmic layer and the proportionality constant is also of the same order as in (3.80). This preferred dune wavelength depends only weakly on the tidal excursion $L = UT/\pi$ (U is the maximum tidal flow velocity) and on the ratio β of frictional adaptation time scale and tidal period.

There is no clear evidence that the logarithmic boundary layer thickness δ_l in tidal flow is proportional to the water depth h. For water depths of 20 m or more it might be related instead to the thickness δ_t of the tidal boundary layer (3.10). In this case the wavelength of strongest growth does not depend on depth. This would be in agreement with dune observations in tidal environments discussed earlier, see for example [331] and Fig. 3.19.

Dune development in the sea is also influenced by waves; it appears that intense wave action tends to inhibit dune development [306]. Fredsøe and Deigaard [157] show that this is due to suspension lag effects, which are particularly important in the case of large tidal flow velocities.

Suspended-load transport

By using (3.75) as the sand transport formula we have implicitly assumed that dune formation proceeds by bedload transport. Observations show that dunes do not grow in very fine sand ($d < 130\mu$) [4]. This may be explained by considering the much stronger delay of the suspended-load response to variations in flow strength, compared to the bedload response. Although the near-bed velocity reaches a maximum upstream of the dune crest, it may happen that the suspended load reaches a maximum downstream, in particular if the flow velocity is high. In that case sediment transport is still increasing when the flow passes the dune crest, thus eroding sediment from the crest. Therefore dune development is inhibited in fine sediment. It should be noted that, according to flume experiments [420], dune height also decreases in very coarse sediment.

3.3.4. *Formation of Bottom Ripples*

Height and wavelength of bottom ripples

The process of ripple formation on a sandy sediment bed starts as soon as the current strength is sufficient for setting sediment grains in motion; the critical velocity for initiation of grain motion is 10–20 cm/s for a cohesionless sediment bed, see Fig. 3.12. Bottom ripples appear on an initially flat seabed as soon as sediment grains start moving. Flume experiments indicate that the ripple wavelength λ is hardly influenced by the shear velocity u_*, but that it mainly depends on the median grain diameter d of the bed sediment according to the relation [501]

$$\lambda \approx (500 - 1000)\, d. \tag{3.84}$$

Flume experiments and field observations also show that ripple formation does not occur in sediment with a grain size larger than about 1000 μm (very coarse sand) or a grain size finer than 100–200 μm. Shortly after emergence the ripple pattern exhibits great regularity with almost straight crests perpendicular to the flow. Ripple heights are consistently observed in the range 10–20 mm, with an average of 15 mm and almost no dependency on grain diameter [370]. When the water depth is less than about one metre or when the flow velocity is below 40–50 cm no seabed structures other than bottom ripples are formed. Dunes appear at greater depth and higher flow velocity; dunes have much greater height and wavelength than the bottom ripples, which are superimposed on the dunes. When the velocity is increased above a value on the order of 1 m/s the ripples disappear and the bed surface layer is entrained by the near bottom flow as a sand sheet; this condition is called 'sheet-flow'.

Ripple formation in coastal environments is often influenced by wave action; waves modify the ripple shape and the thresholds for generation and decay. For instance, field observations on Sable Island Bank (Nova Scotian Shelf, depth 30 m) showed the following sequence of seabed forms during a storm event [278]: (1) relict wave-dominant ripples with worm tubes and animal tracks formed during the preceding fairweather period; (2) irregular, sinuous, asymmetrical current-dominant and intermediate wave-current ripples under bed-load transport; (3) regular, nearly straight or sinuous asymmetrical to slightly asymmetrical wave-dominant ripples under saltation/suspension; (4) upper-plane bed under sheet-flow conditions; (5) small, crest-reversing, transitory ripples at the peak of the storm; and (6) large-scale lunate megaripples which developed when the storm decayed.

Mechanisms for ripple formation

Several processes contribute to the genesis of ripple patterns. When sediment grains start moving, small scattered sediment pileups will first appear on the seabed; these pileups grow by trapping other moving sediment grains [81]. The random distribution of the pileups suggests that they originate from turbulent bursting processes; this is consistent with the observation that ripples are not formed in laminar flow [491, 369]. When the pileups reach the roughness height their spacing becomes more regular; the term sand wavelet is used to designate this initial regular pattern [81]. The height of the sand wavelets is much smaller than the final ripple height and so is the wavelength. The observed wavelengths can be represented by the relationship

$$\lambda \approx 250\,d. \tag{3.85}$$

Upon further growth the spacing of the wavelets increases, probably as a result of coalescence. Small bed features migrate with a speed inversely proportional to their height; differences in wavelet height yield different migration speeds, which lead to merging of wavelets [369]. Finally, a ripple pattern develops with an average wavelength given by (3.84). One may conclude that the final ripple pattern is produced by mutual interaction between ripples, and therefore cannot be described adequately by the initial, linear instability mechanism.

There are several hypotheses about the regularity of the ripple spacing. Observations indicate that new sediment pileups are produced downstream of existing pileups. Flow separation in the lee of these small sediment mounds may produce a vortex which causes erosion downstream of the initial mound and formation of a following mound. In this way a train of bottom irregularities is built up from which a ripple pattern may subsequently develop [419]. This process classifies ripple formation as a forced seabed instability rather than a free seabed instability.

Ripple formation as an inherent seabed instability

The initial development of a wavelet pattern may also be attributed to an unforced instability, similar to the initial development of dunes. Such an

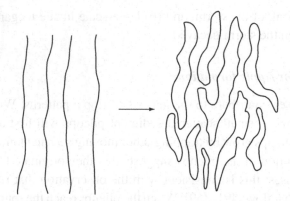

Fig. 3.26. Current ripples evolve from an initial straight-crested sinusoidal pattern to a complex three-dimensional pattern.

assumption is supported by other similarities between ripples and dunes, such as initial shape and migration. Dune formation is related to the response of the outer flow layer to perturbation of the sediment bed, while ripple formation may be related to the response of the near-bottom flow to seabed perturbation. This would explain why dune dimensions are related to the flow depth, while ripple dimensions are not. Observations show that streamlines in the outer flow layer follow the dune bed waves, but not the ripple bed waves. The flow separates at the ripple crest and the trough between the ripple crests is 'filled' with a lee vortex [370], see Fig. 3.27. In contrast, the flow in the roughness layer strongly responds to the presence of ripple bed waves. In the Engelund–Fredsøe model the wavelength of the fastest growing dune instabilities scales with the thickness of the perturbed flow layer. By analogy we may expect initiation of bed waves with fastest growing wavelength proportional to the thickness of the roughness layer. As the layer thickness δ_r is related to the grain diameter, the ripple wavelength should also exhibit a dependency on grain diameter (3.84), as found in flume experiments.

Perturbation of the near-bottom flow layer

For the stability problem of the outer layer a simple analytical solution can be found by assuming in the unperturbed state a flow velocity with little depth variation. This may be a reasonable assumption for the outer layer, but for the near-bottom flow layers, where the velocity first has a linear and then a log-arithmic shape, this is certainly not the case. A solution can be obtained by a

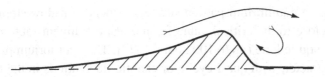

Fig. 3.27. Flow separation and vortex generation at the lee of a ripple. The outer flow layer does not respond to individual ripples, because it experiences the trough between the ripples as being filled with a vortex.

semi-numerical method, as shown by Richards [375], who used a model similar to that of Engelund–Fredsøe, with a more refined description of turbulence, including advection, diffusion and dissipation processes. This investigation showed that in addition to the dune instability, a second instability exists, corresponding to much smaller bedforms with a wavelength related to the height δ_r of the roughness layer. The fastest growing wavelength of these small scale instabilities corresponds well with the wavelength observed for sand wavelets (3.85) [81]. For water depths smaller than about one metre the dune instability disappears and only the ripple instability remains.

Similarity between dunes and ripples

The analysis by Richards [375] supports the hypothesis that a similar instability mechanism is responsible for the initiation of ripples and dunes. Both bedforms induce an asymmetry in the flow pattern between the upstream and downstream flanks of the bedform. This asymmetry is caused by frictional delay of the flow response to acceleration/deceleration over the ripple field and, in particular, the increase in response delay with distance from the bottom. The asymmetry produces a gradient in velocity and in sediment transport at the ripple crest. In other words, the flow response has a spatial phase shift relative to the ripple field. Therefore the ripples will not only migrate with the flow, but they will also grow or decay. Close to the bottom, the flow reaches its maximum value before arriving at the ripple crest. The flow deceleration at the crest and the corresponding decrease of sediment transport then amplifies the initial ripple bed wave.

Fully grown ripples and dunes

The ripple pattern loses its initial regularity in the course of time; the crest lines first become sinuous and later linguoid, see Fig. 3.26. The time scale for reaching an equilibrium pattern strongly depends on flow strength; if the velocity is close to the critical value for incipient sediment motion, the time scale may be quite long (up to 100 hours) [370]. The equilibrium pattern has a dynamic character, with quasi-cyclic or chaotic fluctuations in the ripple configuration. However, the average wavelength and height remain approximately unchanged [193]. The previous linear stability models only describe the initial stage of pattern formation. Ripple steepness increases very soon after initial formation, causing flow separation and vortex shedding at the lee of the ripple. In an early phase of ripple formation, flow separation yields a positive feedback to ripple growth. The spatial coherence of the vortices is weak, however; a vortex which is initially uniform along the crest line, is easily broken up by small disturbances. The vortex structure becomes three-dimensional and perturbs the linearity of the ripple crest [370]. Fully grown current ripples generally exhibit a complex pattern, with linguoid, cuspate or honeycomb structures, see Fig. 3.28. Crest lines are at varying angles to the main flow direction and only the average trend of the crests lines is perpendicular to the flow, see Fig. 3.26. Understanding this behaviour requires modelling of the interaction between

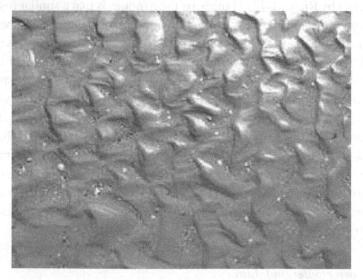

Fig. 3.28. Linguoid ripple pattern, formed in a rip channel on the beach which has fallen dry at low water. The gentle slopes at the right (stoss-side) and the steep slopes at the left (leeside) indicate that the flow was coming from the right.

flow over seabed topography and the structure of turbulent motion. The characteristics of fully grown ripples and dunes has been explored extensively in laboratory flumes. A theoretical interpretation of the main characteristics of the equilibrium shape of ripples and dunes is due to Fredsøe and Deigaard [157].

Formation of megaripples

Megaripples are bottom structures intermediate between ripples and dunes; their wavelength is of the order of a few metres, without strong correlation with water depth. They develop under the same conditions as dunes and ripples and often occur simultaneously. This suggests that their formation mechanism might be related to the formation mechanism of ripples and dunes. When ripples are formed on the sediment bed, the scale of turbulent fluid motion near the bottom is influenced not only by the scale of sediment grains, but also by the wake produced in the lee of bottom ripples. The thickness of the roughness layer is greatly increased by the presence of bottom ripples. The instability mechanism related to frictional flow delay in the near-bottom layer produces bed waves with a wavelength proportional to the thickness of the roughness layer. The increased thickness of this layer due to bottom ripples will therefore trigger the development of bed waves with a greater wavelength than bottom ripples. This wavelength is not directly related to water depth, as is the case for dunes. It has been suggested by Richards [375] that seabed instabilities produced in this way correspond to megaripples.

Megaripples are found both in current-dominated and in wave-dominated environments [167]. In the latter case the presence of wave ripples also produces an increase of the thickness of the near-bottom roughness layer; the thickness of this layer is given by (3.20). Observations in the surf zone of a tidal beach (depth between 0.3 and 1.8 m) show that the formation of megaripples starts at small wavelengths of typically one metre or less [76]. In a few days they grow to bed waves with wavelengths of several metres and finally they reach a size typical of dunes. It seems that in this case dune formation is due to mutual interaction and merging of smaller bedforms, which is basically different from the linear instability mechanism described earlier in this section. The same observations also show that the formation of megaripples is inhibited when flow conditions are strongly variable [76]. This suggests that the wavelength of the fastest growing megaripples depends on flow strength, unlike dunes and ripples.

3.3.5. Genesis of Antidunes

Supercritical flow

When the flow velocity in a flume with sandy bed is gradually increased, bottom ripples become very irregular and then disappear; the upper bed layer starts behaving like a fluid sand sheet which is entrained by the near bottom flow. Sheet-flow conditions appear when the Shields parameter exceeds the value (3.35) $\vartheta \approx 0.5$; this corresponds to flow velocities close to 1 m/s for medium sand. When the flow velocity is still further increased the sand sheet becomes unstable; the sheet layer starts developing bed waves, which tend to move against the flow direction. These bed waves are called antidunes. The water surface is undulated in a similar way as the bed wave, but the water surface amplitude is larger: The water depth is greater at the dune crest than at the dune trough. This happens when the flow becomes near-critical or supercritical; the Froude number $F = u/\sqrt{gh}$ is close to or greater than 1. The surface wave is also unstable and may break; when it breaks the bed-wave is also destroyed, but will soon reappear. Conditions of near-critical or supercritical flow correspond to fast flowing water in very small water depths. Such conditions are rather exceptional in natural coastal environments, but they sometimes occur in gullies on tidal flats and on the beach. For example, antidunes may be observed in the backwash of breaking waves or in the outflow of beach runnels during ebb, see Fig. 3.29.

Difference with subcritical flow

The flow response to an initial bed perturbation is totally different for near-critical conditions then for subcritical conditions. Water is piled up at the dune crest, in contrast to the water level dip occurring in subcritical flow; the flow is not accelerated at the upstream dune flank, but decelerated. The reason is that acceleration of the flow beyond its critical value requires an increase of water surface inclination; however, altering the water level inclination requires upstream propagation of water level change, which cannot occur in critical flow. As a result, the flow velocity is lowest at the crest of the perturbation and highest at the trough; the phase of the longitudinal velocity distribution is, to a first approximation, at 180° to the bottom perturbation.

Suspended-load transport

The impact of a bed perturbation on the flow distribution is described by the Engelund–Fredsøe model, Eqs. (3.58–3.60). We now neglect the

Fig. 3.29. Antidunes appearing in near-critical ebb flow from a beach runnel. The surface-wave amplitude is greater than the bed-wave amplitude. The antidunes are formed in fluidised sand and are highly unstable. Photograph by U. Mooss, reproduced with permission of the Wadden-Vereniging.

friction-related flow perturbation which is on the order of $\sqrt{N/hu_0} \ll 1$; the result is, to first order in ϵ,

$$u(z) = u_0\left(1 + \frac{\cosh kz + hkF^2\sinh kz}{\sinh kh - hkF^2\cosh kh}kz_b\right). \qquad (3.86)$$

For values of the Froude number in the range

$$\sqrt{\tanh(kh)/kh} < F < \sqrt{\coth(kh)/kh}$$

the phase of the flow velocity is exactly opposite to the phase of the bed perturbation z_b. If we assume bedload transport, then the sediment flux and the near-bed velocity have an identical phase; the bedload transport is thus at its minimum

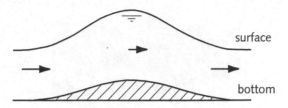

Fig. 3.30. The general pattern of near-critical flow over an antidune. Water depth is greater at the crest than at the trough and the opposite holds for the flow velocity.

right at the crest of the perturbation. Therefore no accretion or erosion of the crest will take place; bed perturbations cannot grow through bedload transport. For the high flow velocities under consideration, the assumption of bedload transport is questionable, however; the flow is strong enough to bring sediment particles in suspension. Suspension transport introduces a delay between transport rate and flow velocity, because deposition and resuspension take some time. Hence, the suspended sediment flux at the crest of the perturbation is not yet at its minimum; this minimum will be reached after the crest. Sediment transport decreases at the crest and the crest will therefore accrete (Fig. 3.30). Bed perturbations may thus grow in supercritical flow because of the transport delay related to deposition and resuspension. Most of the upstream flank of the antidune experiences a sediment-flux decrease, while most of the downstream flank experiences a sediment-flux increase. This implies that the upstream flank accretes while the downstream flank erodes; the perturbation will thus move upstream. If the phase shift of sediment deposition and resuspension is very large the zones of erosion and accretion may be shifted so far that the upstream flank is eroded more than the downstream flank; in that case the antidunes will move downstream.

3.3.6. *Wave-Induced Ripples*

In the coastal environment bed ripples are generated not only by currents, but also by waves. Wave ripples are formed when wave-orbital velocities are not too high, typically below 0.5 m/s, for a fine to medium sandy seabed. Often two distinct regimes of wave ripples are observed [52, 278, 190, 444]: (1) small ripples (also called anorbital ripples) with a height of typically one or two centimetres and a wavelength in the range given by (3.84) and (2) large ripples with a height of typically 5 cm and a wavelength comparable to or somewhat

Fig. 3.31. Typical wave-ripple pattern on the beach. Height and spacing are greater than for current ripples and the pattern is more regular. The ripples are symmetric with rounded crest, due to symmetric wave-orbital motion.

larger than the wave-orbital amplitude [259],

$$\lambda \approx (1-2)U_1/\omega \tag{3.87}$$

where U_1 is the amplitude of the wave-orbital velocity and ω is the wave radial frequency. Large wave ripples generally occur in coarser sediment and at higher wave-orbital velocities than small wave ripples. Wave ripples with characteristics intermediate between (1) and (2) are also sometimes observed. Compared to current ripples, wave ripples have a more symmetric shape, the crest line exhibits greater continuity and the overall ripple pattern has greater regularity, see Fig. 3.31. According to laboratory experiments and field observations, two- and three-dimensional ripple patterns both occur.

Difference with current-induced ripples

Wave-induced ripples differ from current-induced ripples in at least two aspects:

- The turbulence characteristics of wave-induced flow are very different from steady flow. Wave-induced flow can be considered frictionless except in a thin layer at the bottom (Stokes layer), of thickness

$$\delta_w \approx \sqrt{2N/\omega}, \tag{3.88}$$

where N is the generalised viscosity (in case of a smooth bottom $N = \nu$, the kinematic viscosity, see (3.19)).
- Wave ripples do not migrate, or migrate only a little in the case of asymmetric waves. There are no substantial differences in migration rate along

the crest line, as for current ripples. This is probably the main reason for the greater symmetry and regularity of wave-induced ripples compared to current-induced ripples.

Wave ripple formation

Wave ripple formation can be triggered by an instability mechanism similar to the mechanism for current ripples. The near-bottom orbital flow profile does not adjust instantaneously to a bottom perturbation throughout the turbulent boundary layer, but at the top of this layer it is delayed compared to the bottom; this results in a spatial shift between the flow perturbation and the bed perturbation. Averaged over the wave period, residual eddies are generated in the boundary layer which direct the near-bed flow to the ripple crests; the picture is qualitatively the same as for tidal flow over dunes (see Fig. 3.25) [409]. For current-induced ripples, the balance of this growth mechanism with down-slope transport selects a wavelength of fastest growth; this wavelength is mainly determined by the thickness of the roughness layer. For wave-orbital flow the ripple wavelength of fastest growth is selected by a similar balance, if we assume that settling and erosion time lags of bedload can be ignored with respect to the time scale of ripple-induced flow acceleration and deceleration [43].

Small-scale wave ripples have a height and wavelength which are similar to those of current ripples. The wavelengths in the small-scale ripple regime for a medium to coarse grained seabed are substantially larger than the wavelengths found for medium to fine sand [444, 190]. This suggests that small-scale wave ripples are related to a near-bottom boundary layer with thickness determined by grain roughness, in the same way as current ripples.

For sufficiently strong wave-orbital motion a turbulent wave boundary layer may exist on top of the grain roughness layer, with thickness given by (3.88). The turbulent eddy viscosity N in the wave boundary layer scales with the layer thickness δ_w and the maximum wave-orbital velocity U_1 (3.18); it follows that the thickness δ_w of the wave boundary is proportional to the wave-orbital excursion $2U_1/\omega$. This may explain the existence of the large-scale ripple regime with typical wavelengths proportional to the wave-orbital excursion.

Rolling grain ripples

The selection of a wavelength of fastest growth proportional to the wave-orbital excursion does not necessarily depend on down-slope gravity-induced transport. The selection also follows from the assumption that sediment particles in the rolling-grain regime are in continuous motion [409]. If the ripple spacing is much smaller than the wave-orbital excursion, sediment grains are entrained back and forth during the wave cycle over more than one residual eddy; the net displacement will be smaller than for a larger ripple spacing, where sediment grains move back and forth in a single residual eddy. Ripple growth will be strongest in the latter case; the preferred ripple wavelength is then of the same order of magnitude as the wave-orbital excursion.

Vortex ripples

Soon after initial formation the mechanism of ripple formation changes. Flow separation occurs at the ripple crest; vortices are generated in the ripple lee, which stimulate sediment motion towards the crest. These vortices are also a very effective mechanism for bringing sediment in suspension. In steady flow, sediment is entrained as bedload over the gently sloping stoss side of each ripple; the vortex generated at the steep lee side remains attached to the ripple crest or will be swept to the next ripple. In wave orbital flow, vortices are generated at both sides of the ripple, depending on the instantaneous flow direction. These vortices are swept back over the ripple crest when the flow changes direction. The sediment transport process is therefore basically different for current ripples and wave ripples. The similarity between current ripples and wave ripples therefore relates mainly to the initial stage of ripple formation, when vortex shedding has not yet developed. When the wave orbital velocities are increased above a certain threshold, the ripples disappear and sheet flow conditions establish. This occurs when the Shields parameter ϑ is close to 1.

3.4. Sandbanks

3.4.1. *Qualitative Description*

Ripples and dunes belong to the small and medium scale structures of the seabed. There are also much larger bedforms: Sandbanks. They did not appear in the previous analysis of bed-flow interaction. Their large dimensions suggest that sandbank dynamics is probably more related to the horizontal flow structure than to the vertical structure. This hypothesis will be examined in this section, using similar considerations as before.

Sandbank categories

Different types of sandbanks may be distinguished [124]:

- Tidal ridges, also called tidal sandbanks. Their length is in the order of several tens of kilometres, their width in the order of several kilometres and their height on the order of tens of metres; the crest line is almost straight and makes a small cyclonic angle to the main flow direction. Tidal ridges migrate very slowly; the profile is asymmetric, with the steepest slope at the lee side (down the propagation direction). Sediment is coarsest around the crest and finest around the troughs (swales). Tidal ridges occur on the open shelf in areas where tidal currents are strong. They sometimes sit on a coarse gravel basement. In wide tidal embayments different flood and ebb dominated channels are often separated by elongate bars; these elongate bars resemble offshore tidal ridges and may result from similar morphodynamic processes [299].
- Shoreface-connected ridges. Scales are comparable to tidal ridges, the height is often somewhat lower. They migrate along the coast and the crest line follows a downdrift shoreward direction (shoreface attached). The downdrift seaward flank has a steeper slope than the landward flank; sediment is coarsest around the trough and finest on the seaward flank. Shoreface-connected ridges occur on the inner shelf, outside the breaker zone.
- Ebb-tidal deltas. Systems of sandbanks and channels at the seaward side of tidal inlets. Their size and morphology is mainly related to the tidal flow through and around the inlet, but wave-driven longshore transport also plays an important role.
- Headland-associated banks. Elongated banks of a few kilometres long, formed in the vicinity of headlands, on fairly steep sloping coasts.

Forced and free seabed patterns

The occurrence and the scale of the last two categories of sandbanks are linked directly to existing local coastal features; in that respect they differ from the first two categories, which are not directly related to local topography. The latter two sandbank categories may be considered as forced seabed patterns, while the former two are more characteristic of free patterns. Another distinctive

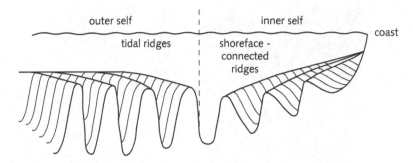

Fig. 3.32. Schematic cross-shore section of the inner and outer shelf with shoreface-connected ridges and tidal ridges. The vertical scale is strongly exaggerated relative to the horizontal scale. The shoreface-connected ridges are situated on the gently sloping inner shelf and curve towards the coast in downdrift direction. The tidal ridges are located on the outer shelf and oriented cyclonically with respect to the main tidal flow axis.

feature of the first two sandbank categories is their great length compared to their width; therefore they are sometimes called linear sandbanks or linear ridges (see Fig. 3.32). In this section we investigate the mechanisms through which these ridges may arise as large-scale free instabilities of the seabed of the inner and mid shelf. Some aspects of ebb-tidal deltas are discussed in the chapter on tide-topography interaction. Beside sandbanks, other large-scale topographic structures exist on the seabed such as isolated mounds or throughs; such structures are related to past geological events, for instance, ice scour or earth crust motions, and will not be considered in the following.

Dynamic equilibrium

Linear sandbanks as described above are found on many continental shelves, for instance, in the southern North Sea, see Fig. 3.33. They satisfy the following conditions:

- they occur in more or less regular patterns,
- they are predominantly composed of the same type of sediments as neighbouring shelf zones,
- they occur in dynamic sedimentary areas, characterised, for instance, by strong currents and cross-flow oriented ripple and dune cover.

These criteria can be considered as an indication that these sandbanks are in dynamic equilibrium with the existing flow and sand transport regimes.

Fig. 3.33. Bathymetric map of the Dutch part of the North Sea (Southern Bight) based on detailed bottom sounding; dark is deep and clear is shallow. Depths range between 10 m (close to the coast) to about 40 m offshore. The North Sea basin has been subsiding for several million years at an average rate on the order of a cm per century [504]. In this period thick sediment layers have been deposited of mainly medium grained sand. The overall topography exhibits an East–West asymmetry, with a smaller depth along the Dutch coast than along the UK coast. The sediment composition has a South–North gradient: Medium-grained sand towards the south and fine-grained sediments towards the north. In the central part of the Southern Bight large ridge-form sandbanks can been seen, with length in the range 25–100 km, width in the range 5–10 km and height in the range 5–25 m. The ridge orientation is generally South–North; the crests make a small cyclonic angle to the major tidal current axis. Close to the coast, ridges of similar dimensions can be observed which have an opposite orientation and which are oriented towards the coast; these are called shoreface-connected ridges. Most of the southern part of the Bight is covered with dunes, which are oriented approximately perpendicular to the main tidal current axis. They have a similar height as the tidal ridges but the wavelength is much smaller (approximately a few hundred metres), see Figure 3.18.

Field observations

The first investigation of ridge fields was made by T. Off in the early 1960s. [336]. Many investigations have followed since, revealing ridge fields as a common characteristic of sandy shelf seas. Sandbank fields have been observed, in particular, on the Atlantic inner shelf of North America [430, 174] and South America [149, 343] the North Sea inner and mid shelf [212, 267, 11] and the Yellow Sea [285, 286]. Similar sandbank fields are found on the same shelves in deeper water, where current-induced sand transport is too weak for maintaining sandbank morphology [33, 174], for instance in the northern North Sea. These relict sandbank fields are in a regressive stage, although there is still evidence of active ripple and dune formation. Seismic soundings in the East China Sea have revealed the existence of extensive sandbank fields buried under Holocene deposits [34].

Relict or modern?

Sandbanks probably developed already at the earliest stages of sea-level rise, but there is still uncertainty about their origin. Their initial formation may be related to the reworking of relict shoreline deposits during the Holocene transgression, since such deposits are found in the core of many present sandbanks [212]. However, the orientation of present sandbanks often does not match ancient shorelines; the evolution of sandbanks far off the present shoreline may have become independent of the initial stage. In the following we will further examine this point of view. It will be shown that sandbank development can be explained by morphodynamic feedback, which does not require any relict structure as an initial trigger. This does not preclude that sandbanks may actually originate from relict deposits; it only means that the same sandbanks could have developed from a flat seabed and that the morphology of mature sandbanks does not necessarily bear a causal relationship to preexisting seabed structures.

3.4.2. Tidal Sand Ridges

Seabed instability

Tidal ridges are immense structures on the shelf sea bottom. A single tidal ridge may contain as much as several billion cubic metres of sand. One may wonder by which forces these submarine giants have been shaped. No geological events have been identified that provide a convincing explanation for their appearance on the outer shelf. But is it conceivable that a structure so huge can generate itself out of a flat bottom, without specific external influences? J. Huthnance [227] was the first to advance such a daring hypothesis and to provide a theoretical underpinning for tidal ridge formation as the result of seabed instability. An important indication for the dynamical origin is the observation that the ridges are always inclined relative to the main flow direction [212, 430, 149, 343, 345, 331, 478, 66]. This implies that the flood and ebb flow have to cross the ridge, thereby creating different flow regimes on the upstream and downstream flanks. The importance of upstream-downstream flow asymmetry for bedform generation has been demonstrated for ripples and dunes. In the case of tidal ridges the asymmetry appears in a different way, but the same factors play a role: Momentum conservation (or its equivalent: Vorticity conservation) and bottom friction. Flow perturbation by tidal ridges is also influenced by earth's rotation, because of the large spatial scale of the perturbation. For that reason, flow perturbations corresponding to cyclonic and anticyclonic orientations of the ridge relative to the main flow direction are not symmetric. In the following we will first give a qualitative description of the process of tidal ridge formation; then we will discuss two more mathematical approaches proposed by Zimmerman [510] and Huthnance [227].

Flow over a tidal ridge

We consider a shallow ridge, at an angle to the undisturbed tidal flow, and examine the path followed by a fluid parcel when crossing the ridge. We will see that it makes no difference whether we choose the flood or the ebb tidal phase; the process of ridge growth is symmetric with respect to the flow direction. The path followed by the fluid parcel is shown in Fig. 3.34. In fact, two paths are shown. The dashed path would be followed by the fluid parcel if we neglected frictional delay. In that case the streamline depends on the following three factors:

- Continuity of fluid discharge across the ridge. As the water depth at the ridge crest is smaller than at the trough, the cross-ridge velocity component increases at the uphill slope and decreases at the downhill slope.
- Bottom-friction torque. When a fluid parcel crosses the ridge, its ridge-parallel velocity component will decrease in response to increased bottom friction. Due to the ridge inclination a cross-flow array of fluid parcels will

not decelerate simultaneously; the parcels closest to the ridge will decelerate first. The fluid array will thus rotate towards the crest line.

- Earth's rotation. When moving uphill the water column is compressed and therefore the potential vorticity (due to earth's rotation) is increased. Vorticity conservation forces the fluid to adopt an opposite rotation; in the Northern Hemisphere this opposite rotation is clockwise (anti-cyclonic). If the ridge is turned cyclonically relative to the flow direction, then the fluid rotates towards the ridge-normal direction; in the opposite case the fluid rotates away from the cross-normal direction.

In Fig. 3.34 the ridge has a cyclonic orientation relative to the undisturbed flow direction. The dashed streamline is the path a fluid parcel would follow if the flow momentum adapted instantaneously to the above three factors. This hypothetic streamline is perfectly symmetric relative to the ridge crest.

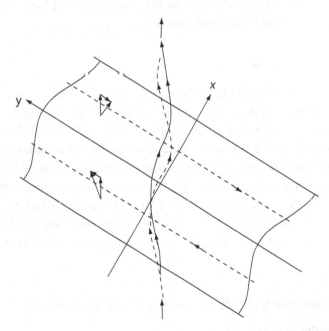

Fig. 3.34. Streamlines crossing a ridge which has a cyclonic orientation relative to the undisturbed flow. The dashed line corresponds to the path a fluid parcel would follow in the absence of inertial delay. The solid line represents the path of a fluid parcel which is delayed by inertia. At the upstream flank the delayed flow has an along-ridge component in positive y-direction and at the downstream flank an along-ridge component in negative y-direction.

Feedback mechanism

The adaptation of the ridge-parallel flow momentum to the frictional torque and to earth's rotation is delayed due to inertia (vorticity conservation). The streamline bends towards the ridge-normal direction with a spatial lag, as represented by the solid particle path in Fig. 3.34. This streamline is therefore not symmetric relative to the ridge-crest. Before crossing the crest, the real streamline is deviated relative to the non-inertial streamline in the ridge-parallel downstream direction; after crossing the ridge the streamline is deviated in ridge-parallel upstream direction. These deviations indicate a secondary circulation around the ridge; this circulation is caused by the flow perturbation due to frictional torques and vorticity conservation. The strength of the flow velocity at the upstream flank of the ridge is greater than at the downstream flank; this is because the circulation contributes asymmetrically to the unperturbed velocity at the upstream and downstream flanks of the ridge (see Fig. 3.34). The flow will therefore carry more sediment towards the ridge crest than away from it and we may expect the ridge amplitude to grow.

Instability of the seabed

As first postulated by Smith [414], the presence of a residual circulation around the bank is essential for its growth and maintenance. Oceanographic surveys at tidal sandbanks confirm the existence of flow circulation around tidal sandbanks [380, 212, 307, 213]. Theoretically, sandbank circulation is initiated already at ridges of very small height. This means that the seabed is unstable; a perturbation of infinitesimal amplitude may grow, provided the horizontal dimensions are such that the flow response to the perturbation is mainly determined by frictional torques and momentum conservation. In the following sections a more mathematical description of the flow response to a ridge perturbation will be presented.

3.4.3. *Zimmerman's Qualitative Ridge Formation Model*

Instant vorticity balance

The vertical structure of the flow was ignored in the previous qualitative description; it does not play a crucial role in the feedback of flow perturbation to ridge growth. This will also be the starting point for the mathematical analysis in which mass and momentum balances will be considered for depth-averaged variables. The analysis is simplified by eliminating pressure gradient terms. This is achieved by considering the vorticity balance instead of the momentum balance (see Appendix A.5). The depth-averaged balance of the potential vorticity $(\zeta + f)/H$ reads (A.19)

$$\frac{\partial}{\partial t}\left(\frac{\zeta + f}{H}\right) + \vec{u}\cdot\vec{\nabla}\left(\frac{\zeta + f}{H}\right) = -\frac{r}{H}\zeta + \left(\vec{u} \times \vec{\nabla}\frac{r}{H}\right)\cdot\vec{e}_z. \qquad (3.89)$$

Here $\vec{u} = (\bar{u}, \bar{v})$ is the depth-averaged flow velocity vector, $\zeta = \bar{v}_x - \bar{u}_y$ the vorticity, f the Coriolis parameter and $\vec{e}_z$ a unit vector in the vertical direction. In this equation a linearised form of the quadratic bottom friction is used, $\vec{\tau}_b = \rho r \vec{u}$, where r is the linear friction parameter with dimension [m/s]. The second term on the r.h.s. of Eq. (3.89) shows that potential vorticity is generated by bottom friction gradients. This happens when the flow crosses isobaths at an angle; the flow is then rotated perpendicular to the isobaths. The first term on the r.h.s. stands for vorticity dissipation by bottom friction. The second term on the l.h.s. of (3.89) represents local increase or decrease of potential vorticity, due to tidal advection.

Tide-averaged vorticity balance

Equation (3.89) shows that vorticity produced by bottom friction has opposite signs for ebb and flood. This implies (a) that ebb-vorticity and flood-vorticity have opposite signs and (b) that tidal vorticity advection has the same sign at ebb and flood. When averaging (3.89) over the tidal period, the tidal vorticity advection term remains as the only tidal contribution in the balance equation of residual vorticity. This tidally averaged balance thus shows that tidal vorticity advection is responsible for the production of residual vorticity at the tidal ridge. In the following it will be shown that the production of residual vorticity is equivalent to the production of a mean flow circulation around the ridge.

Sandbank circulation

Residual circulation can be evaluated by integrating the depth-averaged momentum balance equation (see A.16)

$$\vec{u}_t + (\vec{u}.\vec{\nabla})\vec{u} + f\vec{e}_z \times \vec{u} + \frac{1}{\rho}\vec{\nabla}p - (N\vec{u}_z)_z = 0, \qquad (3.90)$$

along a depth contour. The circulation $C(t)$ along the depth contour can be expressed as

$$C(t) = \oint \vec{u}(t) \cdot \vec{dl} = \int\int_{\Sigma} \zeta \, dx \, dy, \qquad (3.91)$$

where Σ is the area enclosed by the depth contour. The last equality follows from Gauss' law. To integrate the momentum balance equation we use the relation

$$(\vec{u}.\vec{\nabla})\vec{u} = (\vec{\nabla} \times \vec{u}) \times \vec{u} + \tfrac{1}{2}\vec{\nabla}.\vec{u}^{\,2}; \qquad (3.92)$$

the result is

$$\frac{\partial C}{\partial t} = -\oint (\zeta + f)(\vec{u} \times \vec{dl}).\vec{e}_z - \frac{r}{H}C. \qquad (3.93)$$

We integrate this equation over the tidal period with the assumption that the tidal amplitude is much smaller than water depth; we find

$$< C > = \oint \langle \vec{u}(t) \rangle \cdot \vec{dl} = -\frac{h}{r} \oint < \zeta \bar{u}_\perp > dl, \qquad (3.94)$$

where $u_\perp$ is the tidal component perpendicular to contour element dl (outward positive). We now consider the area Σ which corresponds to the tidal ridge above mean bed level. Then it follows from Eq. (3.94) that tidal vorticity advection over the ridge produces a residual circulation around the ridge, because the sign of the product $\zeta u_\perp$ does not change around the contour and is the same for ebb and flood (see Fig. 3.35). The strength of the residual circulation is thus given by the amount of vorticity advected through the ridge contour during a tidal period.

Ridge inclination

The sign of the vorticity produced by the bottom friction torque depends on the ridge orientation relative to the flow direction. It appears that the residual circulation is clockwise if the sand ridge crest is oriented anticlockwise relative to the tidal current; in the opposite case the residual circulation is anticlockwise, if we assume that the bottom friction torque is large enough to overcome the effect of earth's rotation. This principle is illustrated in Fig. 3.36. In addition

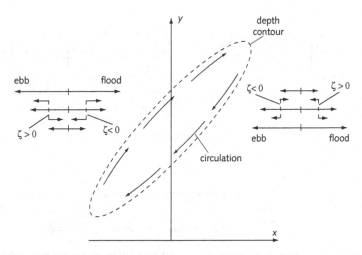

Fig. 3.35. Schematic representation of the tidally averaged vorticity balance, producing residual circulation. Lateral shear of the tidal velocity at both sides of the sandbank is due to the combined effect of sandbank friction and bank inclination. The corresponding vorticity has opposite sign at both sides of the sandbank and also changes sign from flood to ebb. The product of the cross-ridge flow component and vorticity is always positive. (The outward normal direction to the contour is taken positive.)

to this residual circulation, a current is generated along the depth contour with a quarter-diurnal (M_4) period. The contribution of this higher harmonic component to ridge growth is smaller than the contribution of the residual circulation.

Sediment transport towards the crest

As already mentioned, the secondary circulation around a tidal ridge produces spatial variations of net sand transport, with important consequences for ridge growth. This is illustrated by a simplified sediment transport formula, where gravity effects are ignored:

$$\vec{q}(x, y, t) = \alpha |\vec{u}|^2 \vec{u}. \tag{3.95}$$

We evaluate the tide-averaged sand transport $\langle q \rangle$ in a direction perpendicular to the sand ridge crest, which we will call the x-direction. The main tidal flow (u_{M2}, v_{M2}) is assumed to be semi-diurnal and ebb-flood symmetric. Interaction of this tidal flow with the sand ridge produces a residual flow component v_{M0} and a M_4 flow component v_{M4} in y-direction, parallel to depth contour. If we

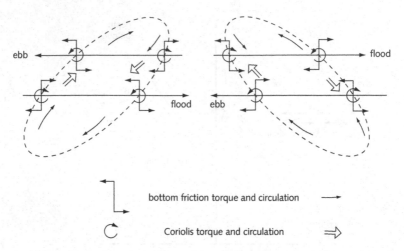

Fig. 3.36. Basic production mechanisms of residual circulation around a tidal sandbank in the presence of earth's rotation [511]. The dashed lines represent the tidal ridge depth contours, the long arrows flood or ebb flow. The frictional torques are indicated as in Fig. 3.35 and produced in the same way. The rotating arrows represent the Coriolis torques. For ridges with cyclonic orientation, frictional torques and Coriolis torques have the same direction; for ridges with anticyclonic orientation the direction is opposite. In the former situation the circulation induced by earth's rotation enhances the circulation induced by friction; in the latter case the circulation induced by friction is decreased by earth's rotation. Tidal ridges observed in the field have a cyclonic orientation relative to the main tidal flow axis; however, the inclination is sometimes quite small.

consider topographic structures of small height compared to the water depth then the secondary flow components are an order of magnitude smaller than the main tidal flow. The first order contribution of the flow perturbation to the tide-averaged sand transport across the sand ridge is given by:

$$\langle q \rangle = 2\alpha \langle u_{M2} v_{M2} (v_{M0} + v_{M4}) \rangle. \tag{3.96}$$

This expression shows that non-zero tide-averaged sediment transport across the ridge requires an angle θ of flow incidence on the ridge different from 0 or $\pi/2$. The residual circulation v_{M0} at the upstream slope of the ridge is aligned with the ridge in downstream direction (v_{M2}, v_{M0} same sign); at the downstream slope the direction is reversed (v_{M2}, v_{M0} opposite sign). This occurs during both ebb and flood. From (3.96) it then follows that the residual sediment transport is always directed towards the ridge crest. Sediment will be deposited at the ridge crest and the crest height will increase. This means that there is a positive feedback between a linear topographic structure which is inclined relative to the flow axis, such as a sand ridge, and tide-induced sediment transport. Such a seabed perturbation will thus grow spontaneously as an instability of the seabed.

Symmetry of cross-ridge flow

An essential difference with the generation mechanism of dunes is the absence of asymmetry in the cross-ridge velocity. In the case of tidal ridges the growth process is not produced by net uphill flow, but by asymmetry between uphill and downhill sediment load. The reason is that the total flow strength (vectorial sum of unperturbed flow and sandbank circulation) is greater at the upstream ridge flank than at the downstream flank.

Influence of gravity

The foregoing suggests that tidal ridge type perturbations of any horizontal dimension (within the validity range of the depth-averaged flow model) will be amplified and thus may develop on the seabed. However, the degree of amplification is not the same for all tidal ridge scales. Equations (3.89) and (3.94) show that the strongest vorticity and the strongest residual circulation are produced by seabed structures of small horizontal scale. Small-scale seabed structures have a steep bottom slope; due to gravity, sediment particles are carried by the flow more easily downhill than uphill. This applies to both bedload and suspended-load transport. The gravity effect implies that small-scale structures will grow less easily than larger scale structures; their height will remain quite small.

Influence of earth's rotation

Flow over large scale topographic structures is influenced by earth's rotation, through the principle of conservation of potential vorticity. When the flow approaches a sand ridge, water depth will decrease and therefore potential vorticity conservation requires that the flow is deflected in anticyclonic direction (to the right on the Northern hemisphere). If the sand ridge crest is oriented cyclonically relative to flow (either ebb flow or flood flow), then the flow will also deflect anticyclonically due to bottom friction: Friction and earth's rotation effects add up. However, if the sand ridge crest is oriented anticyclonically relative to the flow, then potential vorticity conservation and bottom friction torque counteract each other, so less vorticity is created. Thus strongest residual circulation is produced at a sand ridge oriented cyclonically relative to the flow direction, see Fig. 3.36. This may explain why tidal ridges observed on the Northern Hemisphere are generally inclined to the left relative to the main flow direction [431, 148, 345] while on the Southern Hemisphere the inclination of tidal ridges is predominantly to the right relative to the main flow direction [343].

3.4.4. *Huthnance's Analytic Ridge Formation Model*

Flow response to a small bed undulation

An analytical derivation of the previous qualitative results was first made by J. Huthnance [227]. The essential steps of his flow-seabed interaction model are reproduced below. In this model we start with a flat sea bottom, at constant water depth h_0. Next, the sea bottom is disturbed with a small ridge-type undulation $z_b(x, t)$ with wavelength $\lambda = 2\pi/k$ in x-direction and amplitude $\hat{z}_b$ uniform in y-direction:

$$\Re z_b(x, t) = \hat{z}_b e^{\sigma_i t} \cos(kx - \sigma_r t) = \hat{z}_b \Re e^{i(kx - \sigma t)}. \tag{3.97}$$

Here $\sigma_r/k = \Re\sigma/k$ is the migration rate of the bottom perturbation and $\sigma_i = \Im\sigma$ is the growth rate (positive or negative). The depth-averaged undisturbed flow $\vec{u}_0 = (u_0, v_0) = U(\sin\theta, \cos\theta)$ makes an angle θ with the ridge-crest direction y, see Fig. 3.37. Bottom topography and fluid flow are uniform in

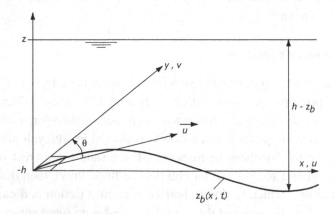

Fig. 3.37. Definition of symbols and axes for fluid flow across a sequence of banks.

y-direction; therefore the flow components in x and y-direction can be derived respectively from the continuity equation,

$$(h_0 - z_b)u = h_0 u_0 = h_0 U \sin \theta \qquad (3.98)$$

and the ridge-parallel momentum balance equation

$$v_t + u v_x + v v_y + f u + g \eta_y + c_D \frac{\sqrt{u^2 + v^2}}{h_0 - z_b} v = 0. \qquad (3.99)$$

The undisturbed flow obeys the equation

$$v_{0t} + f u_0 + g \eta_{0y} + c_D \frac{U v_0}{h_0} = 0, \qquad (3.100)$$

where $U = \sqrt{u_0^2 + v_0^2}$.

Infinitesimal steady flow perturbation

We will first consider the case of steady flow instead of tidal flow, i.e., u_0 and v_0 are constants and $\vec{u}_t = 0$. The bottom perturbation induces a small perturbation of the flow velocity $\vec{u}_1 = (u_1, v_1)$, which is on the order of ϵU, with $\epsilon - \hat{z}_b / h_0$. Development of the continuity equation (3.98) to order ϵ gives

$$u_1 = \frac{z_b}{h_0} U \sin \theta. \qquad (3.101)$$

The flow perturbation u_1 in cross-ridge direction is in phase with the perturbation z_b; this means that there is no upstream-downstream flow asymmetry in cross-ridge direction, as we already expected from the qualitative analysis. Development of the ridge-parallel momentum balance equation to order ϵ gives

$$u_0 v_{1x} + f u_1 + \frac{c_D}{h_0} \left[U v_0 \frac{z_b}{h_0} + U v_1 + v_0(u_1 \sin \theta + v_1 \cos \theta) \right] = 0. \quad (3.102)$$

The linearity of this equation implies that the along-ridge flow perturbation v_1 can be written in the form:

$$v_1 = -\chi \frac{z_b}{h_0} U = -|\chi| \frac{z_b}{h_0} U e^{-i\phi}, \qquad (3.103)$$

where χ is a yet unknown complex function of x and t; ϕ is the phase lag of the flow component v_1 relative to the bottom perturbation z_b. If $\phi = 0$ the alongshore flow perturbation is symmetric around the ridge crest; in that case there will be no sediment transport gradient at the ridge crest and no accretion or erosion will occur. If $\phi = \pi/2$ the alongshore flow perturbation is perfectly asymmetric around the ridge crest; the along-ridge flow perturbation has opposite signs at the upstream and downstream ridge flanks. This corresponds to a flow circulation around the ridge; it has been shown before that such a flow circulation causes asymmetry of sediment transport at the ridge crest leading to ridge growth or ridge decay.

Flow adaptation length scale

Substitution of (3.101 and 3.103) in (3.102) yields for $\chi(x, t)$ the equation

$$\chi\left[ikl\sin\theta + (1 + \cos^2\theta)\right] - \frac{fl}{U}\sin\theta - \cos\theta(1 + \sin^2\theta) = 0, \quad (3.104)$$

where the length scale l is given by

$$l = h/c_D. \tag{3.105}$$

The along-ridge velocity perturbation is found from substitution in (3.103),

$$v_1 = -\frac{\cos\theta(1 + \sin^2\theta + p\tan\theta)}{1 + \cos^2\theta} U \frac{z_b}{h_0} \cos\phi e^{-i\phi}, \tag{3.106}$$

where

$$p = \frac{fl}{U}, \quad \tan\phi = \frac{kl\sin\theta}{1 + \cos^2\theta}. \tag{3.107}$$

This velocity perturbation depends crucially on the length scale l. This length scale corresponds to the spatial adaptation lag of the flow to the ridge-related bottom friction; this can be seen by comparing the scales of the inertial (third) and frictional (last) terms in the momentum balance equation (3.99). As discussed earlier, it is precisely this spatial adaptation lag that produces flow asymmetry relative to the ridge crest. Maximum flow circulation around the ridge crest occurs when the phase lag $\phi = \pi/2$. According to (3.107), this corresponds to a bottom perturbation of infinitely small wavelength.

Convergence of the sediment flux

However, growth of ridges with very small wavelength cannot occur in reality, because the bottom slopes are too steep. This can be seen by using a sediment transport formula which takes into account the stabilising influence of gravity on bottom slope increase (3.37),

$$\vec{q}_b = \alpha |\vec{u}|^{n-1}(\vec{u} - \gamma |u_\perp|\vec{\nabla}z_b), \quad u_\perp = \vec{u}.\vec{\nabla}z_b/|\vec{\nabla}z_b|. \tag{3.108}$$

The critical velocity for sediment motion, u_{cr}, has been ignored for simplicity, considering that significant sediment transport relates to conditions of much stronger flow velocity, $|u| \gg u_{cr}$. The expression (3.108) can be evaluated by substitution of $\vec{u} = \vec{u}_0 + \vec{u}_1$, with $\vec{u}_1$ given by (3.101) and (3.106). Ridge growth is related to the sediment transport gradient,

$$z_{bt} + \vec{\nabla}.\vec{q}_b = 0. \tag{3.109}$$

We only need to consider the x-component of the sediment flux, because there is no gradient in the y-direction. The growth rate can now be calculated using

$$\Im\sigma = \Re[z_{bt}/z_b] = -\Re[\vec{\nabla}.\vec{q}_b/z_b]. \tag{3.110}$$

To first order in ϵ we find

$$\vec{\nabla} \cdot \vec{q}_b = \alpha U^{n-1}\Big[(\cos^2\theta + n\sin^2\theta)u_{1x}$$
$$+(n-1)\sin\theta\cos\theta\, v_{1x} - \gamma U|\sin\theta|z_{bxx}\Big], \tag{3.111}$$

and after substituting (3.101), (3.106) and (3.107),

$$\Im\sigma = \alpha\gamma U^n \frac{(1+\cos^2\theta)^2}{l^2|\sin\theta|}\left(\xi\sin^2\phi - \tan^2\phi\right), \tag{3.112}$$

$$\text{with}\quad \xi = \frac{n-1}{\gamma c_D}\left(\frac{\cos\theta}{1+\cos^2\theta}\right)^2 |\sin\theta|(1+\sin^2\theta + p\tan\theta).$$

As $c_D \ll 1$ and $\xi \gg 1$, growth rates are positive, except for ϕ close to $\pi/2$. According to (3.107), this implies that only very small wavelengths are completely suppressed by the bottom-slope term in the sediment transport formula.

Tidal ridge growth rate

The maximum growth rate $\Im\sigma$ is found for $\Im\sigma_\phi = 0$, or

$$\tan^2\phi = \xi^{1/2} - 1. \tag{3.113}$$

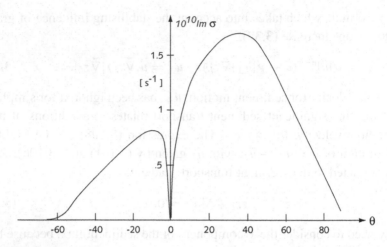

Fig. 3.38. Growth rate as function of ridge inclination θ relative to the flow axis, for $p = 1$ (northern hemisphere), friction coefficient $c_D = 0.003$ and sediment transport parameters $n = 4$ and $\gamma = 1$. The depth ($h = 20$ m) and velocity ($U = 0.5$ m/s) only influence the scale of the curve. The growth rate is positive because for each θ-value a particular wavelength is chosen which corresponds to maximum growth, according to (3.113). For positive angles, corresponding to cyclonic ridge orientation relative to the flow axis, the growth rate is larger than for negative angles.

This yields

$$\Im\sigma = \alpha U^n \frac{\gamma}{l^2} \left[\sqrt{\frac{n-1}{\gamma c_D}} \cos\theta \sqrt{1 + \sin^2\theta + p\tan\theta} - \frac{1 + \cos^2\theta}{\sqrt{|\sin\theta|}} \right]^2.$$

(3.114)

Because $p > 0$ growth rates are larger for positive values of θ than for negative values, see Fig. 3.38. This implies that low ridges with cyclonic orientation grow faster than ridges with anticyclonic orientation, confirming what we already found. The angle θ at which growth is maximum follows from $\Im\sigma_\theta = 0$, yielding (for $\theta > 0$)

$$\sin^{9/2}\theta = p\sin^{3/2}\theta \frac{1 - 2\sin^2\theta}{4\cos\theta} + \sqrt{\frac{\gamma c_D}{4(n-1)}}$$
$$\cos\theta\left(1 + \frac{3}{2}\sin^2\theta\right)\sqrt{1 + \sin^2\theta + p\tan\theta}.$$

Because $\gamma c_D/(n-1) \ll 1$, we have in the case $f = 0$,

$$\sin\theta \approx \left(\frac{\gamma c_D}{4(n-1)}\right)^{1/9}.$$

In this case the ridge inclination θ depends only on the friction coefficient c_D and the bottom-slope coefficient γ; however, the sensitivity to these parameters is weak.

Ridge inclination θ

Taking $n = 4$ and $\gamma = 1$ in the sediment transport formula and taking for the friction coefficient the value $c_D = 0.003$ we find $\theta \approx 25°$. Substitution in (3.113) with $f = 0$ gives $\phi = 73°$ for the phase lag of the secondary ridge-parallel current v_1 relative to the bottom perturbation z_b; this implies that we are not very far from maximum flow asymmetry ($\phi = 90°$). At the ridge-crest v_1 is negative; the flow over the ridge is thus always deflected towards the cross-ridge direction. At 50° latitude we have $f \approx 10^{-4}$ and $p = 1$, assuming 20 m water depth and an undisturbed flow velocity of 0.67 m/s (see 3.107). In that case we find a larger angle, $\theta \approx 35°$.

Ridge wavelength

The wavelength λ is given by (3.107, 3.113)

$$\lambda = 2\pi/k = \frac{\sin\theta}{(1 + \cos^2\theta)\sqrt{\xi^{1/2} - 1}} \frac{2\mu h}{c_D}. \tag{3.115}$$

For $p = 0$ and $\theta = 30°$ we find $\lambda \approx 0.5\,h/c_D \approx 150\,h_0$ and for $p = 1$ we find a slighter larger wavelength, $\lambda \approx 180\,h_0$. In water depth of 20 m the model predicts an initial tidal ridge wavelength on the order of three to four kilometers. If the wavelength is increased above the value of maximum growth (ϕ smaller than (3.113)), growth rate (3.112) remains positive and decreases only slowly to zero, while for smaller wavelengths the growth rate becomes rapidly negative. This implies that perturbations with larger wavelength than (3.115) will also experience substantial growth. Therefore one may expect that in natural situations ridges may occur with wavelengths larger than (3.115).

Fully developed tidal ridges

The foregoing analysis only applies to the initial phase of tidal ridge development. Field observations show that tidal ridges may grow quite high, attaining a large fraction of the water depth. The time scale $1/3\sigma$ for ridge growth follows from (3.114). For the previous example with $p = 1$, $h = 20$ m, $c_D = 0.003$, maximum tidal velocity of 1 m/s and sediment transport parameter $\alpha = 10^{-4}$

(3.34), we find a growth time scale (amplification by a factor e) on the order of a thousand years. This is close to time scales which have been estimated from measurements of sand transport over tidal sandbanks [478]. The following two notes need to be made, however. In the first place, growth of mature tidal ridges from a flat bottom will take much longer than the e-folding time scale. This implies that ancient seabed perturbations, such as relict coastal barriers from an initial stage of transgression, may have played a significant role as kernel for tidal ridge development. This is consistent with the finding by Houbolt [212] that a relict core is present in many tidal sandbanks. In the second place, higher-order terms in the flow equations, which have been ignored in the linear analysis, become important during the growth process. These terms may favour ridge growth at other wavelengths than the initial exponential growth. Correspondence between the wavelength of fully grown ridges and the wavelength of initial ridges may even be considered as a coincidence. Finally it has been suggested that factors such as wave action, ridge curvature and tidal ellipticity [216] need to be taken into account.

Comparison with observations

Fields of tidal ridges occur in the Southern Bight of the North Sea at depths of typically 30 m, see Fig. 3.39. These ridges are rotated at an angle of 20°–30° in cyclonic direction relative to the main flow axis. According to Huthnance's model, the wavelength corresponding to maximum initial growth is roughly 5 km. The observed wavelengths are somewhat larger, but fairly close to this estimate.

Tidal flow

So far we have ignored the tidal nature of the flow. The time dependence of tidal flow can be included in the analysis by considering

$$\vec{u} = (u_0 I(t), v_0 I(t)),$$

with, for instance $I(t) = \cos \omega t$. The mathematical treatment is similar, but Eq. (3.104) now changes into

$$\frac{l}{U}\chi_t + \chi \left[ikII \sin \theta + |I|(1 + \cos^2 \theta) \right] - \frac{fl}{U} I \sin \theta - I|I| \cos \theta (1 + \sin^2 \theta) = 0.$$

$$(3.116)$$

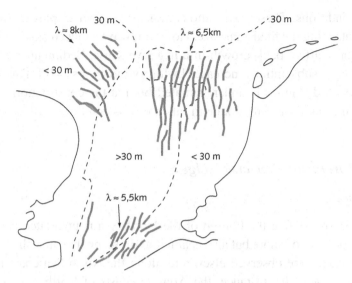

Fig. 3.39. Tidal ridge fields in the Southern Bight of the North Sea. The dashed line is the 30 m depth contour. The height of the ridges is typically 10 m, but some ridges attain almost 30 m. The orientation is cyclonic relative to the main flow axis with an angle between 20° and 30°. The wavelength λ corresponds to the average spacing between the ridges in each ridge field. The ridge fields do not cover the entire Southern Bight; however, their location does not bear any obvious relationship with the distribution of seabed grain diameter, strength of tidal currents or wave action.

An analytical solution can be found only in a few special cases, for instance, when representing the tide by a square-wave function,

$$I = 1 \ \text{ for } \ 0 < t < T/2, \ \ I = -1 \ \text{ for } \ T/2 < t < T.$$

The results are similar to the case of stationary flow except for a correction factor multiplying ξ in Eq. (3.112). The correction factor is close to 1, especially if the ridge wavelength is much smaller than the tidal excursion, i.e., $h/c_D \ll UT/\pi$. The conclusion is, that in case of sufficient flow strength (U on the order of 1 m/s), the wavelength of ridges in tidal flow and in stationary flow depends linearly on water depth with approximately the same proportionality factor.

Wave influence

In periods of high waves a fully developed sandbank will regress; some 20% of the bank volume may be lost [267], while the bank is restored during calm weather conditions. These observations indicate that waves play an important role in controlling the final bank size. However, wave-induced sand suspension does not only affect bank growth in a negative sense. Sediment transport by tidal currents is substantially increased by high wind waves [478]. This suggests that sandbank dynamics is underestimated both at the constructional and the destructional stages, if only tidal currents are considered.

3.4.5. *Shoreface-Connected Ridges*

Occurrence

Figure 3.33 shows that the bottom of the North Sea is undulated, with large ridges not only far offshore but also near to the shore, on the inner shelf. Similar nearshore ridges are observed elsewhere along the North Sea coast [12] and in other shelf seas, for instance, the Atlantic coasts of North America [430, 431] and South America [149, 343] and the coasts of the Yellow Sea and East China Sea [286]. The dimensions of the nearshore and offshore ridges are similar; the nearshore ridges have a wavelength of a few kilometres up to 10 kilometres and their height is in the order of ten metres. They extend from the shoreface in offshore direction over the inner shelf, in water depths of 10 to 30 metres and their slope is not much larger than the inner shelf slope. However, a marked difference with the offshore tidal ridges is the orientation of these ridges. Nearshore ridges often have no anticyclonic orientation relative to the main flow; they always bend to the coast with an angle of 10°–50° and they extend up till the shoreface; for that reason they are called shoreface-connected ridges.

Instability of the seabed slope

In spite of the apparent similarity of tidal ridges and shoreface-connected ridges they are thought to be formed by a completely different mechanism. Tidal flow, bottom friction and gravity-induced downslope transport, which are essential factors for tidal ridges, only play a minor role for the dynamics of shoreface-connected ridges. In contrast to most other seabed structures, the delayed flow response to gradients in bottom friction is not the primary instability factor. On the inner shelf, seabed instability is related to the average cross-shore slope. The

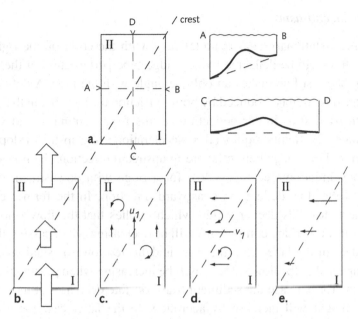

Fig. 3.40. Development of shoreface-connected ridges as an instability inherent to a sloping shoreface with longshore flow. We consider a rectangular domain situated around the ridge crest (a), which divides the domain in an upstream part I at the right lower half, and a downstream part II at the upper left half. The right boundary of the domain is parallel to the shoreline (x-direction); without the ridge, the depth increases linearly from right to left (y-direction). The unperturbed longshore flow enters at the bottom and leaves at the top (b); if the flow strength is not altered by the ridge, the water flux (open arrows) over the ridge is smaller than the incoming and outgoing water fluxes. Continuity of the water flux can be restored by an additional longshore velocity component u_1 at the ridge crest (c). However, this additional velocity component corresponds to a change of vorticity over the crest, because continuity requires $u_{1y} \neq 0$. Without frictional effects, such a change of vorticity cannot exist; besides which friction-induced vorticity would have the opposite sign. Continuity of the longshore water flux over the ridge can also be achieved by an offshore flow component v_1 at the ridge crest (d). Vorticity is conserved if the longshore variation v_{1x} is equal to u_{1y}; this requires that the longshore velocity perturbation u_1 is small compared to the cross-shore perturbation v_1. The offshore water flux over the ridge should be approximately the same in the shallow nearshore zone as in the deep offshore zone. This implies that the offshore velocity has to decrease in offshore direction. As a result, the flow will be decelerating at the ridge crest (e); this flow deceleration entails convergence of the sediment flux at the ridge crest. Hence, sediment is deposited at the crest and the ridge-perturbation will grow.

presence of a seabed slope induces flow asymmetry between the upstream and downstream flanks of an oblique seabed perturbation and a cross-ridge transport gradient at the perturbation-crest. The principle is illustrated in Fig. 3.40. A qualitative description of this generation mechanism is given below.

Feedback mechanism

We assume a stationary coast-parallel flow, with the coast on the right-hand side. The flow will be deflected by any ridge-type perturbation of the seabed, provided the crest line makes an oblique angle with the flow. A deflection in cross-ridge direction is produced by bottom friction torques, but in the absence of bottom friction a similar deflection results from continuity and vorticity conservation. Continuity requires flow acceleration at the upstream slope of the perturbation. For a ridge bent offshore in upstream direction, the flow deflects in offshore direction (see Fig. 3.40); for a ridge with downstream offshore orientation the flow deflects in coastward direction. In the former case the flow deflects towards deeper water, which implies that the flow velocity (in particular its downridge component) will be decreasing when passing the crest of the ridge. In the latter case the flow is deflected towards shallower water, which implies that the flow velocity will be increasing when passing the ridge-crest. In the former case the sediment transport rate decreases at the crest and the ridge height will increase by accretion. In the latter case the crest will erode. The conclusion is that a sloping seabed is unstable only for ridge-type perturbations with upstream offshore orientation.

Unidirectional storm-driven flow

We will now consider an initial ridge-perturbation inclined offshore towards the flood direction. In this case the flood flow will be deflected seaward over the ridge; during flood the perturbation will grow. The ebb-flow, however, will be deflected landward over the same ridge-perturbation; during ebb the perturbation will decay. Therefore shoreface-connected ridges will not develop under symmetric tidal conditions. They can only be formed if one flow direction along the coast is dominant over the other. Along the Dutch coast northward flow dominates over southward flow, mainly because the wind climate is dominated by southwesterly winds. This may explain why south-west running shoreface-connected ridges have developed along the Dutch coast (see Fig. 3.41) as predicted by the previously described instability mechanism. We may expect that shoreface-connected ridges are best developed on coasts where storm winds coming from a preferred oblique direction are the major cause for longshore sediment transport. Such storm dominated coasts are, for instance,

Fig. 3.41. Shoreface-connected ridges along the Dutch coast. The ridges occur in a zone where storm-driven net sediment transport is greater than tide-induced net sediment transport, see Fig. 3.42.

the Atlantic coast of North and South America; along these coasts shoreface-connected ridges are frequent and well developed. Along the North Sea coast, tidal influence is stronger than along the American Atlantic coast, see Fig. 3.42. This may explain why shoreface-connected ridges are less prominent features (lower height, sometimes absent) along the North Sea coast. The stratigraphy of shoreface-connected ridges along the Dutch coast reveals not only storm deposits, but also tidal deposits [310]. A tidally induced growth mechanism might contribute to the growth of these ridges, even if the ridge inclination relative to the flow axis is opposite to the ridge inclination which is favoured by earth's rotation, see Fig. 3.38.

Sediment distribution

As discussed above, shoreface-connected ridges may develop as a result of uni-directional flow along the slope of the inner shelf. This flow is responsible not only for ridge growth, but also for ridge migration. Observed migration rates are on the order of only a few (generally less than 10) metres per year. The

Fig. 3.42. Southern bight of the North Sea and the zone (shaded) where net tide-induced sediment transport is greater than storm-induced net sediment transport [459]. Along most of the Dutch North Sea coast and along the coast of the Wadden Sea storm events contribute more to net sediment transport than tides.

downstream slope of the ridges is steeper than the upstream slope, which is a normally observed feature of migrating bedforms. Observations also show that at the downstream slope the seabed sediment is finer than at the upstream slope [432]. This is consistent with ridge migration by storm currents. During these conditions fine sediment is winnowed from the seabed, especially when the flow is accelerated. This produces a coarsening of the upstream flank of the ridge. The flow strongly decelerates at the downstream flank and part of the previously suspended load will settle. As this load contains relatively more fine than coarse material, the seabed sediment on the downstream flank of the ridge (which is seaward oriented) will be finer than on the upstream flank.

Initial formation

Instability of the inner-shelf slope is not the only explanation for the occurrence of shoreface-connected ridges. Other phenomena may play a role as well. Shoreface-connected ridges are often found at locations where, during the Holocene transgression, the coast has received substantial sediment input from rivers or from erosion of Pleistocene deposits. In these regions coastal barriers have probably developed, which may not have kept pace with sea-level rise. Remnants of these coastal barriers on the inner shelf might form the origin of the present shoreface-connected ridges [212, 27]. There are also indications that shoreface-connected ridges are related to former inlets. Abandoned ebb-tidal deltas along a retreating coast may have provided the sediment source for these ridges [305]. These hypotheses do not necessarily conflict with the theory of formation through instability of the inner shelf-slope. Relict seabed structures might have acted as an initial seabed perturbation which has been remodelled to a shoreface-ridge pattern by the described earlier instability mechanism. Without this morphodynamic feedback relict seabed structures would probably not have persisted.

3.4.6. Trowbridge's Model for Shoreface-Connected Ridges

Flow response to an infinitesimal bed undulation

The first model of seabed instability leading to shoreface-connected ridges was proposed by J. H. Trowbridge [446]. Using a similar approach we will show that shoreface-connected ridges can be generated as an instability of a sloping seabed in the case of uniform frictionless stationary longshore flow. The present model is greatly simplified and far from realistic; many processes influencing the dynamics of ridge generation are left out. The model only aims to illustrate the salient features of pattern generation on a sloping seabed. A justification of several simplifying assumptions, such as ignoring bottom friction, is given by Trowbridge [446] and Calvete [63]. The symbol conventions are shown in Fig. 3.43. The x-axis is in longshore direction, the y-axis points in offshore direction. We will consider a uniform basic flow u_0 in longshore direction and a sinusoidal bottom perturbation z_b parameterised as

$$\Re z_b(x, y, t) = \epsilon h_0 \Re e^{i(kx \sin\theta + ky \cos\theta - \sigma t)}. \tag{3.117}$$

Here k is the wavenumber of the perturbation and θ is the angle between the crest line and the x-axis. The time evolution depends on the complex number σ with $\Re\sigma/k$ being the migration rate and $\Im\sigma$ the exponent for growth or decay. We will consider an emerging ridge for which ϵ is a constant, $\epsilon \ll 1$.

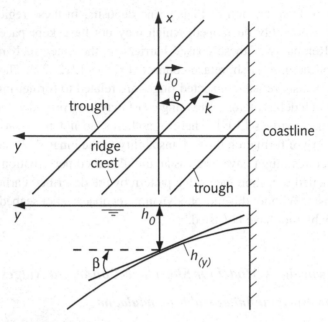

Fig. 3.43. Definition of symbols and axes for longshore flow over a ridge with oblique orientation to the shoreline. Upper part plan view and lower part vertical cross-shore section.

Concave equilibrium shore profile

It will be assumed that the depth h only depends on the cross-shore coordinate y and that it can be approximated by an exponential function

$$h(y) = h_0 \exp(\beta y / h_0). \tag{3.118}$$

This expression represents a convex depth profile, while in reality the shoreface profile is more often concave. If the width of the shoreface h_0/β is much greater than the wavelength $\lambda = 2\pi/k$, we may consider $h(y)$ as a fair approximation of a locally linear depth increase. It will be assumed that the cross-shore depth profile (3.118) is in equilibrium and that the processes responsible for maintaining this equilibrium are not dependent on the flow perturbations caused by the emerging ridge (3.117). For the inner-shelf slope this is a plausible assumption; for the lateral slopes of a flow channel it is probably not. Bank-connected ridges in flow channels do exist and are designated as alternating bars or point bars. However, these bars are thought to arise from a completely different type of current-topography interaction, as will be discussed in the next section.

Frictionless flow perturbation

The flow velocity $\vec{u}$ is to first order in ϵ given by

$$\vec{u} = (u_0 + \epsilon u_1, \epsilon v_1), \tag{3.119}$$

where $\vec{u}_0$ is the unperturbed longshore flow, and $\epsilon\vec{u}_1$ the flow perturbation due to the emerging ridge. We ignore bottom friction and earth-rotation effects. In this case there are no vorticity producing terms in the momentum balance equation. Together with the assumed uniformity of the basic flow u_0 this yields

$$\zeta = v_{1x} - u_{1y} = 0. \tag{3.120}$$

This equation is solved by defining a potential function ψ such that

$$\psi_x = u_1, \quad \psi_y = v_1. \tag{3.121}$$

The continuity equation reads

$$\vec{\nabla}.(D\vec{u}) = \vec{u}.\vec{\nabla}D + D\vec{\nabla}.\vec{u} = 0, \quad D = h - z_b. \tag{3.122}$$

To first order in ϵ this gives

$$h(u_{1x} + v_{1y}) + h_y v_1 = u_0 z_{bx}/\epsilon,$$

and after substitution of (3.121),

$$\psi_{xx} + \psi_{yy} + \frac{\beta}{h_0}\psi_y = iku_0\frac{z_b}{\epsilon h_0}\sin\theta e^{-\beta y/h_0}. \tag{3.123}$$

The solution has the form $\psi = \chi z_b \exp(-\beta y/h_0)$ and the complex number χ is found by substitution in (3.123), yielding

$$\psi = -u_0\frac{z_b}{\epsilon}\sin\theta\frac{\beta\cos\theta + ikh_0}{(kh_0)^2 + \beta^2\cos^2\theta}e^{-\beta y/h_0}. \tag{3.124}$$

Using (3.121) we find for the along-ridge velocity perturbation $v_{//ridge} = -u_1\cos\theta + v_1\sin\theta$ the expression

$$v_{//ridge} \propto z_b e^{i\phi}, \quad \tan\phi = \frac{kh_0\beta(1 + \cos^2\theta - \cos\theta\sin\theta)}{k^2 h_0^2(\sin\theta - \cos\theta) + \beta^2\cos\theta}. \tag{3.125}$$

For large ridge wavelengths (small k) the phase ϕ is always positive, showing that the along-ridge velocity perturbation is directed seaward and that it reaches a maximum upstream of the ridge crest.

Sediment flux convergence

We evaluate the erosion-sedimentation pattern resulting from sediment flux gradients over the perturbed seabed from

$$z_{bt} + \vec{\nabla}.\vec{q} = 0, \tag{3.126}$$

where the seabed porosity factor has been incorporated in the sediment flux $\vec{q}$. From (3.117) it follows that the (positive or negative) growth rate of the perturbation, $\Im\sigma$ is given by

$$\Im\sigma = \Re(z_{bt}/z_b) = -\Re(\vec{\nabla}.\vec{q}/z_b). \tag{3.127}$$

For the sediment transport $\vec{q}$ we will use formula (3.32)

$$\vec{q} = C\vec{u}, \tag{3.128}$$

where C and $\vec{u}$ represent the volumetric sediment load (load divided by sediment density) and the average current velocity in a near-bottom layer in which the bulk of the sediment flux is concentrated. Gravity-induced down-slope transport is ignored.

Wave-induced sediment suspension

As shoreface-connected ridges are observed almost exclusively on storm dominated shelves, we will assume that they are generated on the inner shelf during storm conditions. Under such conditions sediment concentrations near the bottom depend more on wave stirring of the bed than on the mean current velocity u_0. Because the wave height and wave orbital velocities depend on water depth, we may expect that C is also a function of depth, $C = C(h)$. The bed perturbation also influences the sediment concentration through water depth, but for emerging ridges this is a higher order effect that will be neglected.

Growth rate

The expression of the sediment flux gradient can be simplified by using (3.122):

$$\vec{\nabla}.C\vec{u} = D\vec{u}.\vec{\nabla}(C/D). \tag{3.129}$$

To first order in ϵ the result is

$$\vec{\nabla}.C\vec{u} = u_0 C z_{bx}/h + \epsilon h v_1 (C/h)_y = u_0 C z_{bx}/h - \epsilon \psi_y (C - h C_h) h_y/h. \tag{3.130}$$

Using (3.127), (3.128) and (3.124) we find for the growth rate

$$\Im\sigma = \beta\frac{u_0}{h_0{}^2}\cos\theta\sin\theta\frac{(kh_0)^2 + \beta^2}{(kh_0)^2 + \beta^2\cos^2\theta}(C - h_0C_h). \tag{3.131}$$

This expression is evaluated at an offshore distance y corresponding to $h(y) = h_0$. The derivative $C_h = \partial C/\partial h$ is negative, because wave stirring decreases with increasing depth; the factor $(C - h_0C_h)$ is thus positive and could be approximated by the sediment load close to the breaker line. The growth rate is positive for positive values of the angle θ, i.e., for tidal-ridge perturbations with an upstream offshore oriented crest line. This is what we already expected from the qualitative description of tidal-ridge growth. We may conclude that a sloping seabed is unstable if it is subject to a slope-parallel current and wave-stirring.

Ridge inclination

The angle θ for which the growth rate is fastest corresponds to vanishing of the derivative of (3.131) with respect to θ. This yields

$$\cos\theta = \left(2 + (\beta/kh_0)^2\right)^{-1/2}. \tag{3.132}$$

For ridges with a wavelength which is on the order of the width of the inner shelf or smaller ($k \approx \beta/h_0$) we find $\theta \approx 45°$. The angle can be larger if the wavelength is much larger than the inner shelf.

Growth and migration rates

From (3.131) an estimate can be derived of the growth rate (e-folding amplification time scale) of emerging shoreface-connected ridges. A typical value of the sediment load under storm conditions (wave height 4 m, wave period 10 s) is on the order of $C \approx 10^{-4}$ m [418]; we suppose that these conditions occur on average 20 days a year. We further assume that under these conditions the sediment load is inversely proportional to the square of depth and that $u_0 \approx$ 0.5 m/s. We take the depth $h_0 = 15$ m and the slope $\beta = 10^{-3}$ and consider a ridge wavelength of the same order as the width of the inner shelf. Then the time scale for ridge growth $1/\Im\sigma$ is found on the order of a few thousand years. The migration rate $c_{ridge} = \Re\sigma/k \approx \Im(u_0Cz_{bx}/z_bkh) \approx u_0C\sin\theta/h$ is found to be on the order of 5 m per year. Shore-face connected ridges are very slowly evolving structures according to the model and this is consistent with observations.

Ridge wavelength

The expression (3.131) shows that the growth rate decreases with increasing wavenumber k. From this we might conclude that the growth rate is strongest for perturbations with infinite wavelength. Observations show that shoreface-connected ridges have a finite wavelength, which is on the order of the width of the inner shelf (5–10 km). The simple model we have presented fails to reproduce this wavelength as fastest growing instability mode. This is due to the choice of the convex depth profile (3.118); this choice is physically less realistic but simplifies the model by producing linear equations. For small wavelengths only the local depth dependence counts, so the choice of a convex or concave profile does not make a great difference. For large wavelengths, comparable to the width of the shoreface, the difference is crucial. The maximum growth rate of the perturbation corresponds to the situation with greatest offshore flow deceleration. For a convex profile the depth continues increasing in offshore direction. As the offshore deflection of the current increases with wavelength, the flow deceleration will be strongest for very large wavelengths. In reality the width of the sloping inner shelf is limited; the offshore flow deceleration is restricted to this limited width. The ridges with the fastest growth rate then correspond to a wavelength which is comparable to the width of the inner shelf [63]. A finite-amplitude analysis with more refined models, including bottom friction and earth-rotation effects, gives predictions not only of the wavelength, but also of the growth and migration rates and of the final amplitude of shoreface-connected ridges, which are consistent with observations [64]. This analysis also shows that the ridge wavelength and migration rate are not strongly influenced by finite-amplitude effects. These results further support the hypothesis that surface-connected ridges are generated as free instabilities of the sloping inner shelf bed under the action of storm-driven flow.

3.5. Channel Bars and Meanders

3.5.1. *Laterally Bounded Flow*

Morphology

Tidal inlets, tidal basins, estuaries and rivers are examples of laterally bounded flow systems. If bed material is mobile, bedforms may develop in bounded flow as well as in unbounded flow; in laterally bounded flow systems bedforms are even more abundant than in the sea. The generation of small bedforms, like ripples or dunes, is not substantially influenced by lateral flow constraints, as the dynamics of these bedforms is essentially related to the vertical flow structure. The dynamics of large bedforms is related to the horizontal flow structure; one may expect that lateral flow constraints play an important role here and in this section we will see that this is indeed the case. Large-scale bedforms in laterally bounded flow will be called 'bars'. The morphology of these bars is different from the marine sand ridges and their generation mechanism is also different. Most bars in tidal basins and rivers likely arise as free instabilities, but their further development is strongly influenced by topographic constraints imposed on the flow. These topographic constraints are different from place to place; hence, the bar patterns observed in bounded flows are less regular than the ridge patterns in the sea. Sometimes it may even be difficult to identify the initial generation mechanism of fully developed bars. Similar as in the sea, small and large scale bedforms coexist; ripples, megaripples and dunes are often superimposed on each other and on the much larger bars. The flow and sand transport patterns related to these bedforms interact; this interaction increases the morphologic complexity of the resulting compound structures.

Alternating bars

Three main categories of bar structures in tidal channels may be distinguished. We will designate the first category as 'alternating bars', see Fig. 3.44. These bars are connected to the channel banks; alternating bars may have the appearance of tidal flats if the bar crest reaches above low water. They are generally situated at the inner bend of a channel meander; the development of these bars and the development of channel meanders are mutually related processes. The bars are alternately located at opposite channel banks, forming a sequence of channels meanders. Their spacing (wavelength λ) is often related to the channel width b; the empirical relationship

$$\lambda = 6b \tag{3.133}$$

provides a reasonable estimate of observed bar spacings, but the scatter is large ($\lambda/b = 2-10$) [91]. Other empirical expressions also include the channel depth h or the tidal excursion L. Alternating bars are observed only if the width-to-depth ratio of the channel is sufficiently large, i.e., greater than 10 [72].

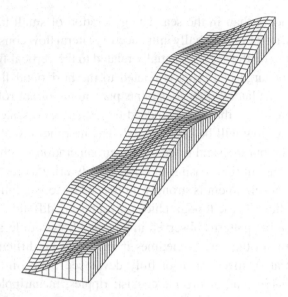

Fig. 3.44. Schematic representation $z_b(x, y) \propto \cos(\pi y/b)\cos(kx)$ of alternating bars in a straight channel. The flow meanders between the bars; opposite to the bars flow depth is maximum and halfway to the bars flow depth is minimum.

Similarity between alternating bars in rivers and tidal basins

In case of strong channel sinuosity alternating bars may have a greater width (cross-channel dimension) than length (along-channel dimension); in that case they appear as flow-transverse structures. Similar structures exist in rivers, where they are called 'point bars', due to the pointed lee side shape. Comparison of alternating bars and meanders in tidal and riverine conditions exhibits strong similarities [22]; this suggests that the basic generation mechanisms of alternating bars and channel meandering are the same for both and that the tidal character of the flow is not a critical factor. A major difference concerns the migration rate of alternating bars; in rivers the bars migrate downstream while in tidal basins almost no migration is observed.

Multiple bars

In tidal conditions the flood flow towards the intertidal portions of the bar is concentrated in channels which end on the tidal flat. Sometimes these channels completely separate all or part of the bar from the bank. This can be considered as a transitional morphology to the second bar category. This second category

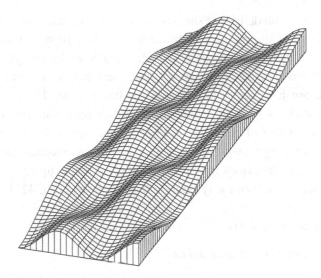

Fig. 3.45. Schematic representation $z_b(x, y) \propto \cos(3\pi y/b) \cos(kx)$ of multiple-row bars in a straight channel.

consists of bars separating different flow channels, see Fig. 3.45. There are different names for these bars: 'multiple-row bars' or 'en-echelon braid bars'; they are similar to the bars observed in braiding rivers. Channel braiding occurs when the width-to-depth ratio is sufficiently large, typically more than 100 [502]; when this ratio increases the number of rows also increases. In regions where tidal currents are strong, for instance close to the inlet, the bars take a more elongated form; the bars form chains, interrupted by swatchways. These elongate bars separate flood and ebb dominated channels. The bars make a small angle with the main flow direction; this suggests that the tidal ridge generation mechanism (Sec. 3.4.2) may contribute to the formation and maintenance of these bars [299]. Another interpretation of the location and the levee-type form of these bars is that they result from the convergence of sediment transport in zones where flood and ebb dominated channels meet.

Delta sand bodies

Tidal channels generally end on shoals where the channel widens; here the flow expands, the flow velocity drops and sediment is deposited. These deposits are called 'deltas' or 'delta sand bodies'; they have a characteristic lobate form, the lobes representing deposition zones. Flood deltas are situated at the interior of tidal basins while ebb deltas are situated seaward of the inlet. In contrast to the rhythmic pattern of alternating and multiple-row bars, the delta bodies are isolated structures; their morphology is strongly related to the overall topography of the tidal basin. The development of these sand bodies influences the tidal flow pattern, not just locally but even at the basin scale. Their development therefore interacts with overall deposition and erosion patterns in the basin and produces shifting of channels and bars. This mutual feedback generates permanent morphologic evolution with a long term quasicyclic character. Morphologic cycles with periods ranging from several tenths up to hundred years have been documented for many tidal delta complexes [133, 425].

3.5.2. *Formation of Bars*

Stability of a straight flat channel bed

Nowadays it is generally accepted that bar formation provides the initial trigger for the development of large-scale channel topography, including channel meandering. This hypothesis will be examined in this section. We will address the following questions:

- which mechanism is responsible for breaking the symmetry of a perfectly straight uniform channel?
- what type of rhythmic perturbation is amplified by channel bed instability, under which conditions?
- which dynamic balance determines the wavelength of the strongest growing perturbation?

The method used is a linear stability analysis, similar to that in the previous sections. Several investigations of this type have been reported in literature, examining different mechanisms and formulations of the bed-flow interaction [62, 342, 155, 256, 42, 399, 394]. We will restrict here to a very simple model, valid only for a limited range of channel flows, which highlights the most essential features of alternating bar dynamics.

Alternating bar perturbation

We consider a straight channel with an initially flat horizontal bottom and a uniform, constant and unidirectional flow. We slightly disturb this uniform channel by imposing a long wave bed undulation of infinitesimal amplitude corresponding to an alternating pattern of a shoal (or 'bar') at one side of the channel, opposite to a pool at the other channel side. The water depth is greater over the pool than over the shoal and therefore more water will flow over the pool than over the shoal. At the upstream slope of the shoal the flow is therefore deviated to the deeper opposite side of the channel and the inverse occurs at the downstream slope. Due to this flow deviation sediment is taken away from the upstream slope of the shoal and brought to the downstream slope. The result is a downstream migration of the shoal-pool pattern, see Fig. 3.46.

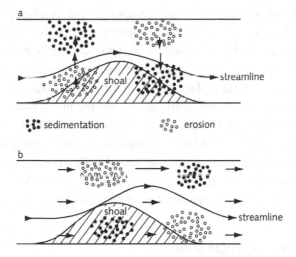

Fig. 3.46. Development of an alternating bar. Figures (a) and (b) show a plan view of a channel with a local perturbation of small amplitude corresponding to a shoal at the right bank. The flow is steady and unidirectional and in both figures the solid line schematically represents a streamline. In (a) the frictional influences on horizontal flow acceleration/deceleration related to shoal topography are ignored, i.e., the distance between streamlines is approximately inversely proportional to depth. At the upstream slope of the perturbation the flow diverges from the shoal and the downstream slope it converges. This causes erosion of the downstream slope and accretion of the upstream slope: The perturbation will move downstream. The streamlines are symmetric with respect to the shoal structure; no net erosion or deposition takes place at the crest of the perturbation. In (b) shoal-induced friction is taken into account. Friction decelerates the flow over the shoal. However, maximum flow deceleration does not occur at the shoal crest, but downstream of the shoal crest. This is caused by inertial lag of the flow response to increased friction. The streamlines are displaced downstream and they are asymmetric with respect to the shoal structure. The flow decelerates not only at the upstream slope, but also around the shoal crest; the lateral flow component around the shoal crest is very small. In the zone around the shoal crest deceleration of the longitudinal flow component causes deposition of sediment and growth of the shoal crest.

Feedback mechanism

But this is only part of the story. Due to friction the flow velocity is decreased over the shoal and increased over the pool. The effect of increasing or decreasing friction is counteracted by inertia; flow deceleration or acceleration lag behind the decrease or increase of depth. Hence, the location of minimum flow velocity is not at the crest of the perturbation, but downstream of the crest. The flow at the crest is still decelerating, implying sediment transport convergence and sediment deposition. Sediment deposition at the crest of the perturbation will cause growth of the amplitude of the perturbation. Growth of the perturbation amplitude will enhance the flow deceleration at the emerging bar crest and therefore stimulate further bar growth. The opposite process of self-enhanced scouring takes place at the other side of the channel; the highest velocity does not occur opposite to the crest of the perturbation, but further downstream. Therefore erosion will take place in the pool region. We conclude that the combined effect of friction and inertia stimulates the initial growth of a channel bed perturbation consisting of a shoal at one side of the channel and a pool at the other side, see Fig. 3.46. When the amplitude of the alternating bar pattern increases the flow will start bending around the shoal, producing centrifugal forces. These centrifugal forces induce secondary circulations directed to the shoal which stimulate development of tidal flats and channel meanders.

Frictional adaptation length scale

The above description suggests that perturbations of any length scale can grow. If this is true we may wonder why channel meanders in tidal basins only occur within a certain range of wavelengths, from one or a few km for small/shallow basins to a few tens of km for large/deep basins. For initiation of channel meandering the channel bed should not be perfectly flat but contain irregularities. These irregularities may be random and cover a broad spectrum of wavelengths. Then, by the process described above, feedback of channel bed irregularities to the flow structure may produce amplification. The point is, however, that not all irregularities will grow as fast. The fastest growing irregularities are those where the flow deceleration at the 'crest' of the irregularity is greatest. These irregularities will dominate the others and become the dominant morphologic features. Flow deceleration at the crest is a lag effect due to inertia. The lag is greatest for irregularities with a small wavelength. At first sight this would imply that small-scale irregularities will grow faster than large scale irregularities. There is a limit, however. The length scale of the perturbation should be large enough to allow adaptation of the flow velocity to the increase/decrease of friction at the slopes of the perturbation. This adaptation length scale follows from the requirement that inertia (represented by the term uu_x in the momentum balance equation) and friction (represented by the term $c_D u^2/h$) are of the same order of magnitude. This yields the length scale

$$\lambda = \pi h/c_D \approx 1000\,h. \tag{3.134}$$

We expect that perturbations with a wavelength (spacing) on the order of λ will experience fastest growth.

3.5.3. *A Simple Model for Alternating-Bar Formation*

Flow response to an alternating bed undulation

We will now investigate the processes that initiate alternating bars in more detail, to clarify the assumptions behind the previous arguments. We consider a channel of constant width b and depth h_0 and uniform depth-averaged flow u_0 in longitudinal direction. We assume $b \gg h_0$ and we look at the response (u, v) of the depth-averaged flow to a small channel bed perturbation. The bottom depth is perturbed by a small longitudinal undulation which has opposite signs at opposing channel banks, see Fig. 3.44; in complex notation:

$$\Re z_b = \epsilon h_0 \cos(\pi y/b)\Re \exp[i(kx - \sigma t)], \tag{3.135}$$

with $\epsilon \ll 1$. The growth rate corresponds to $\Im \sigma$ and the migration rate to $\Re \sigma/k$. The perturbed depth $D(x, y, t) = h_0 + \eta'(x, y, t) - z_b(x, y, t)$. The depth-averaged momentum balance equations read

$$uu_x + vu_y + g\eta_x + c_D \frac{u\sqrt{u^2 + v^2}}{D} = 0,$$

$$uv_x + vv_y + g\eta_y + c_D \frac{v\sqrt{u^2 + v^2}}{D} = 0,$$

and the mass balance equation:

$$(uD)_x + (vD)_y = 0.$$

All variables are scaled with respect to the undisturbed situation as follows:

$$u \to u_0 u, \quad v \to u_0 v, \quad \eta \to h_0 \eta, \quad D \to h_0 D, \quad z_b \to \epsilon h_0 z_b,$$
$$x \to h_0 x/c_D, \quad y \to by.$$

The x-scaling is derived from (3.134); the scaling of x and y is chosen such that the order of magnitude of flow properties and their gradients are the same. The scaled equations read

$$uu_x + Gvu_y + F^{-2}\eta_x + \frac{u\sqrt{u^2 + v^2}}{D} = 0, \tag{3.136}$$

$$uv_x + Gvv_y + F^{-2}G\eta_y + \frac{v\sqrt{u^2 + v^2}}{D} = 0, \tag{3.137}$$

$$(uD)_x + G(vD)_y = 0, \tag{3.138}$$

where

$$F^2 = u_0^2/gh_0, \quad G = h_0/bc_D. \tag{3.139}$$

Infinitesimal perturbation and linearisation

The velocity u, v and the surface slope η_x contain a small perturbation component u', v' and η'_x, which can be expressed as a power series in ϵ,

$$u' = \epsilon u_1 + \epsilon^2 u_2 + \cdots, \quad v' = \epsilon v_1 + \epsilon^2 v_2 + \cdots, \quad \eta' = \epsilon \eta_1 + \epsilon^2 \eta_2 + \cdots$$

The possible dependence of the friction coefficient c_D on depth and velocity is ignored. Now we only retain terms which are of first order in ϵ,

$$u_{1x} + F^{-2}\eta_{1x} + 2u_1 + z_b = 0, \tag{3.140}$$

$$v_{1x} + F^{-2}G\eta_{1y} + v_1 = 0, \tag{3.141}$$

$$-z_{bx} + u_{1x} + Gv_{1y} = 0. \tag{3.142}$$

Scaling analysis

In the frictional terms we have neglected the contribution of η_1 compared to z_b; this approximation is equivalent to the assumption of small Froude number F. The perturbation of the sediment transport mainly depends on u_1; an estimate of u_1 can be derived directly from the first equation (3.140) if we assume that the pressure gradient term $F^{-2}\eta_{1x}$ is small compared to the other terms.

Intuitively this seems a reasonable assumption, since the perturbation of the surface slope at one side of the channel is partly offset by the perturbation at the other side, i.e., $\eta_{1x} \approx 0$. A more quantitative argument is obtained by estimating the order of magnitude of the longitudinal pressure gradient term $g\eta_{1x}$ from the last two equations. From (3.142) we estimate the order of magnitude of $v_1 : O[v_1] = G^{-1}O[z_b]$. Next, we note that in (3.141) the first term changes sign somewhere halfway the bar slope (where $|v_1|$ is maximum) while the second and third term do not. The last two terms therefore must be of equal order of magnitude, i.e., $O[\eta_1] = F^2 G^{-2} O[z_b]$. From this we derive that the condition $O[F^{-2}\eta_{1x}] \ll O[z_b]$ is equivalent to

$$G^2 \gg 1, \quad \text{or} \quad b^2 \ll h_0{}^2/c_D{}^2. \tag{3.143}$$

Taking $c_D \approx 0.003$, we find that for channels with a width which is not too large, say $b/h_0 \leq 100$, the pressure gradient term in (3.140) can be ignored.

Effect of centrifugal surface slope

In the following section the same equations will be analysed for fully developed channel meanders. It will be shown that the transverse surface slope η_{1y} is induced mainly by the centrifugal force term v_{1x} in Eq. (3.141). The water level is raised at the pool (outer channel bend) and lowered at the shoal (inner channel bend). This implies that the flow towards the shoal experiences two opposing forces: An acceleration, due to dipping of the water level at the shoal and a deceleration, due to increased bottom friction. The acceleration term is large for short meanders, i.e., if the ratio of channel width to bar wavelength is large. The deceleration term is large if the bar wavelength is large. In the following we will assume that deceleration due to friction is much larger than acceleration due to the centrifugal surface slope.

Phase lag

We now go back to the original dimensional variables and look for a solution of the flow response u' to the perturbation z_b of the form

$$u' = -\chi \frac{u_0}{h_0} z_b = -|\chi| \frac{u_0}{h_0} z_b e^{-i\phi}, \tag{3.144}$$

where ϕ is the spatial phase lag between the velocity perturbation u' and the topography z_b. Substitution in Eq. (3.140) yields

$$\cos \phi = 2|\chi|, \quad k = \frac{2c_D}{h_0} \tan \phi. \tag{3.145}$$

The greatest frictional lag corresponds to $\phi = \pi/2$. The length scale k^{-1} of the alternating bar then equals zero, i.e., the bed perturbations with the fastest growth rate would be those with infinitely small wavelength. However, the scaling analysis is not valid for wavelengths much smaller than h_0/c_D. Moreover, such bars would have very steep slopes and we may expect that bar growth in that case will be strongly counteracted by gravity effects. Strongest growth of alternating bed perturbations is therefore more likely to occur if the frictional lag is substantial, but not as large as $\pi/2$. This is the case, for instance, if $\phi \approx \pi/4$. Substitution of this phase-lag in (3.145) yields a wavenumber k which is identical to the earlier estimate (3.134).

Sediment transport including gravity effects

The influence of gravity effects on the growth rate of bed perturbations can be estimated from the sediment transport formula (3.38)

$$\vec{q}_b = \alpha |\vec{u}|^{n-1} (\vec{u} - \gamma |\vec{u}| \vec{\nabla} z_b). \tag{3.146}$$

The γ-term in this expression represents gravity-induced downslope sediment transport. The critical velocity for incipient sediment motion, u_{cr}, is ignored, following the assumption that morphologic evolution mainly takes place in periods that $|u| \gg u_{cr}$. We evaluate the gradient of the bedload sediment flux to first order in ϵ,

$$\vec{\nabla}.\vec{q}_b = \alpha u_0{}^n \left(nu'_x/u_0 + v'_y/u_0 - \gamma(z_{bxx} + z_{byy}) \right). \tag{3.147}$$

Now we substitute v'_y from (3.142); x and y-derivatives follow from (3.135). The result is

$$\vec{\nabla}.\vec{q}_b = \alpha u_0{}^n \left[ik \left((n-1)\frac{u'}{u_0} + \frac{z_b}{h_0} \right) + \gamma z_b (k^2 + (\pi/b)^2) \right]. \tag{3.148}$$

Convergence of sediment transport

The growth of the bed perturbation follows from the sediment balance equation

$$z_{bt} + \vec{\nabla}.\vec{q}_b = 0, \tag{3.149}$$

which expresses the fact that for each channel segment a difference between incoming and outgoing sediment fluxes produces a change of bed elevation. The bed porosity factor is left out as it is not essential for our discussion. By substituting (3.135) we find from (3.149) expressions for the time dependency of the bed perturbation, and in particular the migration rate $\Re\sigma/k$ and the growthrate $\Im\sigma$

$$\Re\sigma/k = \Im\frac{\vec{\nabla}.\vec{q}_b}{kz_b} = \alpha u_0^n \left[(n-1)\Re\frac{u'}{u_0 z_b} + \frac{1}{h_0} \right],$$

$$\Im\sigma = -\Re\frac{\vec{\nabla}.\vec{q}_b}{z_b} = \alpha u_0^n \left[(n-1)k\Im\frac{u'}{u_0 z_b} - \gamma(k^2 + (\pi/b)^2) \right]. \tag{3.150}$$

After substitution of (3.144) and (3.145),

$$\Re\sigma/k = \alpha\frac{u_0^n}{h_0} \left[1 - \frac{n-1}{2}\cos^2\phi \right], \tag{3.151}$$

$$\Im\sigma = \alpha u_0^n \frac{c_D}{h_0^2} \left[(n-1)\sin^2\phi - 4\gamma c_D \tan^2\phi - \gamma\frac{\pi^2}{c_D}\left(\frac{h_0}{b}\right)^2 \right]. \tag{3.152}$$

Wavelength

The greatest growth rate $\Im\sigma$ corresponds to

$$\cos^2\phi = 2\sqrt{\gamma c_D/(n-1)}, \tag{3.153}$$

or to a bar wavelength (see 3.145)

$$\lambda = \frac{2\pi}{k} \approx 2\pi \left(\frac{\gamma c_D}{4(n-1)}\right)^{1/4} \frac{h}{c_D}. \tag{3.154}$$

Substituting 'regular' values $\gamma = 1, n = 4$ and $c_D = 0.003$, we find for the wavelength of strongest bar growth the estimate

$$\lambda \approx 0.8h/c_D \approx 270\,h. \tag{3.155}$$

The dependency of the growth rate on the wavelength λ is shown in Fig. 3.47. The growth rate decreases sharply for perturbations with shorter wavelength,

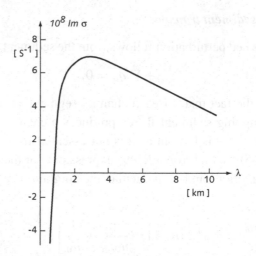

Fig. 3.47. Growth rate $\Im\sigma$ as a function of bar wavelength λ assuming an unperturbed flow depth h_0 of 10 m. Other parameters are $u_0 = 1$ m/s, $\alpha = 10^{-4}$, $\gamma = 1$, $n = 4$ and $c_D = 0.003$. Perturbations with a wavelength corresponding to the greatest growth rate are amplified first. Perturbations with larger wavelength also experience substantial growth. Perturbations with smaller wavelength are suppressed by gravity-induced downslope sediment transport.

but not for perturbations with greater wavelength. This implies that in natural flows, bars with a spacing larger than $270h_0$ should occur more frequently than bars with a smaller spacing.

Growth and migration rates

The maximum growth rate is obtained by substituting (3.153) in (3.151),

$$\Im\sigma = \alpha u_0{}^n \frac{c_D}{h_0{}^2}\left[(n-1) + 2\sqrt{(n-1)\gamma c_D} - \gamma\frac{\pi^2}{c_D}\left(\frac{h_0}{b}\right)^2\right]. \qquad (3.156)$$

Substituting the previous estimates of γ, n, c_D we find

$$b/h_0 > \sqrt{\frac{\gamma\pi^2}{(n-1)c_D}} \approx 33 \qquad (3.157)$$

as a necessary condition for a positive growth rate. This implies that alternating bars cannot develop in narrow channels. The migration rate $\Re\sigma/k$ (3.151) is always positive because $\cos^2\phi \ll 1$; assuming an average flow velocity $u_0 = 1$ m/s, a depth $h_0 = 10$ m and $\alpha = 10^{-4}$ (3.34) the migration rate is about one metre a day. For the growth rate we find an e-folding amplification time scale $1/3\sigma$ of half a year.

Influence of channel width

Neglecting the surface slope in the momentum balance of the perturbed flow (Eq. (3.140)) is justified only for channels with a depth-to-width ratio of approximately 100 or less. For wider channels, the surface slope contribution plays a significant role, by opposing frictional deceleration at the shoal. With increasing channel width, a greater shoal wavelength is required for flow deceleration at the shoal. On the other hand, the shoal wavelength should be short enough for producing frictional delay and transport convergence at the shoal crest. This implies that the wavelength of maximum growth will be close to the estimate (3.155) for small width-to-depth ratios, but that the bar wavelength will take larger values than predicted by (3.155) for larger width-to-depth ratios or for larger friction coefficients. An analytical analysis, similar to the one presented above, but including the surface-slope terms has been carried out by Parker [342]. In this case a wavelength of maximum perturbation growth can be found even if gravity-induced downslope transport is ignored,

$$\lambda \propto \sqrt{F/Gh_0/c_D}, \tag{3.158}$$

where F and G are defined in (3.139); the proportionality constant equals about 4. This yields a comparable value for the wavelength as (3.155) when taking, for instance, $F = 0.1$, $b/h_0 = 140$ and $c_D = 0.003$.

Comparison with data

Data on the wavelength of alternating bars are available mainly for rivers, if it is assumed that meander spacing is a good indicator of bar spacing (this point will be discussed in Sec. 3.5.4). When comparing observed meander length scales [396] with the predictions (3.155) and (3.158), it appears that both expressions underestimate the observed values by a factor 3 to 7, but the scatter is large (up to a factor 10). The trend in the data is better represented by (3.158) than by (3.155) [228]. A comparison with observed spacings between inner-bend tidal flats in Dutch tidal basins is shown in Fig. 3.48. Again both expressions (3.158) and (3.155) underestimate the observed tidal-flat spacings by a factor 3 to 7. The data are fairly well represented by the first estimate (3.134). Figure 3.48(b) suggests that a best fit to the data would involve the square of $\sqrt{F/Gh_0/c_D}$.

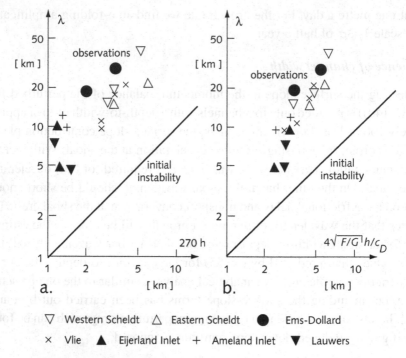

Fig. 3.48. Observed meander length scales λ in Dutch tidal basins; the meander length scales correspond to twice the spacing of successive (opposed) inner-bend tidal flats. In all cases the ratio of channel width to channel depth is greater than 100, and may reach 500 for the Wadden Sea basins. The observed length scales are plotted against the predictions of expression (3.155) in Figure (a) and against the predictions of formula (3.158) in Figure (b). The observed values are a factor 3 to 7 larger than the predicted values. The general trend in the data is better reproduced in (a) than in (b); however, the scatter in the data is smaller in (b), suggesting that $\sqrt{F/G}h_0$ is a better indicator for meander length than h_0.

Results of a numerical analysis of alternating bar growth indicate that the discrepancy between the observed wavelengths and the wavelengths of fastest growth in the linear stability analysis may be explained by higher order non-linearity in the morphodynamics of alternating bars at finite amplitude; this will be discussed later.

Influence of tidal flow alternation

The tidal case differs from the river case in several respects, for instance, the influence of intratidal areas, density currents and waves. Here we will discuss only the differences related to the oscillating nature of tidal flow, the greater strength of the currents and the finer sediment.

The process of bar formation in the previous simple model is independent from the flow direction; change of flow direction does not make any difference for bar growth, if it takes place simultaneously over the whole cross-section. However, in most tidal basins flow direction does not change simultaneously over the whole cross-section. Frictional dissipation of momentum is greater at the bar than at the pool; therefore the flow will respond faster to the tidal variation in water surface inclination at the bar than at the pool, where a time lag is due to inertia. The flow reversal time lag between bar and pool in an estuary is typically of the order of one hour. However, at the initial stage of bar emergence this time lag is much smaller. Numerical simulations indicate that tidal flow alternation does not strongly affect the growth rate of alternating bars [399].

The major difference between the river case and the tidal case is related to bar migration. If the strength of ebb flow and flood flow are similar then no bar migration will occur under tidal conditions.

Suspended load instead of bedload

Sediment is transported in tidal basins mainly as suspended load, due to the strength of the flow and the high fraction of fine sediment. Transport gradients are shifted with respect to flow gradients due to settling and suspension lag. If this shift amounts to a substantial fraction of the bar wavelength, then the transport maximum will occur further upstream of the bar crest and the growth rate will be increased. This is confirmed by the numerical analysis of Fredsøe [155], who investigated the influence of the ratio suspended load to bedload on the generation of alternating bars. The analysis shows that in situations where no alternating bars develop with bedload transport alone, bar growth may become possible when suspended-load transport is taken into consideration.

Settling-resuspension lag

In the case of suspended-load transport the sediment balance equations (3.148) and (3.149) may be replaced by (3.49):

$$z_{bt} = \frac{1}{T_{s,e}} \left(C - C_{eq}(u) \right), \tag{3.159}$$

where C is the local suspended load and C_{eq} the equilibrium load for velocity u. Underlying (3.159) is the assumption that the suspended load converges exponentially to its equilibrium value with time scale $T_{s,e} = T_s$ for sedimentation and $T_{s,e} = T_e$ for erosion; transverse and gravity-induced downslope transport components have been left out. We consider $C_{eq} = \alpha |\vec{u}|^{n-1}$, yielding

$$C_{eq} = C_0 + C'_{eq}, \quad C_0 = \alpha u_0^{n-1}, \quad C'_{eq} = \alpha(n-1)u_0^{n-2}u'. \tag{3.160}$$

We assume that around the bar crest settling dominates over erosion; we therefore take $T_{s,e} = T_s$ as the settling time scale; this time scale is inversely proportional to the settling velocity and proportional to the height of the water column over which significant suspension takes place. If we also assume T_s much smaller than the tidal period and $l = u_0 T_s$ much smaller than the bar wavelength ($kl \ll 1$) we find from (3.39) that the spatial variation of the local suspended load C lags a distance l behind the equilibrium concentration, $C = C_0 + C'_{eq} \exp(-ikl)$. (There is no phase lag in the unperturbed state because C_0 is constant.) Substitution in (3.160) and (3.159) yields the growth rate

$$\Im\sigma = \Re\frac{z_{bt}}{z_b} = \Re\frac{C'_{eq}(e^{-ikl} - 1)}{T_s z_b} = \alpha u_0{}^{n-1}(n-1)k\left(\Im\frac{u'}{z_b} - \frac{kl}{2}\Re\frac{u'}{z_b}\right). \tag{3.161}$$

Using (3.144) and (3.145) we find

$$\Im\sigma = \alpha u_0{}^n \frac{c_D}{h_0{}^2}(n-1)\sin^2\phi(1 + lc_D/h_0). \tag{3.162}$$

This is identical to the expression (3.150) without settling/resuspension lag, except for the last factor. If we consider for the settling/resuspension time lag the approximation $T_s \approx h_0/w_s$ then the last factor in (3.162) equals $1 + c_D u_0/w_s$. The conclusion is that settling/resuspension lag enhances alternating bar growth substantially only for sediment with a settling velocity w_s on the order of $c_D u_0$ or smaller. If $u_0 \approx 1\,\text{m/s}$ and $c_D = 0.003$, then the effect of settling/resuspension lag is substantial only for sediment particles smaller than $100\,\mu\text{m}$.

Multichannel system

Instead of a single row of alternating bars, a second or possibly a third row of alternating bars may divide the cross-section into two or more channels, see Fig. 3.45. This situation occurs in wide estuarine mouths or in tidal bays.

These multiple-row bars are generated by an instability mechanism similar to a single row of alternating bars. At the upstream slope of each bar the flow is decelerated due to frictional effects and due to inertial effects the minimum flow strength is not located at the crest of the bar but downstream of it. The resulting convergence of sediment transport at the crest produces amplification of an initial bar perturbation. The bar slopes are steeper than in the case of simple alternating bars. If the width-to-depth ratio is not very large — less than about 100 — the down-slope gravity-induced transport prevents growth of multiple-row bars. For values of $b/h \gg 100$ multiple-row bars will develop in addition to alternating bars [342, 200]; the wavelength of maximum growth is about the same for both bar types [394].

Similarity with alternating bars

The simple model for alternating bars can easily be extended to the case of multiple bar channel systems. We then have to replace (3.135) by

$$z_b = \epsilon h_0 \cos(m\pi y/b) \exp[i(kx - \sigma t)], \qquad (3.163)$$

with $m = 3$ for a two-channel system, $m = 5$ for a three-channel system, etc. Introducing the same approximations as for the alternating bar case we find an identical expression (3.151) for the growth rate with b replaced by b/m. This does not affect the expression for the preferred (fastest growing) wavelength (3.154). This conclusion holds only in a qualitative sense, however, as the perturbation of the surface slope (the terms η_{1x} in 3.140 and η_{1y} in 3.141) cannot be ignored in this case.

What happens after initial bar formation?

The initially rectilinear tidal flow gets more and more disturbed during the growth of alternating bars and multiple-row bars. Centrifugal forces, represented in particular by the term Gvu_y in (3.136), come into play when the tidal flow is substantially deviated by the bars. This term is quadratic in the perturbation and its influence therefore cannot be investigated by linear stability analysis. Centrifugal forces disrupt the symmetry between ebb and flood flow and create additional lag effects, which affect the erosion/sedimentation pattern of the initial alternating bars. Numerical simulations of alternating bar development in an elongated tidal basin at finite amplitude indicate a complete reorganisation of the initial bar pattern when the bar amplitude increases [200]. The wavelength of the resulting fully developed bar pattern is about 4 times larger than that of the initial bar pattern. This factor 4 matches fairly well the observed discrepancy between observed wavelengths of alternating bars in tidal basins and the wavelength predicted by linear stability analysis.

Flood and ebb flow will follow different streamlines and the bars will no longer be aligned with the flood and ebb flow. The spatial velocity distribution in meandering flow is very different from rectilinear flow and the same goes for the sediment transport pattern. This will be discussed more in detail in Sec. 3.5.4.

Elongate estuarine bars

Flood and ebb-dominated channels are often separated by elongate bars; these bars are sometimes designated as channel margin bars. Examples of such bars are shown in Fig. 3.49. These bars usually make a small angle to the main flow axis and therefore may respond to the same growth mechanism as tidal ridges [299]. The combination of cross-bar tidal flow with along-bar residual circulation produces convergence of sediment transport at the bar crest both during ebb and flood, see Fig. 3.36. On the open shelf the maximum growth rate of tidal sandbanks corresponds to wavelengths (bank spacing) of several kilometres. The elongate estuarine bars have a much smaller width; this may be due to topographic constraints imposed on the flow. Another possibility is that elongate bars develop from multiple-row banks. The generation process of multiple-row banks and tidal sandbanks is fundamentally different; it is possible, however, that both mechanisms are acting simultaneously.

3.5.4. Channel Meandering

Similarity between tidal flats and alternating bars

The initial development of alternating bars stimulates the formation of channel meanders. The flow in channel meanders is far more complex than the flow in a rectilinear channel at the onset of bar formation. The flow has a three-dimensional character, with fluid parcels following a spiral pattern. Channel meanders in tidal basins bend around tidal flats, situated at the inner channel bends. The typical wavelength of tidal flats measured along the channel axis is

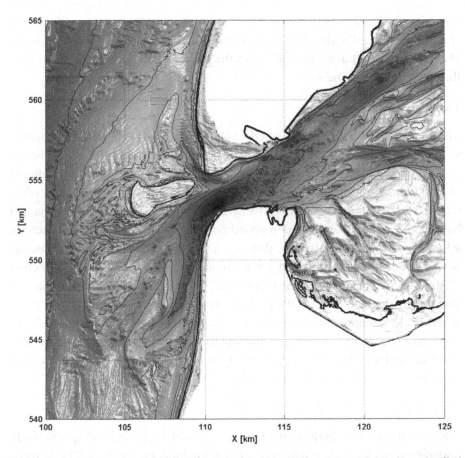

Fig. 3.49. Seabed topography of the inflow and outflow zones of Texel inlet, from detailed bottom soundings (Survey Department Rijkswaterstaat North-Holland, 2002). Increasing darkness corresponds to increasing water depth; depth is greatest at the inlet and attains 40 m. The x and y-axes correspond to the Paris coordinate system. The large ebb delta shoal Noorderhaaks ($x = 106$–109, $y = 553$–555) remains partly dry at mean high water. Tidal flow in the ebb delta is generally ebb-dominated, except in channels close to the North-Holland coast ($x = 110$, $y = 540$–552) and the Texel coast ($x = 110$, $y = 557$–560). The maximum depth-averaged ebb-tidal velocity at the inlet is almost 1.5 m/s, for an average tide. Besides the large intertidal shoals, elongate structures are visible both landward and seaward from the inlet. The elongate ebb delta bar at ($x = 108$, $y = 547$–550) makes a cyclonic angle to the neighbouring channel axes. The general orientation of the elongate shoal in the flood delta ($x = 116$–122, $y = 554$–561) also makes a small cyclonic angle to the main channel axes.

of the same order as the wavelength of alternating bars in rivers with similar channel depth and width. This may be considered as a coincidence, but it can also be interpreted as an indication for the physical relationship between tidal flats and alternating bars.

Tidal basin morphology

The characteristic morphology of tidal basins consists of a meandering channel system with tidal flats located mainly at the inner bends and of channels and at the landward boundary, see Figs. 4.4 and 3.52. Channel branches split off in flood direction towards the tidal flats; the bifurcation points are most often situated at the channel bends or just downstream in flood direction. The channel has flood dominated and ebb dominated portions, often separated by a shoal or a bar. Flood domination occurs downstream in flood direction of the channel bend along the outer bank and ebb domination downstream in ebb direction of the channel bend along the outer bank, see Fig. 3.52. The flood and ebb dominated channels meet halfway along the meander; a shoal is formed at the junction. In general, flood and ebb dominated channels are not aligned, but seem to avoid each other [476]. The shoals or bars separating flood and ebb dominated channels often obstruct navigation; the shoals have to be dredged to maintain sufficient navigation depth.

Flow in a channel bend

Many of the morphologic characteristics of tidal basins can be qualitatively understood from the flow pattern in channel meanders. Figure 3.50 schematically shows a typical cross-section at a channel bend with corresponding horizontal view. The flow in a channel bend is quite complicated, in particular due to the transient nature of the flow structure. However, the picture is much simpler if we assume that the length of the channel bend is sufficient for adaptation of the flow structure to frictional forces (velocity independent of along-channel coordinate x). We assume that the flow velocity at the shallow inner bend is lower than at the deeper outer bend; if the along-channel velocity component increases linearly across the channel we may write the cross-channel momentum balance in cylindrical coordinates ($x = R\theta$ is the along-channel coordinate and y is the radial coordinate, see Fig. 3.50),

$$-\frac{u^2}{R} + fu + \frac{1}{\rho}p_y - (Nv_z)_z = 0. \tag{3.164}$$

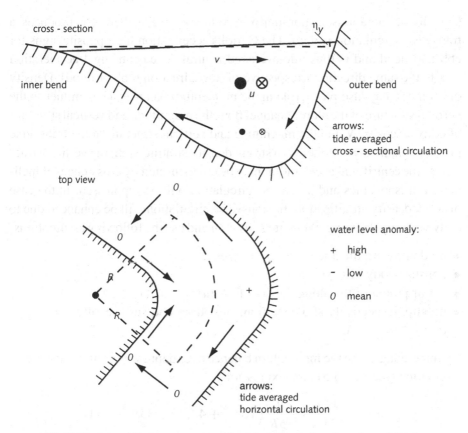

Fig. 3.50. Schematic representation of residual circulations in a channel bend.

The symbols have the following meaning: u velocity along the channel axis, v transverse velocity, $R = u/u_y =$ channel-bend radius, z vertical coordinate (bottom: $z = -h$, surface: $z = \eta$). The first term in the equation is the centrifugal acceleration, the second term is the Coriolis acceleration, the third term is the pressure gradient and the last term represents the momentum diffusion by turbulent viscosity. The cross-channel pressure gradient is related to the cross-channel water surface inclination η_y and to transverse density differences ρ_y (inclination of isopycnics) in the case of stratification.

Transverse circulation

Equation (3.164) shows that a non-uniform vertical distribution of the along-channel flow velocity $u(y, z)$ implies vertically non-uniform centrifugal and

Coriolis accelerations; a non-uniform velocity profile therefore produces a transverse circulation $v(y, z)$. The Coriolis acceleration has opposite signs for ebb and flood and is thus tide-averaged neutral. The centrifugal acceleration acts in the same direction irrespective of flow direction (ebb or flood). Density differences may also play a role in the momentum balance: They influence the vertical structure of the along-channel flow, the magnitude and vertical structure of eddy viscosity N and the magnitude and vertical structure of the transverse pressure gradient p_y. If there is a strong density stratification, the vertical structure of the centrifugal acceleration may be compensated by cross-channel inclination of isopycnics and transverse circulation will be suppressed. In the case of weak density stratification the transverse circulation will be enhanced due to eddy viscosity damping. We write $z^* = z/h$ and use the following assumptions:

- no density stratification (i.e., $p_y = g\rho\eta_y$),
- constant eddy viscosity N,
- linear profile of the along-channel flow ($u(z) = u_s(1 + z^*)$),
- no-slip, respectively slip conditions at bottom and surface ($v(z^* = -1) = 0$, $v_z(z^* = 0) = 0$).

By integrating twice the momentum balance over depth we obtain for the cross-circulation (see Fig. 3.51) the expression

$$v(z) = \frac{h^2 u_s}{24N} \left[-\frac{u_s}{5R} \left(10z^{*4} + 40z^{*3} + 33z^{*2} - 3 \right) \right.$$
$$\left. + \frac{f}{2} \left(8z^{*3} + 9z^{*2} - 1 \right) \right]. \tag{3.165}$$

Spiral flow

In the lower part of the water column the centrifugal force is too small to balance the surface-slope pressure gradient. Therefore, near the bottom, the transverse flow component is directed to the inner bend and carries sediment from the outer channel bank to the inner bank. The outer bank will erode, while the inner bank will accrete. At the surface the transverse flow component is directed to the outer bend. Due to the transverse circulation the tidal flow in a channel bend will have a spiraling character. In the northern hemisphere the Coriolis acceleration enhances the transverse circulation if the channel bend lies to the right of the main flow (opposite bends for ebb and flood). In wide

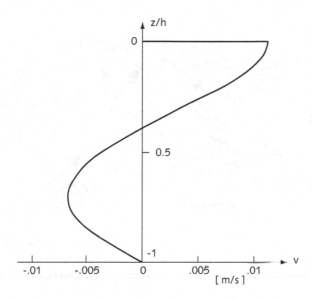

Fig. 3.51. Transverse circulation in a channel bend, for depth $h = 10$ m, longitudinal surface velocity $u_s = 1$ m/s, bend radius $R = 5000$ m, Coriolis parameter $f = 10^{-4}$ s^{-1} and eddy viscosity $N = c_D h \bar{u} = 0.0025\, h u_s$.

basins the earth's rotation tends to concentrate the flood flow along the flood right-hand bank, while the ebb flow will be concentrated along the ebb right-hand bank. Averaged over the water column the centrifugal force is balanced by a pressure gradient corresponding to a transverse surface slope. From (3.164) and (3.165) we find

$$\eta_y = \frac{9}{20} \frac{u_s^2}{g R} - \frac{5}{8} \frac{f u_s}{g}.$$

Headland eddy

The water level at the outer bank of a channel meander is higher than at the inner bank. The corresponding transverse water level inclination also causes an along-channel surface slope. At the location of strongest channel curvature the water level has a maximum along the outer bank and a minimum along the inner bank, see Fig. 3.50. The corresponding along-channel pressure gradient, $\rho g \eta_x$, adds to the inertial pressure gradient, $\rho u u_x$, generated by frictional deceleration/acceleration of the flow at the upstream/downstream regions of the shallow inner bend and frictional acceleration/deceleration of the flow at

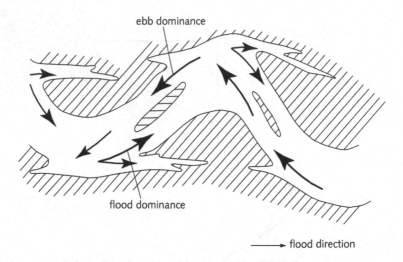

Fig. 3.52. Meandering channel system with flood and ebb dominant channels and tidal flats.

the upstream/downstream regions of the deep outer bend. These along-channel forces act together in driving a horizontal circulation, directed away from the bend along the outer bank and directed towards the bend along the inner bank. This horizontal circulation is called a headland eddy. This residual eddy is superimposed on the main tidal flow. Ebb and flood streamlines are therefore asymmetric; the flow is concentrated at the outer bank downstream of the location of strongest channel curvature. Flood and ebb flow therefore do not follow identical curves in the channel bend and, as a consequence, distinct flood and ebb channels will develop. The headland eddy has the same effect on sediment transport as the previously discussed transverse circulation: A net sediment transport from the outer bank to the inner bank. The centrifugal acceleration forces the flood or ebb flow towards shoals landward or seaward from the channel bend. These shoals are nourished with sediment carried by the flood and ebb flow and therefore tend to grow. As the water level rises during flood, flood flow is more effective for shoal nourishment than ebb flow.

Repulsion of flood and ebb channels

The tide-averaged migration of sediment along the bed in flood-dominated channels follows the flood direction; in ebb-dominated channels bed sediment migrates in the ebb direction. At the junction of flood and ebb dominated channels sediment transport converges. This convergence is responsible for the development of shoals separating flood and ebb channels and it explains the tendency of flood and ebb channels to 'avoid each other'. In some situations flood and ebb dominated channels are separated by elongate linear bars; there is some evidence that sediment transport convergence also contributes to the development of these elongate linear bars [336].

Tidal flat development

Channel meandering induces both vertical and horizontal residual circulations which contribute to transferring sediment from the outer channel bend to the inner channel bend. In this way tidal currents build up tidal flats. The strength of this process is related to the strength of tidal currents. Topographic surveys in the Eastern Scheldt estuary in the Netherlands show a strong morphologic response to long term increases and decreases of the tidal amplitude. Tidal flats build up and channels are deepened when tidal currents are strong, while tidal flats recede and channels silt up when tidal currents are weak, see Figs. 4.9 and 4.10. There are many other indications that tidal currents are responsible for tidal flat development. For instance, spring-tide sediment deposits are much thicker than the neap-tide deposits [257]. The topographic relief of tidal basins is thus increased or decreased depending on increase or decrease of tidal current strength.

From alternating bars to channel meandering

In most natural rivers alternating bars are present at the inner bend of channel meanders. This suggests that the formation of alternating bars and channel meanders are closely linked. In the previous section the formation of alternating bars was explained as a positive feedback of the flow response to an alternating-bar perturbation of the channel bed. However, the channel banks were restricted to straight lines, preventing the development of true channel meanders. One may therefore ask: What would happen if the channel banks were allowed to shift by accretion or erosion? This question has been investigated by several authors, in particular by Ikeda, Parker and Sawai [228]. They considered a simplified model for depth-averaged bend flow to which they added a bank erosion formula. With this model they performed a linear stability analysis, to investigate the stability of a small initial bank undulation and to evaluate the wavelength of maximum growth. The feedback mechanism in their model is different from the mechanism which is responsible for the growth of alternating bars; in the present case the inertial delay in the adaptation of the flow distribution to the channel curvature plays a crucial role. The greatest flow strength along the outer bend occurs downstream from the bend centre; opposite along the inner bend the flow strength has a minimum. The flow accelerates at the centre of the outer bend, causing erosion, and decelerates at the centre of the inner bend, causing accretion. Hence, the perturbation of the channel bank will grow. The wavelength of fastest growth is predicted as

$$\lambda \approx 4h/c_D, \tag{3.166}$$

which is approximately a factor 4 larger than (3.155) and matches fairly well observed meander wavelengths. A similar result was obtained by Kitikandis and Kennedy [256], who investigated bend instability through transverse bend circulation.

These results confirm that the feedback process initiating the formation of alternating bars (inertial delay in the flow adaptation to longitudinal and lateral variations in friction) is not the same as the feedback process which dominates the formation of channel meanders (inertial delay in the flow adaptation to channel curvature). The emergence of alternating bars may trigger channel meandering (see Fig. 3.53); this will be the case in wide channels with fixed banks. Afterwards channel meandering will become the dominating process which determines the longitudinal morphologic scale. The initial alternating bars are rearranged to become point bars at the inner meander bends in the river case and to become tidal flats at the inner meander bends in the tidal case.

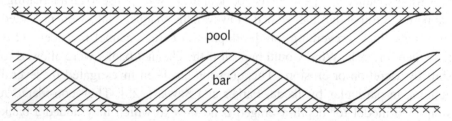

Fig. 3.53. A meandering channel can be interpreted as a straight channel with a sequence of alternating bars.

Meander evolution

Channel meandering is a continuing process in natural tidal basins. Meanders tend to grow if space for growth is available; during the growth process the channel length increases and frictional energy losses become more important. The water level difference along the meander increases, favouring tidal flow over the inner-bend shoal at the expense of tidal flow through the main channel. A bypass channel through the shoal may develop, shunting the main channel; the main channel will silt up and may even be abandoned, see Figs. 2.21 and 2.22. This process is an illustration of the instability of a two-channel system discussed in Chapter 2. The new main channel, which at first cuts straight through the tidal flat, starts developing a new meander and the whole process may repeat. In this scenario, tidal morphology is continuously evolving.

Morphological cycle in ebb-tidal deltas

Cyclic morphologic evolution is a common phenomenon in coastal systems. Cyclic behaviour is observed in particular when different opposing processes compete, as in the ebb-tidal delta of barrier tidal inlets [413, 188, 232]. The channel-shoal configuration is permanently modified due to competition between, on the one hand, channel obstruction due to wave-induced spit building and bar migration, and on the other hand, channel scouring and channel meandering due to tidal inflow and outflow. This competition leads to cyclic morphologic evolution between a situation with multiple inlets or multiple-channel inlets and a situation with a single-channel inlet. A typical example of these two alternating inlet configurations is shown in Fig. 3.54. The cycle period ranges from around 5 years for small inlets up to 100 years for large inlets. If the inlet configuration is not significantly modified after completion of a morphologic cycle the system is said to be in a state of dynamic equilibrium.

Important ignored processes

Channel meandering and tidal flat development are in reality more complex processes than described above. For instance, the simple models ignore the influence of waves and the influence of biotic processes, which play an essential role in marsh development. Yet the spatial patterns and temporal scales derived from the above idealised description of current-topography interaction correspond reasonably well with observations in natural systems, see Table 3.2.

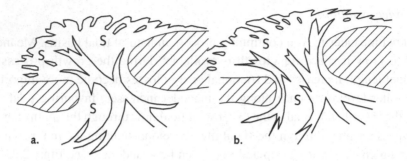

Fig. 3.54. Morphologic cycles are a common phenomenon in barrier tidal inlets (top = sea, bottom = basin). The inlet channel system is generally characterised by a central ebb-dominated channel (outflow jet) and marginal flood channels at the updrift side of the inlet (inflow from the left after LW) and at the downdrift side (inflow from the right around HW). Shoal locations are indicated by a capital S. (a) At the updrift side of the inlet wave-induced spit grow pushes the updrift marginal flood channel towards the centre of the inlet throat. At the downdrift side wave-induced shoreward bar migration blocks the downdrift marginal flood channel. (b) The marginal flood channels are restored by breaching of the updrift spit and shore attachment of the downdrift shoal. This leads to a morphological cycle between the two extreme situations shown in the figure.

This correspondence might nourish the impression that the idealised models reflect the major dynamics of pattern formation. This impression is in general misleading; it has to be kept in mind that the dynamic balances for fully developed morphology are different from the dynamic balances at the onset of pattern formation. Nevertheless we may conclude that the strongly simplified descriptions are relevant for understanding how pattern formation by current-topography interaction comes about.

3.6. Will Seabed Perturbations Always Grow?

We end this chapter with a few remarks on the response of coastal morphology to man-induced perturbations. Many different types of bedforms can be generated by current-topography interaction, see Table 3.2; the overview presented in this table is not exhaustive. The reader might have the impression that the seabed is unstable for almost any type of perturbation. In some ways this is true; at many length scales positive feedback may occur between seabed perturbations and the resulting flow perturbation. However, the most common response of the seabed to perturbation is decay, i.e., a return to the original situation. If the seabed is locally raised the most common flow response will be a local increase

Table 3.2. Bedforms which can develop as free instabilities with characteristic length scales, found from linear stability analysis discussed in this chapter and from observations. c_D is the friction coefficient.

Seabed instability	Scale parameters	Length scale relationships stability analysis	Length scale relationships observations
Current ripple	Grain diameter d	200–$4000\ d$	250–$1000\ d$
	Roughness height δ_r	2–$30\ \delta_r$	
Megaripple	Roughness height δ_r		50–$500\ \delta_r$
Dune	Water depth h	$11\ h$	1–$16\ h$
Alternating bar/	Water depth h	$0.8\ h/c_D$	3–$7\ h/c_D$
tidal flat	Channel width b		3–$9\ b$
	h, b, flow velocity u	$4(ub\sqrt{h}/c_D\sqrt{g})^{1/2}$	$16(ub\sqrt{h}/c_D\sqrt{g})^{1/2}$
Channel meander	Water depth h	$4\ h/c_D$	
	Width b		3–$9\ h$
Tidal ridge	Water depth h	0.5–$0.6\ h/c_D$	0.5–$1.5\ h/c_D$
Shoreface-connected ridge	Inner shelf width b	b	b

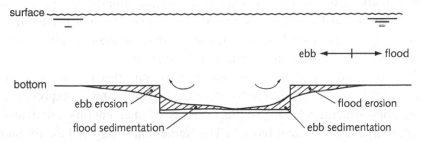

Fig. 3.55. Decay of a local seabed perturbation. The pit is filled by a diffusive process involving erosion and settling time lags and dissipation of local morphologic structures.

of flow velocity which causes scouring of the initial mound; if the seabed is locally deepened the most common flow response will be a local decrease of flow velocity with subsequent sedimentation in the initial pit, see Fig. 3.55.

The explanation of this apparent contradiction is that the seabed is seldom in a state of unstable equilibrium. Unstable seabed states may occur shortly after extreme events wiping out existing bedform patterns or shortly after sudden changes in external conditions. For instance, on a tidal beach the external conditions are strongly variable at the tidal timescale; after passage of the swash zone, the flattened submerged beach may be unstable for small scale perturbations, leading to ripple formation. However, the state of seabed instability will not last very long, compared to the growth time scale of perturbations.

The most common situation is a seabed morphology close to equilibrium or evolving towards equilibrium. Complete (dynamic) equilibrium is seldom achieved, because external conditions fluctuate faster than the time scale for reaching a state of equilibrium. The morphology is closest to equilibrium at small spatial scales, where the adaptation time scale is very short, and at very large scales, where the external conditions are slowly varying (time scale of sea-level rise). If the coastal system is in a state close to equilibrium a small perturbation (small with respect to the scales characteristic for the equilibrium morphology) will decay; otherwise the system would not evolve towards this equilibrium. In a state evolving towards equilibrium a small perturbation will also decay or be incorporated in the ongoing morphologic evolution. Sometimes this evolution can be accelerated, but this requires generally a perturbation of substantial magnitude. One may think, for example, of interventions which strengthen the imbalance between two competing channels, see Sec. 2.5.

Human intervention is not needed to trigger the evolution of an unstable seabed morphology towards a stable equilibrium; an infinitesimal perturbation is sufficient and such triggers exist in nature. One may conclude that in practice small changes in the seabed morphology will seldom generate positive feedback and be amplified as a natural response of the system.

This conclusion does not hold for interventions that modify in a structural way the external conditions to which the coastal system is exposed. Hard engineering structures belong to this category of interventions. For example, the (partial) closure of tidal basin in The Netherlands triggered the formation of sandbanks in the adjacent coastal zone and the growth of sand spits. The construction of training walls in several estuaries in the UK, to concentrate the flow in the main channel, triggered the infilling of secondary channels, see Sec. 4.6.3. Hard engineering interventions may in principle move external conditions beyond a critical threshold for the formation of new morphologic patterns. However, most thresholds are already explored by the strong natural variability generally present in the external conditions. Hard interventions that trigger the development of new types of morphologic patterns are therefore more the exception than the rule.

Chapter 4

Tide-Topography Interaction

4.1. Abstract

This chapter deals with the physical principles of tidal inlet morphodynamics. In our terminology tidal inlets include back-barrier tidal basins, estuaries and tidal rivers. Two types of inlets systems will be distinguished: Barrier tidal inlets, which have developed after breaching of a coastal barrier and river tidal inlets, which have developed after drowning of a river valley. The morphology of tidal inlet systems is highly complex; yet the analysis of field data indicates that several gross features of tidal inlet morphology can be fairly well described by simple relationships. These relationships depend primarily on tidal characteristics, but also on river influences for river tidal inlets and on wave influences for barrier tidal inlets. Before discussing the underlying tide-topography interaction we first present a general description of tidal inlet morphology and an idealised schematisation of the typical geometry of barrier tidal inlets and river tidal inlets. We review several empirical relationships; comparison with observations leads to the important concept of critical flow strength. We then discuss in a semi-qualitative way the characteristics of tidal propagation in shallow basins; a formal derivation is given in Appendix B. The findings are illustrated and further detailed for a few typical tidal basins, based on field evidence. These results are the starting point for a discussion of the major morphodynamic balances which are steering the morphology of river tidal basins and barrier tidal basins. From these balances we derive a few relationships for the equilibrium morphology, which are analysed with reference to the observed geometry of a large number of tidal inlet systems. These relationships provide some clues for the response of tidal inlets to human interventions or to sea-level rise. At the end of the chapter some consideration is given to the role and the transport dynamics of fine sediments in tidal basins.

197

4.2. Description of Tidal Inlets

4.2.1. *Origin and Evolution*

Occurrence

Tidal inlets are present all along the world coastline. Often they intrude far inland and thereby lengthen the coastline substantially. Most tidal inlets are found in the large coastal plains along the American Atlantic coast, the north-west coast of Europe, the southeast coast of Africa, the southeast coast of Asia and the northern Siberian coast, see Fig. 4.1 [45]. Tidal inlets also exist but are less frequent in coastal areas with more pronounced topographic relief [405]. In this chapter we will focus on tidal basins in low relief coastal plains, where the tide may interact with the sedimentary environment and create its own morphology. Tidal basin morphology in high relief coasts is constrained by the existing topography which results from motions of the earth's crust or glaciation processes.

Sea-level rise

Bedforms arising from tide-topography interaction span the whole water col-umn; for instance, tidal flats often build up so high that they are inundated only occasionally. The dependence of tidal basin morphodynamics on these inter-tidal zones implies a great sensitivity to sea-level rise. Most present tidal basins

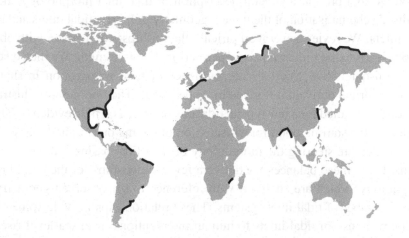

Fig. 4.1. Continental coastal plains where most of world's large tidal inlets are situated, after [45]. Most of the large continental plains are situated at subsiding coasts [36].

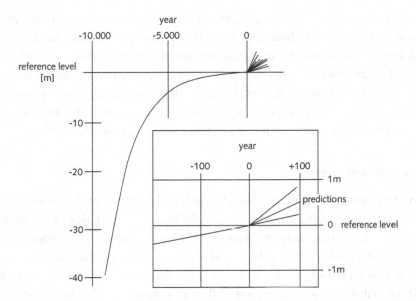

Fig. 4.2. Sea-level curve for the past 10 000 years, derived from geological reconstructions for coastal areas in the temperate climate zone of the Northern hemisphere. Expected sea-level curves for the coming century according to IPCC [484].

are recent geological structures; their development is related to the decrease of the rate of sea-level rise, starting about 5000 years ago, see Fig. 4.2. At that time, tidal currents could take control over the sediment infill in shallow inland coastal areas and morphologic features characteristic of tidal basins have started developing. The stability of tidal basin morphology has often been sufficient for keeping pace with sea-level rise and for survival of the basin up to the present. Observations of long-term average sedimentation in tidal lagoons along the US Atlantic and Gulf coast and in tidal basins of the Dutch Wadden Sea point to a net accretion rate which is comparable with the rate of sea-level rise [362, 325, 293]. It appears, however, that not all tidal lagoons on the US Atlantic and Gulf coast follow sea-level rise; sediment availability is a major limiting factor.

Marine sediment supply

Strong tidal currents are capable to import sediment from the sea, so increasing sediment availability. Along the southern North Sea coast and Wadden Sea coast, the coastal zone is the major sand source for the adjacent tidal basins

[28]. These basins developed during the holocene transgression in coastal plains behind wave-built sand barriers. We will call this type of tidal basin 'barrier tidal inlets'. These basins are nourished mainly with marine sand, which is brought to the inlet by wave-induced transport and tidal currents. Along the coast of Holland several large barrier tidal inlets have disappeared by sediment infilling in excess of sea-level rise in the period between 7000 BP and 3500 BP, in spite of a higher rate of sea-level rise than at the present [29].

River sediment supply

During the holocene transgression tidal basins also formed by drowning of river valleys. We will call these basins 'river tidal inlets' or 'estuaries'. Fine-grained terrigenous materials carried by river inflow are often the major sediment source for these tidal basins. River inflow is also responsible for estuarine circulation, with a landward along-bottom flow component which contributes, together with tidal currents, to the retention of fluvial sediments within the basin and to gradual infilling. In response to this infilling the upstream basin cross-section is narrowed, producing the funnel shape characteristic of many estuaries. The funnel shape develops if fluvial sediment input is large, but not larger than the transport capacity of tidal currents; otherwise the estuary will fill in and seaward progradation will create a subaerial river delta, from which the tide is excluded. Examples are the Yellow River, the Yukon and the Mahakam delta; yet other rivers with high sediment load are funnel-shaped, for instance, the Amazon, Fly, Ganges-Bramaputra and Hooghly (see Fig. 4.3).

Role of biogeochemical processes

The infill of tidal basins is not only due to marine and fluvial sediment supply; biogeochemical processes also play an important role [349]. Deposits of fine sediment would easily be washed away by currents and waves if they were not stabilised by geochemical and microbiological processes. The resistance against surface stress is greatly increased by the production of biofilms and algal mats and by exposure to sunlight [8]. Biotic activity stimulates the formation of large particle aggregates, especially during summer; this leads to increased sedimentation, also because viscosity is decreased at high water temperature. At the same time, erodibility is increased by reworking of sediment through bioturbation [9]. Colonisation of tidal flats by pioneer vegetation strongly enhances stabilisation and retention of sediment [320]. Once a bare

Fig. 4.3. Average annual sediment yield (10^9 kg/year) of world's major river basins. For the large continental plains the sediment yield is relatively small, except for the south-east Asian plain and the South-American Atlantic plain. Redrawn after [26].

tidal mud flat is colonised by plants, the sediment bed rapidly accretes till high-water level. After that, the accretion rate decreases until the marsh finally reaches an equilibrium level close to the highest astronomical tide [436]. At that stage the marsh has ceased to serve as flood storage area; in fact, salt marsh development excludes these areas from tidal influence and therefore affects tidal hydrodynamics in a way comparable to land reclamation.

Tidal basin evolution

Tidal basin infilling proceeds in several distinct stages. The following sequence has been established for the Dutch coastal zone [460, 31]. After the sea intruded in low lying coastal plains, marine and fluvial sediments were initially reworked by tidal currents into a pattern of channels and tidal flats. In the second stage, biogeochemical processes contributed to transforming tidal flats into marshes where fine sediments were easily trapped, raising the sediment bed above the high water level. The tidal discharges decreased; the decrease was in the next stage enhanced by wave-induced sedimentation at the inlet. The fluvial influence increased, turning increasing parts of the lagoon into fresh water marshes with vigorous vegetation. Successive generations of marsh vegetation grew into peat bogs, which expanded vertically and horizontally. The tidal influence further decreased and finally the basin was closed by longshore sand transport. In reality this process had to compete with sea-level rise, and complete infilling of tidal basins was not always achieved. Several Wadden Sea tidal basins, for instance, have persisted without great alteration since at least a few thousand years. Later in this chapter we will see that tide-topography interaction contributes to the stability of tidal basins by steering their response to sea-level rise.

4.2.2. *Barrier Tidal Inlets*

Barrier tidal inlets are typically sandy systems. They develop in coastal plains, after barrier breaching by storms and transgression of the coastal plain due to sea-level rise. Examples of this type of tidal basins include the Wadden Sea basins, the Eastern Scheldt, the Baie d'Arcachon and smaller inlets along the Portuguese Algarve coast, the US Atlantic coast and coasts of Australia and New Zealand. Most sediment in these basins is derived from the sandy near-shore zone. Inflowing rivers have a small width compared to the back-barrier basin and fluvial sedimentation is of minor importance compared to marine sedimentation. Figure 4.4 gives an impression of the morphology of barrier tidal inlets in the Wadden Sea and along the Southern Bight of the North Sea.

Tidal basin morphology

Several characteristics of tidal basin morphology in a sandy coastal environment have already been described in the chapter on current-topography interaction. Tidal flow over a broad, flat channel bed generates instabilities at different spatial scales, producing bedforms such as ripples, megaripples, dunes, bars and ridges. Some of these bedforms may grow into large structures at the basin scale; an alternating shoal-pool instability, for instance, may grow into a meandering and braided channel system. Tidal flow through channel meanders stimulates the development of distinct flood and ebb dominated channels, separated by shoals and elongate bars; flow divergence and sediment deposition at the end of these channels contributes to creating inner flood and ebb deltas. The outer-bend channel banks are steeper than the inner-bend banks. Channel bifurcations are situated at the outer channel bends, while tidal flats are located at the inner bends. The channel width converges in the landward direction; the channel depth also decreases, but not as strongly as the width. The

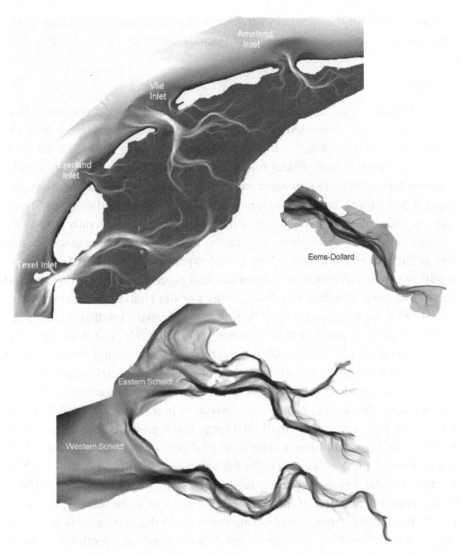

Fig. 4.4. Tidal basins along the Dutch coast; the basin length (in km) and tidal range (in m) are indicated in brackets. The Western Scheldt (estuarine part 90, 4.5; dark is deep) and Eastern Scheldt (40, 3.5; dark is deep) are part of the Rhine–Meuse–Scheldt Delta, the Texel Inlet (45, 3; dark is shallow), Eijerland Inlet (15, 3; dark is shallow), Vlie Inlet (35, 3; dark is shallow), Ameland Inlet (25, 3; dark is shallow) and Ems-Dollard (35, 3.5; dark is deep) are situated in the Wadden Sea. The Eastern Scheldt was greatly affected by the construction of a storm surge barrier at the inlet and compartment dams at the landward end in 1985; the Texel Inlet is the main inlet of the former Zuyderzee, which was dammed off in 1930. The main channels of the Western Scheldt and Ems-Dollard are permanently dredged for navigation.

meander length of the channel decreases with decreasing depth. Many of the characteristic large-scale morphologic features can be recognised in Fig. 4.4.

Tidal flats

Strong tidal flow prevents accumulation of fine sediments on the channel bed; fine sediments are deposited on tidal flats, in wave-sheltered areas near the boundaries of the basin and at the divide of tidal basins in the Wadden Sea. In the seaward part of barrier tidal inlets, tidal flats are more sandy than muddy, because fine sediments deposited under calm summer conditions are resuspended by currents and wave action during winter. In the landward part, mudflats may occupy a substantial part of the intertidal zone, especially when they become colonised by halophytic vegetation; this enhances their capacity to trap fine sediments and stimulates their lateral expansion [196]. The dynamics of mudflats strongly depends on biogeochemical processes, but also on tides and waves; the gross morphologic characteristics of mud flats appear to be primarily related to the tidal range [126]. Sediment supply to the mudflat is stimulated by the flood dominance of tidal currents [354]. The tide also determines the maximum height to which mudflats can grow and it determines the duration of exposure to wave action for different parts of a mudflat. Several studies indicate that the depth profile of tidal flats is consistent with the requirement of uniform bottom shear stress, so as to minimise gradients in residual sediment transport [508, 163]. In a situation of small tidal range and high wave activity, the depth profile is typically concave, minimising gradients in the wave-induced shear stress. Similar profiles are found for the shoreface of an open coast (see 5.5). In this case the sediment grain size decreases with decreasing depth [508]. In the opposite situation of large tidal range and low wave activity, the depth profile is typically convex, minimising gradients in the maximum tidal current bottom stress; here the grain size increases with decreasing depth [163]. In the Wadden Sea both situations occur [508]. Other factors also influence tidal flat profiles, such as the shoreline curvature [163]. The great sensitivity to wave action causes strong seasonal variations in mud flat sedimentation and erosion.

Outer deltas

Wave-induced longshore transport brings marine sediment to the tidal inlet; this sand input may nourish the basin, but it may also lead to constriction of the inlet or in some cases even to closure [457]. The actual inlet width is a dynamic balance between longshore sand transport and tidal flow through the inlet [170]. If longshore sand transport is strong the inlet is narrow; otherwise the inlet will be wide. The cross-sectional area A_C is primarily related to the tidal prism P (the water volume passing through the inlet during ebb or flood); it is affected by wave-driven sand import only to a lesser extent. Tidal inlet dynamics in the presence of strong tides and waves is very complex. At the inlet an outer delta system of shoals is formed, also known as ebb-tidal delta. The outer delta shoals act as swash platforms for breaking waves; swash-bar migration plays an important role in allowing longshore wave-induced sand transport to bypass the inlet [47]. Often this is not a continuous process, but involves cyclic channel and shoal migration at the inlet. These inlet cycles may take from over a decade up to almost one hundred years for large inlet systems [334, 402, 413, 232], see also Sec. 3.5.4. The geometry of tidal inlets may strongly differ, depending in particular on tidal range, strength and direction of wave-induced sediment transport [403]. Nevertheless, certain gross characteristics of outer deltas seem to be directly related to the tidal prism P. Walton and Adams [480] have established an empirical relationship by comparing the tidal prism with the sand volume V stored in the outer tidal delta of inlet systems along the US Pacific, Atlantic and Gulf coasts. They have found that these two quantities are almost linearly related,

$$V = c_e P^{n_e}, \qquad (4.1)$$

where c_e is a constant which primarily depends on wave exposure and n_e an exponent with average value 1.23. A similar relationship has been found for the outer deltas of the sandy tidal lagoons along the coast of Florida [357]; this study shows a considerable scatter in the constant c_e and the exponent n_e. Average values are $c_e = 0.2$ (S.I. units) and $n_e = 1$.

4.2.3. *River Tidal Inlets*

River tidal inlets have developed when the sea entered existing river valleys. Often, the drowned river valley was scoured into ancient sediments with considerable resistance to erosion; the existing topography therefore places an important constraint on morphologic development related to tidal intrusion. Bed sediments in river tidal inlets consist dominantly of silt and clay, especially at the inner estuarine zone. Coarse-grained bed sediments, such as sand and gravel, prevail in the outer estuary. Many of the large river tidal inlets in the world belong to this category of tidal systems, for instance, the Thames, Humber, Severn, Seine, Gironde, Elbe, Weser, Delaware, St. Lawrence, Hooghly, Ord, Fly, etc. The term estuary is often used for these tidal inlets; however, this term refers usually to the gradual seaward transition from fresh water to sea water and to the influence of this transition on water motion, due to density gradients. The term river tidal inlet refers to morphologic characteristics; these characteristics are only partly related to the influence of density gradients between fresh water and sea water. When we want to emphasise morpho-logic aspects we will use the term river tidal inlet instead of the more common appellation estuary.

Sedimentation and turbidity maximum

A major proportion of the sediment input is of fluvial origin; these sediments contain a large fine fraction of clay and silt. Several strong mechanisms operate to retain fine fluvial sediments within the estuary. Flood and ebb currents nor-mally have different strengths. The tidal current component in tidal river inlets is asymmetric with typically stronger flood than ebb currents; the tide therefore contributes to a net upstream sediment transport. Fresh water inflow produces a density gradient along the estuary; this density gradient drives a circulation which, near the bottom, is directed upstream. Both mechanism contribute to trapping fine sediments in a zone near to the limit of seawater intrusion (or even upstream of it, in the case of low river discharge [6]). In this zone the turbidity reaches a maximum and a fluid mud layer may develop at the channel bed, in particular during neap tidal conditions (the Severn and Gironde are well documented examples [5, 254, 83]). The turbidity maximum moves up and down the estuary with tide and river discharge, over distances comparable to the displacement of the seawater intrusion limit, see Fig. 4.5. The turbid-ity maximum constitutes a pool of mobile sediment that, unlike sand, cannot develop into long-lasting bedforms. Most of the stable bedforms in the estuary consist of sand imported from the inlet region; these sand bodies may develop into sandbanks overlying the estuarine mud bed [90]. In a mature stage they may be capped with mud, develop vegetation and even grow into islands (for instance in the Fly and Ord rivers, [195]). Mud is mainly deposited along the channel banks, where the shear stress exerted by currents is less strong than

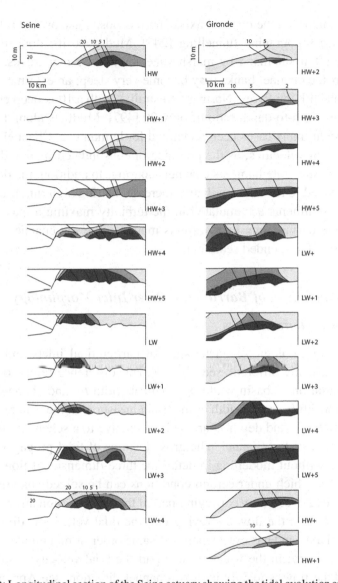

Fig. 4.5. Left: Longitudinal section of the Seine estuary showing the tidal evolution of a turbidity maximum at spring tide and low river discharge (October 1978). Numbers indicate hours after high or low water at Le Havre. Seawater intrusion is indicated by the salinity isolines for 20, 10, 5 and 1 ppt. Redrawn after [18]. Right: Longitudinal section of the south channel of the Gironde estuary showing the tidal evolution of a turbidity maximum at spring tide and high river discharge (May 1974). Seawater intrusion is indicated by the salinity isolines for 10, 5 and 2 ppt. Redrawn after [5] with permission from Elsevier. The turbidity maximum is best developed shortly before HW and LW, i.e., close to peak-flood and peak-ebb velocities. The turbidity maximum extends upstream from the limit of seawater intrusion.

at the bottom along the channel axis. These deposits narrow the tidal channel and produce topographic funnelling [247]. Mud deposits along the channel banks are often stabilised by marsh vegetation or by mangroves. In narrow tidal rivers the channel banks may become very steep; an extreme example is the macrotidal King River (Northern Australia), with a 10 m deep channel and an average width-to-depth ratio of only 20 [497]. Mudflats along the channel banks have in many cases been reclaimed for human use. Channel narrowing increases tidal velocities, as the presence of consolidated mud and underlaying resistant substrate hampers channel scouring. In addition, the downstream velocity related to river flow will also increase. During high runoff events, part of the fine sediments accumulated in the turbidity maximum may be flushed out of the estuary; strong tidal currents may also prevent channel obstruction by dispersal of suspended sediment.

4.2.4. *Comparison of Barrier and River Inlet Morphology*

Morphologic characteristics

The gross morphologic characteristics of barrier tidal inlets and river tidal inlets are basically different, see Fig. 4.7. This difference refers to the along-channel variation of basin width b_S, channel width b_C and channel depth D. The basin width b_S is the width of the basin measured at the water surface; the channel width b_C and depth D are defined relative to a schematic rectangular representation of the channel. The gross features of tidal propagation can be established without modelling in detail the three-dimensional flow field. In a crude model, which under certain conditions can be solved analytically, tidal flow is divided over a flow conveying part of the cross-section and a water storage part [115]. In the flow conveying part the tidal velocity is directed along the channel axis and has everywhere the same order of magnitude as the average flow velocity; in the water storage part the tidal velocity is substantially lower and the flow is mainly in cross-channel direction, see Appendix B and Fig. B.2. The former part of the cross-section is defined as the channel; D is the representative depth of the channel. The width b_C is chosen such that the product Db_C is equal to the cross-sectional area A_C of the channel, see Fig. 4.6. The tidal velocity u is defined such that the product $A_C u$ equals the total discharge Q. Another related characteristic morphologic quantity is the propagation depth D_S, defined as $D_S = A_C/b_S = Db_C/b_S$; it is called the propagation depth because the tidal propagation speed c strongly depends on this quantity. The quantities D, D_S, b_S and b_C depend on the tidal phase; however, a simple analytical solution of the tidal equations can be obtained only if we consider tide-averaged values: h instead of D and h_S instead of D_S.

Typical barrier inlet morphology

The morphology of barrier tidal inlets is typically represented by a channel cross-section A_C decreasing in landward direction and a more or less uniform basin width b_S, see Fig. 4.7. Towards the head of the basin, tidal flats constitute an increasing part of the cross-section. This type of morphology produces significant tidal wave reflection at the landward boundary.

Typical river inlet morphology

The morphology of a river tidal inlet is typically funnel-shaped. The basin width b_S strongly decreases upstream and the channel cross-sectional area A_C decreases at approximately the same rate, see Fig. 4.7. The landward decrease of channel depth is much less, however, and the same applies to the propagation depth h_S. In idealised models of river tidal inlets we will initially neglect the landward decrease rate of propagation depth $|h_{Sx}/h_S|$ relative to the decrease rate of basin width $|b_{Sx}/b_S|$. The landward width convergence of river inlets strongly reduces the upstream propagation of tidal energy. No strong tidal wave reflection occurs in this type basin.

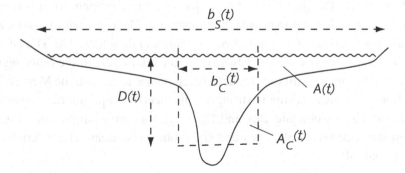

Fig. 4.6. Schematisation of the basin cross-section into a rectangular flow-conveyance section and a storage section. The schematisation depends on the tidal phase.

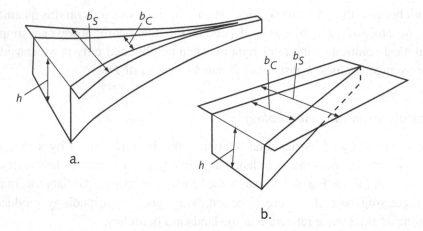

Fig. 4.7. Sketch of the typical geometry of (a) river tidal basin and (b) barrier tidal basin. The morphology of river tidal basins is characterised by landward convergence of channel and basin width, the morphology of barrier tidal basins by non-converging basin width.

4.2.5. *Other Inlet Types*

The previous description of barrier inlets and river inlets is a very crude simplification of reality. Most tidal basins deviate substantially from these simple descriptions, for several reasons. Topographic constraints may play a role, as also human interventions in the basin morphology. Many tidal basins have characteristics belonging to both categories. The outer region can be characterised as a barrier tidal inlet, while the inner region corresponds to a drowned river valley. Typical examples of such mixed inlets are the Western Scheldt and the Weser-Jade. The inner part of these basins is river influenced; it is funnel-shaped with a well developed turbidity maximum. The outer part is sandy, with meandering flood and ebb dominated channels and tidal flats. The Thames and the Elbe also possess characteristics of barrier tidal inlets in the outer region. Other tidal basins with mixed characteristics are the Tagus and the Mersey. The morphology of these basins is strongly influenced by topographic constraints at the inlet zone. Cobequid Bay and Delaware Bay are examples of a funnel-shaped drowned river valley which exhibit otherwise many characteristics of barrier tidal inlets.

Tidal systems not in pace with sea-level

There is also a category of tidal basins with morphologic characteristics differing both from barrier inlets and river inlets. In these basins sediment infill does not keep pace with sea-level rise [325] because of insufficient sediment supply. This is the case for several large microtidal basins along the US Atlantic coast, in particular Pamlico Sound and Chesapeake Bay [322]. The inner parts of these basins have no pronounced channel and tidal flat structures. Tidal currents are weak, except near the entrance, and the shallow nearshore zone is wave-dominated. Due to the weakness of tidal currents in the inner parts of both excess-supply and deficit-supply basins one should not expect a strong sedimentation feedback to tide-induced sediment transport.

4.2.6. Critical Flow Strength

Equilibrium inlet cross-section

One may expect tidal basin geometries to be very different, according to tidal basin origin, substrate, prevailing hydrodynamic conditions, sediment supply and climate. In spite of this great diversity, certain geometric features seem to be shared by most tidal basins. A well established common feature of tidal inlets is the almost universal relationship between the tidal inlet cross-sectional area A_C and the tidal prism P; this relationship was first empirically established by O'Brien [330],

$$A_C = C_A P^{n_A}, \tag{4.2}$$

where P is the tidal prism for an average spring tide and where the constants C_A and n_A are found by fitting (4.2) to data on A_C and P for different coastal inlets. A study including data of 162 inlets along the US coast indicates that the value of n_A ranges between 0.84 and 1.1 [237]. This relationship should be considered as a long-term average, ignoring the short-term variations in the cross-sectional area of the channel. Equation (4.2) can also be written as a relationship between the tidal prism and the maximum tidal current speed U at the inlet, averaged over the channel cross-section. Assuming a sinusoidal tide at the inlet we have

$$A_C = \omega P/(2U). \tag{4.3}$$

Combining this expression with (4.2) yields

$$U = \frac{\omega}{2C_A} P^{n_A - 1}. \tag{4.4}$$

The values of C_A and n_A are not the same for all sedimentary coastal environments; they are influenced by factors such as the magnitude of annual mean littoral drift, by flood-ebb asymmetry and by river discharge [168]. Other studies have also revealed a significant influence of jetties or headlands on the parameter C_A [237, 219]. Yet many inlets are fairly well represented by (4.2) with the values $n_A = 1$ and $C_A = \omega/2$. This implies that for these inlets the maximum tidal current speed U for an average spring tide is in the order of 1 m/s (averaged over the inlet cross-section).

Critical shear stress

From several studies it appears that the relationship between cross-sectional area and tidal prism not only applies to inlets, but also to the channels within a tidal basin [73, 60, 245]. Friedrichs [162] explains this relationship by comparing the peak spring tidal velocity U with the critical shear stress for channel bed scouring. Therefore he investigates the empirical equilibrium relationship

$$U \propto h^{-1/6} Q_m^{1-n}, \tag{4.5}$$

where Q_m is the peak-tidal discharge, $Q_m = A_C U$. The factor $h^{-1/6}$ enters equation (4.5) through the friction coefficient c_D which has been assumed proportional to $h^{-1/3}$ (3.16). Comparison with observations shows that, on the average, the coefficient n is close to 1; in flood dominated estuaries n is slightly larger than 1 and in ebb dominated estuaries n is slightly smaller. Flood dominance ($n > 1$) relates to a landward increasing equilibrium velocity, because Q_m decreases in upstream direction. Friedrichs proposes that this is due to upstream sediment transport driven by tidal asymmetry; this upstream transport produces a turbidity maximum and tends to constrict the channel cross-section. The channel will not get clogged, however, because channel constriction raises the tidally induced shear stress above the critical equilibrium value at which the channel bed is scoured and sediment is dispersed.

Channel convergence

Uniformity of the cross-sectionally averaged, maximum ebb/flood velocity U throughout the tidal basin is related to the convergence rate of the channel cross-section. An expression for the cross-sectional convergence rate can be derived from the assumption of constant U for idealised schematisations; these rates can be compared with convergence rates of existing tidal basins. This comparison provides a consistency check for the constancy of the maximum tidal velocity. For a sinusoidal tide the landward decrease of tidal prism per unit length is given by $-(2/\omega)dQ_m/dx$, where Q_m is the maximum tidal discharge, $Q_m = A_C U$, and $\omega = 2\pi/T$ is the radial tidal frequency. We assume that the variation of tidal level is almost synchronous in the basin; this is equivalent to the condition that the tidal basin length (or the convergence length for strongly converging basins) is much smaller than a quarter tidal wavelength. The mass balance then implies that the landward decrease of tidal prism is also equal to the tidal volume stored in the basin per unit length, $2ab_S$, where a is the tidal amplitude and b_S the average surface width of the basin. We call h and $h_S = A_C/b_S = hb_C/b_S$ the tidally averaged channel depth and propagation depth, we assume $h, h_s \gg a$ and assume that the maximum tidal velocity U is constant throughout the basin. We then may write

$$dQ_m/dx = h_S U db_S/dx + b_S U dh_S/dx = -\omega b_S a. \qquad (4.6)$$

In the following we will derive from this equation an estimate of the tidal basin length scale (or inverse cross-sectional convergence rate) for three idealised situations, the first corresponding to the characteristic geometry of river tidal inlets, the second corresponding to a geometry with constant width-to-depth ratio and the third corresponding to the characteristic geometry of barrier tidal inlets.

Tidal rivers with decreasing width-to-depth ratio

The characteristic geometry of river tidal inlets is the funnel shape. In these estuaries the cross-sectional area decreases in the upstream direction mainly through a decrease of basin width b_S and channel width b_C. The depth h decreases too, but at a lesser rate; $|db_S/dx|/b_S$ is typically much greater than $|dh_S/dx|/h_S$. If we ignore, as a first approximation, the second term at the r.h.s. of (4.6), then the surface width b_S should obey the equation

$$\frac{1}{b_S}\frac{db_S}{dx} = -\frac{a\omega}{h_S U}. \tag{4.7}$$

The solution of this differential equation reads

$$b_S(x) = b_S(0)e^{-x/L_b}, \tag{4.8}$$

where $L_b = -b_S/b_{Sx}$ is the width convergence length of the basin. This length scale is given by

$$L_b^{(o)} = \frac{h_S U}{a\omega} = U\left(\omega\frac{a}{h}\frac{b_S}{b_C}\right)^{-1}. \tag{4.9}$$

A better approximation $L_b^{(1)}$ is obtained by taking into account the landward decrease of h_S; therefore we substitute (4.8) and (4.9) in (4.6) and find

$$L_b^{(1)} = L_b^{(o)}\left(1 + L_b^{(o)}\frac{h_{Sx}}{h_S}\right)^{-1}. \tag{4.10}$$

Length of tidal rivers with constant width-to-depth ratio

An approximate expression for the length of tidal rivers with triangular channel cross-section and constant width-to-depth ratio has been derived by Prandle [359], based on the assumption of constant (x-independent) tidal amplitude. In the case of friction-dominated tidal flow he found the following expression for the tidal velocity amplitude U

$$U \approx 0.024 c_D^{-1/2} a^{1/2} h^{1/4} \approx 0.5 a^{1/2} h^{1/4}, \tag{4.11}$$

where h is the average cross-sectional depth (half the maximum channel depth) and where we have used for the friction coefficient the value $c_D = 0.0025$. According to this expression the tidal velocity amplitude decreases slightly in landward direction with decreasing depth. The tidal river length L_h, defined by $h(0) = -\int_0^{L_h}(dh/dx)dx$, is given by [359]

$$L_h \approx 5.6\,10^3\,a^{1/2}h(0)^{1/4} \approx 1.6\frac{h(0)U(0)}{a\omega}. \tag{4.12}$$

This expression is similar to (4.9), but the length L_h is greater than $L_b^{(o)}$. This is consistent with the higher-order approximation (4.10), which accounts for the landward decrease of h_S.

Length of barrier tidal inlets

The characteristic geometry of barrier tidal basins corresponds to an almost constant basin width b_S and a linearly decreasing channel cross-section A_C in landward direction

$$A_C(x) = A_C(0)(1 - x/L_A), \tag{4.13}$$

where $L_A = -A_C/A_{Cx}$ is approximately equal to the basin length. Constancy of b_S implies that h_S varies as A_C, $h_S(x) = h_S(0)(1 - x/L_A)$. Substitution of (4.13) in (4.6) with the assumption of constant U yields

$$L_A^{(0)} = \frac{h_S(0)U}{a\omega}, \quad L_A^{(1)} = L_A^{(0)}\left(1 + L_A^{(0)}\frac{b_{Sx}}{b_S}\right)^{-1}. \tag{4.14}$$

It appears that the expressions for $L_b^{(0)}$ and $L_A^{(0)}$ are essentially the same.

Comparison with observations

In Fig. 4.8 the observed convergence length scales L_b, L_A of existing tidal basins (see Table 4.2) are compared with the length scale

$$l = hb_S^- U_c/ab_S^+\omega, \tag{4.15}$$

where $U_c = 1\,\text{m/s}$ and where b_S^+ and b_S^- represent the total basin width at resp. high water and low water. We use l instead of the estimates (4.10), (4.14), because it can be computed more easily from existing data (Table 4.2). The condition that the basin length (or convergence length) is much smaller than a quarter tidal wavelength is not well satisfied for the longest tidal basins, in particular the Hudson River and the Bay of Fundy. The comparison shows that the observed length scales L_b, L_A are sometimes close, and in general of the same order of magnitude as the estimate l. The relationship between l and the estimates (4.10), (4.14) is discussed below.

(a) The channel width b_C is generally smaller than the low water width b_S^-. In this respect l is an overestimate of L_b, L_A.

(b) The cross-sectionally averaged, maximum tidal velocity U is not always close to 1. In some basins U is smaller, for instance in the Tamar, Conwy and Potomac; these basins have also a higher ratio l/L_b than other basins of comparable length. For these rivers l is an overestimate of L_b, L_A. In other

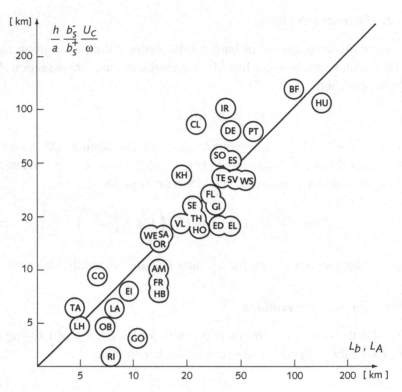

Fig. 4.8. The observed convergence length scales L_b for river tidal basins and L_A for barrier tidal basins are compared to the estimate $l = hb_S^- U_C/ab_S^+ \omega$. A good correspondence (solid line) is expected if (1) the maximum tidal velocity is on the order of 1 m/s and constant throughout the basin, (2) the channel width is well represented by the basin width at low water and (3) propagation depth is uniform for river tidal basins and basin width is uniform for barrier tidal basins. Taking these conditions into consideration it may be concluded that the observed length scales are consistent with the assumption of uniform maximum tidal velocity throughout the tidal basins (see the text).

basins U is larger than 1, for instance, in the Thames, Humber, Severn, Seine, Gironde, Columbia, Ord, Hooghly and Fly.

(c) At small tidal basins the propagation depth and the width both decrease in the landward direction. This holds, for instance, for the Ribble estuary, Gomso Bay and Humber. In that case l is an underestimate of the convergence length scale, because $L_b^{(1)}$, $L_A^{(1)}$ are significantly larger than $L_b^{(0)}$, $L_A^{(0)}$. For large tidal basins the ratios $|db_S/dx|/b_S$, $|dh_S/dx|/h_S$ are smaller than for small tidal basins and can more easily be ignored in (4.10) and (4.14).

(d) The average propagation depth h_S used for the computation of l is smaller than the propagation depth $h_S(x = 0)$ at the mouth. The difference between h_S and $h_S(x = 0)$ is greatest for small tidal basins; correcting for this difference will increase the value of the convergence length for small barrier inlet systems.

Consistency with uniform tidal velocity

The four considerations above indicate that most larger basins will be above the line $l = L_b, L_A$ if we correct for the maximum tidal velocity. They are brought downward, however, by taking into account the smaller channel width relative to low water width. Correcting for the maximum tidal velocity brings all the smaller tidal basins below the line $l = L_b, L_A$; they are brought upward, however, by taking into account the landward decrease of both basin width and propagation depth and by correcting for $h_S(0)/h_S$. The considerations (a)–(d) together suggest that the convergence length scales L_b, L_A of all tidal basins are fairly well represented by the expressions (4.10) and (4.14). This implies that the observed convergence lengths are consistent with the assumption of almost constant maximum tidal velocity U throughout the tidal basins.

Tidal flat accretion and erosion

Sufficient strength of tidal currents is not only necessary for initiating transport of channel bed sediments. It is also a prerequisite for the development of tidal flats. This is illustrated by the evolution of tidal flats in the Eastern Scheldt basin, a large meso-tidal basin in the southwestern part of The Netherlands. The tidal dynamics of the Eastern Scheldt and its morphologic evolution are described in Sec. 4.4.4. During two successive periods of several decades each, the tidal currents in the Eastern Scheldt at first increased by about 10% and later decreased by about 30%. The first intervention increased the area of tidal influence, producing an increase of tidal prism and tidal current strength. A significant increase of the tidal flat area was observed afterwards and a corresponding increase of the channel depth, see Fig. 4.9. The second intervention, about 20 years later, decreased the tidal amplitude (by the construction of a storm surge barrier) and the area of tidal influence (by the construction of compartment dams). This produced a decrease of the tidal current strength, which was followed by a decrease of the tidal flat area and a corresponding

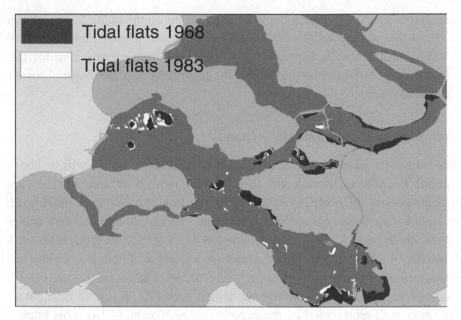

Fig. 4.9. Tidal flats in the Eastern Scheldt in 1968 (black) compared to tidal flats in 1983 (black and white), after [103]. In 1965 the tidal prism of the Eastern Scheldt was increased by about 10 % due to the construction of a compartment dam between the Eastern Scheldt and the Grevelingen inlet further to the north. By 1983 the tidal flat area in the Eastern Scheldt had significantly increased compared to 1968. Most of the sediment was delivered by channel scour.

decrease of the channel depth [294], see Fig. 4.10. During the second period tidal flat erosion was mainly due to wave action and eroded sediments were deposited on the subtidal channel flanks. Before the first intervention, the Eastern Scheldt basin had evolved for many decades without major interventions; we may therefore assume that the basin morphology was close to equilibrium. The observed morphologic changes may therefore be interpreted as a natural morphologic response to the external interventions of first an increase and later a decrease of tidal prism and current strength.

Morphodynamic balance

These observations illustrate that sand is moved from the channel to the intertidal shoals when tidal currents are above a critical strength; this process tends to reduce the tidal current strength by deepening the channels and increasing their cross-section. Conversely, when tidal currents are below this critical strength

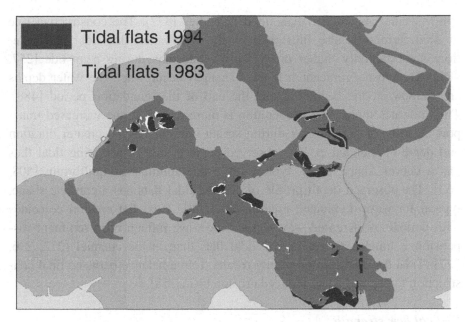

Fig. 4.10. Tidal flats in the Eastern Scheldt in 1994 (black) compared to tidal flats in 1983 (black and white), after [103]. The construction of a storm surge barrier at the inlet and compartment dams in the seaward part of the basin in 1985 reduced the tidal prism of the Eastern Scheldt by about 30%. By 1994 the tidal flat area had significantly diminished compared to 1983. Most of the sediment was deposited in the channels below the LW line.

the sedimentation-erosion balance of the tidal flats is shifted towards net erosion and the channel cross-section is reduced. This process tends to increase the tidal current strength and to restore the sedimentation-erosion balance of the intertidal areas. It appears that tidal flats play a crucial role in tidal basin morphodynamics. They serve as sediment buffers for adjusting tidal currents to a critical strength at which constructive and destructive forces acting on the intertidal shoals are in balance.

Tide-wave competition

Many other observations confirm the existence of a sensitive balance between accretion and erosion of tidal flats. Tidal flats generally grow under calm weather conditions, even though suspended sediment concentrations are much lower than under wave action. Tidal flat accretion largely depends on sediment import from the channel; accretion is observed even in situations where ebb velocities are higher than flood velocities [377]. The constructive role of tidal currents is also illustrated by the observation that tidal flat accretion is substantially higher during spring tide than during neap tide [257, 488]. The most significant bed level changes occur when the water depths are shallow, at the beginning and the end of the inundation period [488]. This indicates that tidal flat accretion is more related to the increased transport capacity of the currents during spring tide than to the greater duration and depth of inundation. The observed depth profile of accreting tidal flats has a convex shape [255], typical for a tide-dominated equilibrium [508, 163]. The observed depth profile of eroding tidal flats has a concave shape, typical for wave-dominated equilibrium [508]. Tidal flat erosion coincides with periods of increased wave action; the wave influence on sediment suspension is much stronger on the tidal flats than in the channel [272, 236, 377]. Tidal flat equilibrium mainly results from a balance between tidal construction forces and wave-induced erosion [362, 258].

Critical flow strength

One may expect that the critical tidal current strength, at which accretion and erosion of tidal flats are in balance, depends on the strength of erosional forces acting on the tidal flats and on the erosional resistance of tidal flat deposits. At barrier tidal inlets, tidal flats consist mainly of sand and erosion is mainly caused by wave action. Maximum tidal velocities throughout these basins are typically on the order of 1 m/s, which appears to be the critical tidal current strength for balancing accretion and erosion of tidal flats. Channel erosion or channel accretion is the major form of morphodynamic feedback for maintaining this maximum tidal velocity.

Tidal flats in river inlets consist mainly of more or less consolidated muddy sediments, which are often stabilised by vegetation cover. When, in addition, the channel bed is resistant to tidal scouring, there is no strong morphodynamic feedback to maintain the tidal velocity at a given critical value. Tables 4.1 and 4.2 shows that maximum tidal current velocities in river tidal inlets often substantially deviate from 1 m/s; observed maximum tidal velocities may be either larger or smaller.

It should be noted that the large-scale morphodynamic equilibrium of tidal inlets does not depend only on cross-sectional sediment redistribution, but also on sediment transport in the along-channel direction. Ebb-flood asymmetry

Table 4.1. Morphologic parameters of tidal systems. All tides are of the semidiurnal type except Columbia River, which is of the mixed-diurnal type.

Identifier	Region	Tidal inlet	L_e (km)	h (m)	a/h	b_S^+/b_S^-	L_b, L_A (km)
WS	Scheldt-	Western Scheldt	160	15.0	0.135	1.4	50
ES	Delta	Eastern Scheldt	45	15.0	0.1	1.6	40
TE	Wadden-	Texel Inlet	45	4.7	0.17	1.2	35
EI	Sea	Eijerland Inlet	15	3.0	0.31	3.0	10
VL		Vlie Inlet	35	3.9	0.24	1.7	20
AM		Ameland Inlet	25	3.75	0.3	2.4	15
FR		Frisian Inlet	20	4.25	0.27	3.0	15
LA		Lauwers Inlet	20	3.3	0.32	3.6	8
ED		Ems-Dollard	42	6.4	0.22	1.8	35
OB		Otzumer Balje	10	3.6	0.39	4.0	7
EL		Elbe	180	10.0	0.2	2.0	40
WE		Weser		8.0	0.22	2.0	13
TH	UK	Thames	95	8.5	0.24	1.4	25
LH		Langstone Harbour	6	4.0	0.38	3.9	5
HB		Humber	120	7.0	0.46	2.0	15
TA		Tamar	31	5.0	0.63	2.0	4.7
SV		Severn	150	15.0	0.2		41
FT		The Fleet	12.5	0.7	0.81	5.3	
RI		Ribble	28	7.0	0.4	6.0	8
CO	Wales	Conwy	22	3.0	0.8		6.3
GI	France	Gironde	160	9.0	0.28		30
SE		Seine	160	11.0	0.32		25
PR	USA-	Price	7.1	3.3	0.21	5.5	
MM	Atlantic	Murrells Main Creek	8	1.9	0.39	2.5	
MO		Murrells Oaks Creek	4.7	1.4	0.55	8.0	
LR		Little River	10	2.0	0.29	3.0	
MA		Manasquan	9.2	1.5	0.4	1.2	
WA		Wachapreague	10	3.6	0.15	3.5	
NA		Nauset	8.2	2.0	0.39	1.5	
SB		Swash Bay	6.4	2.5	0.23	2.3	
SH		Stony Brook Harbor	5.2	1.7	0.5	1.6	
FG		Fort George	8	2.6	0.28	1.6	
CH		Chatham	14	2.4	0.44	1.56	
SR		Shark River	4.4	1.9	0.32	1.67	
NO		North inlet	6.5	2.6	0.28	2.5	
SA		Satilla River	50	6.0	0.22	2.0	14
HU		Hudson	245	9.2	0.075		140
PT		Potomac	184	6.0	0.11		54

Table 4.1. Contd.

Identifier	Region	Tidal inlet	L_e (km)	h (m)	a/h	b_S^+/b_S^-	L_b, L_A (km)
DE		Delaware	215	5.8	0.11		40
BF	Canada-	Bay of Fundy	190	50.0	0.06		100
SL	Atlantic	St. Lawrence	330	7.0	0.36		
CL	USA Pacific	Columbia River	240	10.0	0.2		50
OR	Australia	Ord	65	5.0	0.5		15
LC		Louisa Creek	6.6	2.0	0.78	5.0	
GO	Korea	Gomso Bay	15.0	6.0	0.37	5.0	11
HO	India	Hooghly	72	5.9	0.35		25
IR	Burma	Irawaddy	124	12.4	0.08		35
KH	Iraq	Khor	90	6.7	0.19		21
SO	Vietnam	Soirap	95	7.9	0.165		34
FL	New Guinea	Fly	150	7.0	0.25	1.2	30

Table 4.2. Morphologic parameters of tidal systems (continued). B stands for Barrier Tidal Basin, R for River Tidal Basin and V (Valley) indicates the presence of rocky constraints.

Identifier	Tidal inlet	$U(m/s)$	Q_R (m³/s)	Inlet type	Ref
WS	Western Scheldt	1.0	100	B/R	[121]
ES	Eastern Scheldt	1.0	50	B	
TE	Texel Inlet	1.0	250	B	
EI	Eijerland Inlet	1.0	0	B	
VL	Vlie Inlet	1.0	0	B	
AM	Ameland Inlet	1.0	0	B	
FR	Frisian Inlet	1.0	0	B	
LA	Lauwers Inlet	1.0	0	B	
ED	Ems-Dollard	1.0	140	B	
OB	Otzumer Balje	1.0	0	B	[132]
EL	Elbe	1.0	700	B/R	[395, 268]
WE	Weser	1.0	324	B/R	[378, 176]
TH	Thames	1.0	70	R/V	[332, 308, 161]
LH	Langstone H.	0.5	45	V	[168]
HB	Humber	1.5	230	R	[244, 353]
TA	Tamar	0.6	34	R/V	[450, 452, 176, 164]
SV	Severn	1.5	100	R/V	[448, 449, 194, 483, 268]
FT	The Fleet	0.3		V	[381, 160]
RI	Ribble	1.0	44	R/B	[463]
CO	Conwy	0.6		R/V	[487, 447, 268]
GI	Gironde	1.5	725	R/V	[5, 6, 69]
SE	Seine	1.5	380	R/V	[18, 55]

Table 4.2. Contd.

Identifier	Tidal inlet	$U(m/s)$	Q_R (m³/s)	Inlet type	Ref
PR	Price			B/R	[159]
MM	Murrells Main C.			B/R	
MO	Murrells Oaks C.			B/R	[158]
LR	Little River	0.8		B/R	
MA	Manasquan			B/R	
WA	Wachapreague			B/R	[159]
NA	Nauset	0.5		B/V	[324]
SB	Swash Bay			B	[160]
SH	Stony Brook H.	1.0		B/V	
FG	Fort George			B/R	[159]
CH	Chatham			B/R	
SR	Shark River			B/R	
NO	North Inlet			B	
SA	Satilla River	0.8	70	R	[41]
HU	Hudson	1.0	400	R/V	[220, 268, 51]
PT	Potomac	0.8	336	R	[136]
DE	Delaware	1.0	333	R	[159]
BF	Bay of Fundy		100	R	[206]
SL	St. Lawrence		8500	R	[268]
CL	Columbia River	1.5	7300	R/V	[191, 238]
OR	Ord	1.5	163	R/V	[497]
LC	Louisa Creek	1.0	0	V	[275]
GO	Gomso Bay			B/V	[71]
HO	Hooghly	1.5	1850	R	[308, 268]
IR	Irawaddy		13500	R	[497]
KH	Khor			R	
SO	Soirap			R	
FL	Fly	1.5	6000	R	[195, 494]

plays a dominant role in along-channel sediment transport in barrier tidal inlets, as will be shown later; for river tidal inlets the role of river discharge and ebb-flood asymmetry are similarly important.

4.3. Tides

4.3.1. *Tide Generation*

Moon and sun

The regular daily upward and downward motion of the water surface along the coastline is the most visible expression of tidal forcing. In ancient Greece it was recognised that tides are in some way related to sun and moon, but this relationship could not be explained. The explanation of tidal motion as a consequence of gravitational forces was given by Newton. The lunar and solar tide generating forces are extremely small; they are a factor $\approx 5.10^{-7}$ smaller than the gravitational force of the earth. It may therefore seem surprising that the attraction forces of moon and sun are capable of producing water level elevations of several metres and tidal currents exceeding 1 m/s. Indeed, the momentum dissipation associated with such strong tidal currents is much larger than the tide generating forces, at least in continental shelf seas. This is precisely the reason why no substantial tidal motion is generated in shallow enclosed seas or lakes.

Ocean basin resonance

Tides are generated in ocean basins where tidal currents are small and the water depth so large that momentum dissipation is smaller than the tide generating forces. Ocean resonance plays an important role; tidal motion is amplified if the frequency of tidal forcing is close to the frequency of free oscillations in the ocean basin. Resonance occurs if the dimensions of the tidal basin match the tidal wavelength. The tides generated on the ocean propagate to the continental shelf, where the tidal range may in some cases increase further due to local resonance phenomena and due to concentration of the tidal energy flux in areas of reduced width and depth. The strong tidal motion occurring in many shelf seas is thus not locally produced by tide generating forces, but results from co-oscillation with ocean tides and from local topographic resonance. Numerical tidal studies for the northwest European shelf [351] and for the East China shelf [251] show that the tide generating force influences the co-oscillating semidiurnal tide in these shelf seas by no more than about 1%.

Semidiurnal periodicity

The gravitational forces exerted by moon and sun on the earth's water masses are not sufficient to explain tidal motion. Tide generation results from the local imbalance at the earth's surface of two opposing forces: The gravitational forces acting between the earth and the moon (and between the earth and the sun), and the centrifugal forces related to the orbital motions of these celestial bodies. In fact only the tangential component of the resulting force is relevant, pointing towards the equator. Because the gravitational force and the centrifugal force cancel at any location on the earth's surface just twice during each diurnal rotation, the major periodicity is approximately semidiurnal.

Diurnal tide and spring-neap cycle

The tide generating force of the moon is about twice as large as that of the sun. Because the solar and lunar orbits do not coincide with the equatorial plane (the angle between orbital plane and equatorial plane is called declination), a daily inequality arises in the semidiurnal cycle. The daily inequality is strongest in basins which resonate at diurnal frequency, leading to a diurnal tide in areas where the semidiurnal tidal wave has a node. Due to the $\approx$ 30-day orbital motion of the moon, the moon-earth and sun-earth axes approximately coincide every 15 days (syzygy). This causes a 15 day cycle of neap-tide and spring tide; spring tide follows (with a short delay) full moon and new moon and neap-tide follows half-moon. In fact, the 15 day period corresponds exactly to the frequency difference of the semidiurnal lunar component (M_2) and the semidiurnal solar component (S_2); the interference of these components produces the neap-spring variation of the tidal amplitude.

Other tidal components

Due to the cyclicity of gravitational forces several mono-periodic tidal waves are generated in the oceans, with the same periods as those present in the relative motions of earth, moon and sun. The lunar semidiurnal tidal component (M_2, period $\approx$ 12.4 h) is generally the largest tidal constituent, followed by the diurnal component which is related to the declination of the lunar and solar orbits relative to the equatorial plane (K_1, period $\approx$ 24 h). Other important components are the diurnal lunar component (O_1, period $\approx$ 26 h) and the semidiurnal and diurnal solar components (S_2, period 12 h and P_1, period $\approx$ 24 h). There are many other tidal components of smaller magnitude and lower frequency related to periodicities in the lunar and terrestrial orbits, for instance, the 18.6 year oscillation in the declination of the lunar orbit. Tides with higher periodicity are generated due to the nonlinearity of tidal propagation. The periods of these higher harmonic components are multiples of the astronomical basic tidal periods. These locally generated higher harmonic components are associated with tidal-wave distortion or tidal asymmetry. They play a crucial role in tide-topography interaction (see Fig. 4.11).

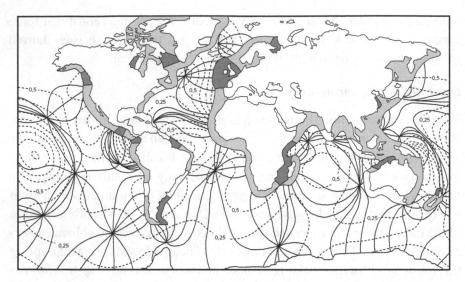

Fig. 4.11. System of semidiurnal lunar tidal waves (M_2) in the oceans represented by lines of equal tidal phase (solid, intervals of $30°$) and lines of equal tidal amplitude (dashed, intervals of 0.25 m). Three ranges are indicated for the tidal amplitude a on the continental shelf: microtidal (white fringe, $a < 1$ m), mesotidal (light grey fringe, $1 < a < 2$ m) and macrotidal (dark grey fringe, $a > 2$ m). Adapted from [26].

4.3.2. *Tide Propagation*

The periodic oscillation produced by tide generating forces propagates over the ocean towards the shallow ocean margins, where most tidal energy is dissipated. The tide can be regarded as an extremely large wave, but it has a sinusoidal variation only in the ocean. In coastal shelf seas it is distorted and flood and ebb become asymmetric; the tidal wave may even break (see the example of Qiantang estuary, Sec. 2.2). The generation of tidal asymmetry can be understood by examining the propagation speeds of high water and low water (both defined here from the viewpoint of a fixed observer); a difference between these propagation speeds causes a distortion of the tidal wave. In this section the tidal propagation process is described in a semi-qualitative way; a mathematical treatment is given in the appendix (Sec. B).

We first discuss tidal propagation for frictionless tidal flow in a rectangular channel. This case is not representative for real tidal inlets, where tidal flow is strongly influenced by friction and where the cross-section is non-rectangular and dependent on the tidal phase. We discuss not only this second case but also a third case where the tide propagates in a strongly converging channel. The qualitative description highlights the fundamental differences between the propagation processes in these three cases.

Frictionless wave propagation

In the ocean, tidal wave propagation is almost frictionless. Frictionless wave propagation corresponds to a local departure of the sea surface from equilibrium which propagates by converting continuously potential energy into kinetic energy and vice versa, without loss of energy. Wave propagation in a single spatial dimension x can be described by the wave equation

$$\eta_{tt} = c^2 \eta_{xx},\qquad(4.16)$$

where $\eta(x, t)$ is the departure of the sea level from equilibrium and c is the wave speed (propagation speed of the tidal wave). We will use the convention that indices x, y, z, t stand for partial differentiation to these variables; for instance, $\eta_{tt} \equiv \partial^2 \eta / \partial t^2$, etc.

Flow acceleration by wave surface slope

The surface slope of the tidal wave accelerates the flow over the entire water column; the magnitude of this slope depends on the tidal phase, the tidal amplitude a and the wavelength $L = 2\pi/k$, where k is the wavenumber, see Fig. 4.12. Around slack water ($u = 0$) the slope is approximately given by ak, during

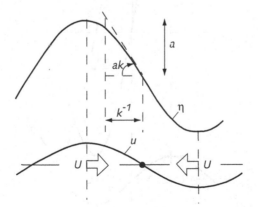

Fig. 4.12. Flow acceleration under a wave front. In the case of frictionless tidal wave propagation the acceleration from negative to positive tidal velocity depends on the slope ak of the wave front.

a period of approximately $\omega^{-1} = T/2\pi$, where T is the tidal period and ω is the radial tidal frequency. The flow velocity thus reaches a maximum given by $U = gak/\omega$, where $g = 9.8\,\text{m/s}$ is the gravitational acceleration. The ratio $\omega/k = L/T$ is equal to the wave speed c. The maximum tidal flow velocity U and the wave speed c are thus related by

$$U = ga/c. \tag{4.17}$$

Progression of the wave front

The wave front advances with speed c; this is the distance travelled by the rising wave front per unit time. The corresponding rise of the wave front requires that per unit time a quantity of water approximately equal to $2ca$ flows towards the wave front, see Fig. 4.13. At the wave crest the flow has accelerated to the velocity U; at the wave trough the flow has accelerated in the opposite direction to a velocity $-U$. The amount of water flowing towards the wave front per unit time is thus approximately given by $2Uh$, where h is the water depth. From the equality $2ac = 2Uh$ it follows that the maximum tidal flow velocity U and

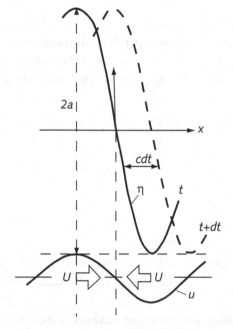

Fig. 4.13. Flow convergence at the wave front and related wave propagation. Solid curve: Wave at time t, dashed line: Wave at time $t + dt$.

the wave speed c are related by:

$$hU = ac. \tag{4.18}$$

Tidal wave propagation speed

By comparing both relationships (4.17, 4.18) we find the expression for the wave speed

$$c = \sqrt{gh}. \tag{4.19}$$

In the case of frictionless flow the tidal wave speed is independent of the tidal amplitude. In ocean basins, where the depth exceeds 1000 m, the wave speed and the tidal wavelength are very large. For a semidiurnal tide the ocean tidal wavelength is in the order of 10 000 km. This is about twice the width of the Atlantic ocean; the semidiurnal tide is therefore close to resonance in this ocean basin.

Friction-dominated wave propagation

In a frictionless progressive wave the wave amplitude a does not change during propagation; it is independent of x. In the case of frictional damping the wave amplitude decreases while propagating; a is a decreasing function of x (assuming propagation in the positive x direction). The wave shape is shown in Fig. 4.14 at two successive time intervals t and $t + dt$. In the case of strong

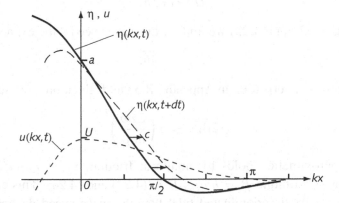

Fig. 4.14. Wave propagation in the case of friction-dominated tidal flow. Solid curve: Wave shape as a function of the spatial phase kx at time t, dashed line: Wave shape at time $t + dt$, dotted line: Flow velocity as a function of kx at time t. The time t corresponds to HW at the mouth $x = 0$. In friction-dominated tidal flow the maximum tidal velocity depends on the steepness ak of the wave crest.

friction the wave amplitude vanishes almost completely over an interval of one wavelength L. This is the case if the momentum dissipation by bottom friction is on average much greater than the inertial acceleration,

$$\langle |\tau_b / h| \rangle \gg \langle |\rho u_t| \rangle. \tag{4.20}$$

If we assume linear friction, $\tau_b = \rho r u$, where r represents the linearised friction coefficient, the condition (4.20) is equivalent to

$$\beta \equiv h\omega / r \ll 1. \tag{4.21}$$

In this case the maximum tidal velocity U (or equivalently, the strongest frictional momentum dissipation, $\tau_b^{max} = \rho r U$) occurs when the pressure gradient p_x is maximum. From Fig. 4.14 we estimate $p_x^{max} \sim \rho g k a$, where a is the tidal amplitude at $x = 0$ and where $k = 2\pi / L = \omega / c$ is the wavenumber. From the approximate equality $\tau_b^{max} / h \approx p_x^{max}$ follows an expression relating U to the wave speed c,

$$U \sim gkha / r = \beta ga / c. \tag{4.22}$$

From Fig. 4.14 we estimate that the flow convergence in the interval $0 < kx < \pi$ during the time dt corresponds approximately to $hU\,dt$. This flow convergence produces an advance $c\,dt$ of the wave front over a height in the order of a. This yields a second relationship between U and c,

$$U \sim ac / h. \tag{4.23}$$

Combining (4.22) and (4.23) we find for the wave speed c the expression,

$$c \sim \sqrt{\beta gh}. \tag{4.24}$$

The mathematical derivation in Appendix B.5 yields the more precise estimate

$$c \approx \sqrt{2\beta gh} \approx h \left(\frac{2g\omega}{ac_D} \right)^{1/3}. \tag{4.25}$$

The last approximation holds for nonlinear friction, $\tau_b = \rho c_D u^2$, and can be found by substituting $\beta = h\omega / c_D U$ in (4.22) and (4.24). This expression shows that for friction-dominated tidal flow the wave speed depends on the tidal amplitude.

Comparing (4.19) and (4.25) it follows that the wave speed for frictional flow is smaller than for frictionless flow, because of the condition (4.21). This condition also implies that only waves with a very long period, $T \gg 2\pi h/r$, are influenced by friction. Tidal waves fall within this category, but for many other wave phenomena (for instance, wind waves) frictional influence on propagation can be neglected.

It should also be noticed that in the case of friction-dominated flow the tidal variation of flow velocity is not in phase with water level. Maximum flow velocity occurs at the time of steepest surface slope and not at the time of high water. The steepest surface slope occurs in advance of high water; therefore the phase of the flow velocity is advanced relative to the frictionless case. The mathematical treatment (Appendix B.5) shows that the phase advance of the flow velocity amounts to $\pi/4$ radians. This phase advance is smaller than $\pi/2$, because the tidal wave has no sinusoidal shape as a function of x, due to frictional damping (see Fig. 4.14).

Tidal diffusion

The change of tidal level per unit time, η_t, is proportional to the amount of water displaced by flow divergence, hu_x. In the case of dominant friction ($\beta \ll 1$) the flow velocity u is proportional to the water level slope η_x, so flow divergence is proportional to the second derivative η_{xx}. It follows that tidal wave propagation can be described by a diffusion-type equation,

$$\eta_t = \Xi \eta_{xx}, \tag{4.26}$$

where Ξ is the diffusion coefficient. The expression of the diffusion coefficient is derived in Appendix B.4,

$$\Xi = c^2/2\omega = \beta gh/\omega = gh^2/r. \tag{4.27}$$

The diffusion character of tidal wave propagation implies damping of the tidal amplitude during propagation. The damping rate is inversely proportional to $\Xi^{1/2}$; a small diffusion coefficient (large friction coefficient r) corresponds to strong tidal wave damping. Equation (4.26) illustrates the fundamentally different nature of tidal wave propagation in the frictional case compared to the frictionless case (4.16).

Wave propagation in a channel with non-rectangular cross-section

In the foregoing sections the width dependence of tidal flow has been ignored; the validity of the results is therefore restricted to tidal channels with rectangular cross-section and constant width. In reality, the conveyance width of tidal flow depends on the tidal phase — in particular on the tidal level — and can be much larger than the width of the deepest channel parts. Some time before high water the flow spreads over tidal flats, while some time before low water the flow is confined to the deepest channel parts. Therefore a distinction has to be made between the channel width, b_C, representative of along-channel flow, and the surface width, b_S, representing the water-covered surface of the tidal basin. Although these quantities depend on the tidal phase, we will assume for simplicity that they may be approximated by their tide-averaged values, see Fig. 4.6. The mass balance equation (4.18) for frictionless flow then becomes

$$b_C hU = b_S ac, \qquad (4.28)$$

and a similar modification applies to the cases of friction-dominated propagation. The equations keep the same form as before by introducing the propagation depth h_S defined by

$$h_S = b_C h / b_S. \qquad (4.29)$$

The expressions for tidal propagation in a non-rectangular channel are similar to the rectangular case, with the depth h replaced by the propagation depth h_S. In the case of frictionless tidal flow the wave propagation velocity is given by

$$c = \sqrt{gh_S}, \qquad (4.30)$$

and in the case of friction-dominated tidal flow by

$$c = \sqrt{2\beta gh_S} = \left(\frac{2ghh_S^2\omega}{ac_D} \right)^{1/3}, \qquad (4.31)$$

where the first equality refers to linearised friction ($\beta = h\omega/r$) and the second equality to quadratic friction. In tidal basins with broad tidal flats we have $b_C \ll b_S$ and therefore $h_S \ll h$; comparison of (4.19) with (4.30) and (4.25) with (4.31) shows that the wave speed decreases when the tidal flat surface increases. The advance of the tidal wave in the channel is slowed down by flow diversion for filling the tidal flats.

Strongly convergent basins

We have assumed so far that the channel geometry is uniform in the tidal propagation direction. The propagation of the tidal wave was therefore related to flow convergence and divergence resulting from longitudinal variation of the flow velocity. If the channel geometry is non-uniform, flow convergence and divergence may also result from longitudinal variations of channel depth and channel width. In the following we will examine the case of a strongly converging channel width but approximately uniform channel depth, as a crude representation of river tidal inlets. Progression of the wave front corresponds to convergence of the tidal discharge Q; we will assume that the convergence of the discharge is mainly due to the convergence of the channel cross-section, as shown in Fig. 4.15. Flow convergence must be maximum when the vertical

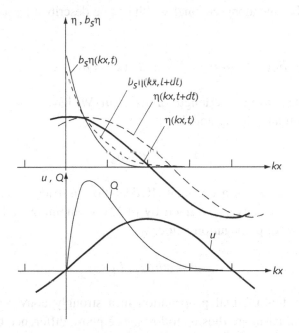

Fig. 4.15. Tidal wave η as a function of the spatial phase kx, in a long, strongly converging basin at time t (solid line) and at time $t + dt$ (dashed line). The time t corresponds to HW and flow reversal (HSW) at the inlet $x = 0$. The water volume displaced by tidal elevation, $b_s\eta$, strongly decreases in the landward direction (solid line for time t and dashed line for time $t + dt$). The tidal velocity u leads the tidal elevation by a quarter tidal period. The longitudinal distribution of tidal discharge Q is mainly due to channel convergence; progression of the wave front η is related to the convergence of discharge $\partial Q/\partial x$. This convergence delivers the water volume increase $b_s\eta_t$ contained in the advancing wave front. Progression of the wave crest (rising water) relates to landward tidal flow (flood) and progression of the wave trough (falling water) to seaward tidal flow (ebb).

water displacement is maximum, i.e., midway between high water and low water. As in the present case flow convergence is maximum when the flow velocity is maximum, the phase of the tidal velocity precedes tidal elevation by a quarter tidal period or by $\pi/2$ radians. The tidal velocity vanishes at the wave crest and at the trough. This implies that channel convergence will produce no flow convergence or divergence at crest and trough and that the tidal wave will not be damped or amplified by channel convergence. In a time interval ω^{-1} the wave front advances over a distance $\omega^{-1}c = k^{-1}$, displacing a volume of water equal to $ab_S\omega^{-1}c$. This volume of water is delivered by flow convergence equal to $h\Delta b_C\omega^{-1}U$, where Δb_C is the decrease of channel width over the distance k^{-1}; the equality of both water volumes yields $U = ab_Sc/h\Delta b_C$. We will assume that the along-channel width can be described by an exponential function:

$$b_C(x) = b_C(0)e^{-x/L_b}, \quad b_S(x) = b_S(0)e^{-x/L_b}, \qquad (4.32)$$

where L_b is the convergence length of the basin. We have $\Delta b_C = b_C/kL_b$, so the maximum tidal velocity equals

$$U = a\omega L_b/h_S, \qquad (4.33)$$

where $hb_C/b_S = h_S$ is a constant. If tidal flow is friction-dominated, the maximum tidal velocity U is given by (4.22). Combining both expressions yields for the wave propagation velocity

$$c = ghh_S/rL_b. \qquad (4.34)$$

The characteristics of tidal propagation in a strongly converging, friction-dominated tidal basin are thus as follows. The phase difference between tidal velocity and tidal elevation equals $\pi/2$, just as for a standing wave. Nevertheless the tidal wave propagates, with a velocity which is inversely proportional to the linearised friction coefficient. In spite of the friction-dominated tidal flow, the tidal wave amplitude is not damped, but constant along the basin. One may conclude that tidal propagation characteristics are strongly dependent on basin geometry and that tidal propagation in a strongly converging, friction-dominated basin combines aspects of both a standing tidal wave and a frictionless propagating wave.

4.3.3. *Tidal Asymmetry*

Role of tidal asymmetry

Tidal asymmetry plays an essential role in producing large-scale residual sediment transport in tidal inlets. This tide-induced residual transport has important consequences for the equilibrium morphology of tidal inlets and for the morphologic response to changing external conditions. Tidal asymmetry depends on the characteristics of tidal propagation and these characteristics depend on the morphology of the tidal inlet, as shown in the previous section. In other words, there is a mutual dependency between tidal asymmetry and inlet morphology. In this section we will examine this mutual dependency and derive explicit expressions. These expressions will be used later to analyse the stability of tidal inlets and to derive relationships for the equilibrium morphology.

Inequality of flood and ebb duration

The tidal wave speeds derived in the previous sections are solutions of the linearised tidal equations, as shown in Appendix B.3. The major difference between the nonlinear and the linear equations relates to ignoring in the latter case the tidal variation of the channel depth D and the propagation depth D_S. The influence of the tidal variation of these quantities on tidal propagation is discussed in Appendix B.5, under the assumption that the tidal variation of D and D_S is substantially smaller than the tide-averaged values h and h_S. The results show that the characteristics of the solution of the nonlinear equations can be fairly well approximated by solving the tidal equations separately for short periods around high water and low water, where D and D_S are taken as constants equal to their value at high and low water. For high water we take

$$h = D^+, \quad h_S = D_S^+ \tag{4.35}$$

and for low water

$$h = D^-, \quad h_S = D_S^-. \tag{4.36}$$

The propagation velocities of the high water and low water tidal phases can be found by substituting (4.35) and (4.36) in the expressions found earlier for the mean wave speed (4.30), (4.31) and (4.34). Once the times of high water (HW) and low water (LW) at the seaward boundary of the tidal basin are known, the

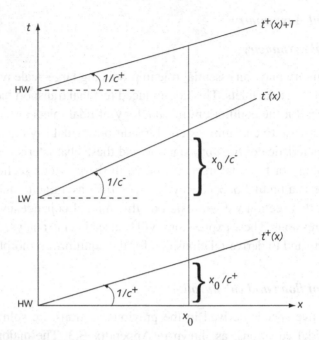

Fig. 4.16. Propagation characteristics of HW and LW in a tidal basin. The wave speed is assumed constant throughout the basin, but different for HW (c^+) and LW (c^-). In this example the propagation speed is greater for HW than for LW ($c^+ > c^-$); the period of rising tide in the basin will then be shorter than the period of falling tide. The corresponding tidal curve is depicted in Fig. 4.17.

times of HW ($t^+(x)$) and LW ($t^-(x)$) within the basin can be derived from the high water propagation velocity c^+ and the low water propagation velocity c^-,

$$t^+(x) = t^+(0) + x/c^+, \quad t^-(x) = t^-(0) + x/c^- \qquad (4.37)$$

(see Fig. 4.16), where x is the distance to the seaward boundary. If the tide is symmetric at the seaward boundary, i.e., the times of HW and LW are equidistant ($|t^+(0) - t^-(0)| = T/2$), then the difference between the periods of falling and rising water within the basin is given by

$$\Delta_{FR} = 2(t^-(x) - t^+(x)) - T = 2x \, (1/c^- - 1/c^+). \qquad (4.38)$$

For instance, if the HW propagation speed c^+ is higher than the LW propagation speed c^-, then the duration of the rising tide is shorter than the duration of the falling tide, see Fig. 4.17. This figure shows that the tidal wave becomes asymmetric when propagating through the tidal basin. In Appendix B.6 it is

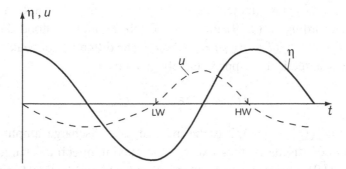

Fig. 4.17. Shape of an asymmetric tidal wave with shorter duration of rising tide than falling tide (solid line), obtained from Fig. 4.16, by interpolating between times of high water and low water. Dashed line: Corresponding velocity curve, assuming $\pi/2$ phase lead relative to tidal elevation (HW coincides with high slack water and LW with low slack water). Flood duration is shorter than ebb; the maximum flood velocity is therefore larger than the maximum ebb velocity.

shown that the difference between ebb and flood duration is given by a similar formula:

$$\Delta_{EF} = 2(t_S^-(x) - t_S^+(x)) - T/ \propto 1/c^- - 1/c^+, \qquad (4.39)$$

where t_S^+ is the tidal phase of flow reversal from flood to ebb and t_S^- is the tidal phase of flow reversal from ebb to flood. The expression (4.39) is approximately valid for tidal basins with weakly varying geometry and short length (basin length much shorter and length scale of geometric variability much longer than the inverse wavenumber k^{-1}). For strongly converging, friction-dominated tidal basins, the times of slack water coincide approximately with HW and LW; in that case we have

$$\Delta_{EF} = \Delta_{FR} = 2x \ (1/c^- - 1/c^+). \qquad (4.40)$$

Inequality of maximum flood and ebb discharge

When excluding the river runoff flow component, tidal inflow must on the average equal tidal outflow. Therefore, if the flood duration is shorter, flood velocities must on the average be higher than ebb velocities; the inverse holds if ebb duration is shorter than flood duration, see Fig. 4.17. If the flood and ebb tidal curves are approximately sine-shaped, then the equality of flood and ebb volumes requires that the ratio of peak-flood discharge Q^+ to peak-ebb discharge Q^- is approximately equal to the ratio of ebb duration to flood duration. According to (4.39) the ratio of ebb duration to flood duration is given by $(T + \Delta_{EF})/(T - \Delta_{EF}) = Q^+/Q^-$. The difference of peak-flood and peak-ebb discharge is thus approximately given by:

$$Q^+ - Q^- \approx \frac{2\Delta_{EF}}{T} Q_m, \qquad (4.41)$$

where $Q_m = (Q^+ + Q^-)/2$ is the average tidal discharge amplitude. The maximum flood discharge thus exceeds the maximum ebb discharge in tidal basins where HW propagates faster than LW and vice versa in tidal basins where LW propagates faster than HW. For strongly converging, friction-dominated tidal basins we may use (4.40); this yields

$$Q^+ - Q^- \approx \frac{2Q_m}{T} \left(\Delta_{FR}^{inlet} + 2x \left(\frac{1}{c^-} - \frac{1}{c^+} \right) \right), \qquad (4.42)$$

where Δ_{FR}^{inlet} is the difference in duration of tidal fall and tidal rise of the offshore tidal wave at the inlet ($x = 0$).

Stokes drift

If the tidal velocity and tidal elevation are almost $\pi/2$ radians out of phase, an expression similar to (4.41) holds for the difference between peak-flood velocity U^+ and peak-ebb velocity U^-,

$$U^+ - U^- \approx \frac{2\Delta_{EF}}{T} U. \qquad (4.43)$$

This is the case for strongly converging friction-dominated tidal basins. For weakly converging basins a correction is required for the influence of the covariance between tidal velocity variation and tidal variation of the channel cross-section. This covariance, the Stokes drift Q_S, is given by

$$Q_S = \int_0^T D(t) b_C(t) u(t) dt. \qquad (4.44)$$

Stokes drift is directed landward, as the phase difference between tidal velocity variation $u(t)$ and tidal elevation $\eta(t)$ is between 0 and $\pi/2$ radians. Mass conservation requires an increase of ebb velocities relative to flood velocities to compensate for the Stokes drift. Therefore an additional term has to be introduced in (4.43), which is given in the Appendix, Eq. (B.72).

Ebb-flood asymmetry and basin geometry

From the expressions for the wave speed, (4.31) and (4.34), it follows that

$$c^+ - c^- \propto D^+ D_S^+ - D^- D_S^- = D^{+2} b_C^+ / b_S^+ - D^{-2} b_C^- / b_S^-. \qquad (4.45)$$

Here we have assumed that $|D^+ - D^-| \ll D^+ + D^-$ and $|D_S^+ - D_S^-| \ll D_S^+ + D_S^-$. Through this equation tidal asymmetry is directly related to the geometric characteristics of the tidal basin. The relevant geometric characteristics are the channel depth at high and low water and the propagation depth at high and low water. For tidal basins without large intertidal areas, the larger depth at HW (D^+) relative to LW (D^-) and the larger channel width at HW (b_C^+) relative to LW (b_C^-) imply a shorter flood duration than ebb duration; for tidal basins with large intertidal areas, the larger basin width at HW (b_S^+) relative to LW (b_S^-) implies a longer flood duration than ebb duration. In the former type of basins the maximum flood velocity exceeds the maximum ebb velocity, while in the latter basins the ebb velocity will be dominant.

Slack water asymmetry

Velocity asymmetry does not arise only from tidal wave propagation. Asymmetry in the duration of high slack water (HSW) and low slack water (LSW) is directly related to local channel geometry. As shown later, this asymmetry is particularly relevant for net transport of fine sediment in a tidal basin. Slack water asymmetry can be demonstrated by relating the tidal discharge to a change in tidal volume in the landward part of the basin:

$$Q(x, t) = A_C(x, t) u(x, t) = \int_x^l b_S(x', t) \eta_t(x', t) dx', \qquad (4.46)$$

where η_t is the change of tide level per unit time. We consider tidal basins where at HW and LW the tide level is almost constant throughout the basin (synchronous tide); this is the case if the length l of the basins (or the convergence length scale L_b for strongly converging basins) is much smaller than the tidal wave length. If the basin cross-section has either a uniform width or an exponentially converging width, Eq. (4.46) is equivalent to

$$u(x, t) \propto b_S(x, t) \eta_t(t) / A_C(x, t) = \eta_t(t) / D_S(x, t). \qquad (4.47)$$

This equation shows that the highest tidal velocities occur when b_S is large, i.e., just after the tidal flats become inundated. For low-lying tidal flats the highest tidal velocities occur close to LW, while in the case of high tidal flats the highest tidal velocities occur close to HW. This is confirmed by many observations, for instance, those reported in [40] for Satilla River.

In the short basins considered here the tidal variation of $\eta(t)$ is $\pi/2$ radians out of phase with $u(t)$; this implies $\eta_t \approx 0$ at HSW and LSW. By differentiating (4.47) we then obtain the following formula for the ratio of the velocity variation around HWS and the velocity variation around LWS:

$$\frac{|u_t|_{HSW}}{|u_t|_{LSW}} \approx \frac{D_S^-}{D_S^+} \frac{|\eta_{tt}|_{HSW}}{|\eta_{tt}|_{LSW}}. \tag{4.48}$$

The last factor on the r.h.s. of (4.48) equals 1 if the offshore tide is symmetric. Equation (4.48) shows that in basins with large tidal flats and deep channels ($D_S^+ < D_S^-$), the tidal velocity changes faster around HSW than around LSW: The period of slack water around HW is shorter than at LW. In basins with small tidal flats and shallow channels ($D_S^+ > D_S^-$), the tidal velocity changes slowly around HSW compared to LSW: The period of slack water around HW is longer than at LW. In the former case sedimentation per unit area will be greater at LSW than at HSW and vice versa in the latter case.

Influence of offshore tides

Tidal asymmetry in a tidal basin is influenced by offshore tidal asymmetry. A faster tidal rise than fall at sea causes higher flood flow velocities than ebb flow velocities in the adjacent tidal basins; slower tidal rise has the opposite effect. In long tidal basins offshore tidal asymmetry can be offset by the influence of basin geometry; in short tidal basins the influence of offshore tidal asymmetry dominates over internally generated asymmetry. A long period of HW at sea causes a long period of slack water, whereas a short period of HW at sea causes a short period of slack water, and the same goes for LW. Tidal wave distortion along a coast influences ebb-flood asymmetry in adjacent basins and therefore the net sediment import or export and morphologic development of these basins. Figure 4.18 shows curves for the Dutch coast and its tidal basins; it appears that the shape of the offshore tide often persists within the adjacent basins. The question "Why is the offshore tide asymmetric?" is discussed in the Appendix (Sec. C.2).

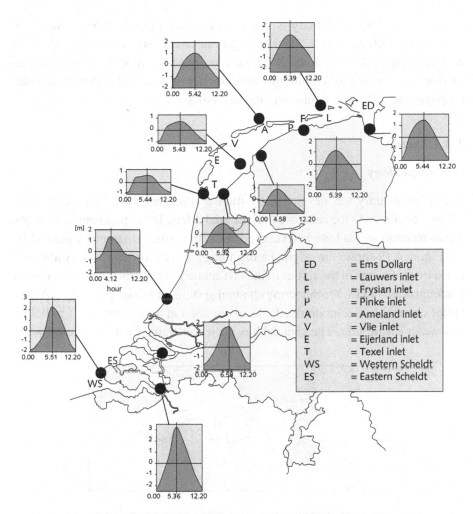

Fig. 4.18. Dutch coastline and tidal basins. Tidal curves averaged over a decade (for spring tide) give an impression of tidal amplitude and tidal asymmetry at various coastal and inland locations. Offshore tidal asymmetry persists at least partly in most tidal basins.

4.4. Tidal Propagation in a Few Selected Basins

In this section we will take a closer look at tidal propagation in a few tidal systems, based on field observations. The selected systems are typical examples of the different tidal basin categories discussed before: The Seine is a macrotidal, funnel-shaped river inlet, the Eastern Scheldt is a mesotidal barrier inlet, the Western Scheldt is a meso/macrotidal funnel-shaped inlet with small river inflow and the Rhine estuary is a microtidal river inlet with high river flow and almost uniform width. Tidal propagation characteristics in these systems will be compared with the idealised wave propagation models of the previous section. The comparison highlights the essential influence of geometry on tidal propagation, and in particular on tidal asymmetry.

4.4.1. *Seine*

Geomorphology

The Seine estuary has developed by marine transgression of the Seine river valley; it belongs to the category of river tidal inlets. In the past centuries major reconstruction works have been carried out in the estuary to improve navigability. Figure 4.19 shows the original morphology and the present estuary after the latest interventions in the sixties of the last century. The former estuarine mouth was characterised by a meandering channel system with broad tidal flats. A new outlet channel has been dredged and maintained at approximately 7 m below low water spring tide. The channel has been constrained within training walls

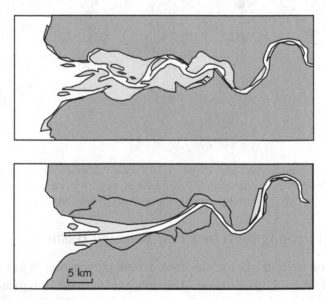

Fig. 4.19. The Seine estuary in the beginning of the 19th century and the estuary today. The intertidal area is shaded light grey. Most tidal flats have been reclaimed; the main channel is constrained by training walls.

and most tidal flats have been reclaimed. In 1834 the intertidal area covered some $130 \, km^2$; since, the intertidal area has been reduced to $30 \, km^2$ and the total estuarine HW cross-section has been reduced from over $20000 \, m^2$ to $5500 \, m^2$. By removing the inlet shoals the famous tidal bore (Fig. 2.6) has been almost suppressed. The present Seine inlet is funnel-shaped with an exponentially decreasing width over a length of $25 \, km$ (with approximately the same convergence length, $L_b \approx 25 \, km$) and an almost constant width further upstream.

Tide and river discharge

The tidal range at spring tide reaches 7 m at the mouth. The tide propagates up to Poses, $160 \, km$ upstream, where the remaining tidal wave is partially reflected by a weir. Figure 4.20 shows tide curves at spring and neap tide in the 19th century and Fig. 4.21 shows tide curves in the present situation at spring tide, for tidal elevation and surface flow velocity. The tidal range is largest at the mouth and decreases landward; the decrease takes place in the uniform reach of the river. At spring tide, damping is much stronger than during neap tide. Seawater could intrude up to $70 \, km$ upstream from the mouth, before the reconstruction of the Seine estuary [18]. At present, seawater intrudes over a distance of $25 \, km$ for an average river discharge of around $400 \, m^3/s$; under these conditions the estuary is partially well mixed. At low discharge the seawater intrusion limit extends some $10 \, km$ further inland. The river discharge may exceed $1000 \, m^3/s$ during a few months a year (with peak discharges exceeding $2000 \, m^3/s$); under such conditions almost all seawater is flushed out of the estuary at the end of each ebb period [273].

Tidal asymmetry

Tidal rise is substantially shorter than tidal fall even at the mouth, especially at spring tide. In the original situation, the tidal wave front became steeper when propagating landward and at spring tide a tidal bore developed at some $40 \, km$ from the mouth. In the present situation, steepening of the tidal wave front is suppressed and no significant tidal bore develops. The maximum flood velocity exceeds the maximum ebb velocity in the first $100 \, km$ of the tidal river, for low to average river discharges. Almost everywhere the low water tidal phase coincides with tide reversal from ebb to flood (designated LSW = low slack water); high water advances tide reversal from flood to ebb (designated

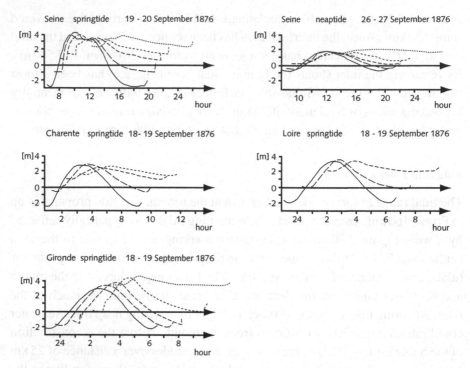

Fig. 4.20. Observed tidal curves in several French rivers in the 19th century, before major dredging interventions [84]. Strong tidal asymmetry develops during upstream tidal propagation; at spring tide HW overruns LW in the Seine and the Gironde rivers, causing a tidal bore ('mascaret'). The tidal curves correspond to different stations along the river; distance in km from the mouth and line characteristic are indicated between brackets. Seine: Honfleur (0, solid line), Quillebeuf (24, long-dashed), Caudebec (46, medium-dashed), Duclair (78, short-dashed) and Elbeuf (136, dots); Charente: Embouchure (0, solid line), Carillon (32, long-dashed), Taillebourg (56, medium-dashed), La Baine (80, short-dashed); Loire: St. Nazaire (0, solid line), Cordemais (24, long-dashed), Nantes (56, medium-dashed); Gironde: Pointe de Grave (0, solid line), La Maréchale (38, long-dashed), Bec d'Ambes (72, medium-dashed), Portets (116, short-dashed), Langon (142, dots).

HSW = high slack water) by approximately one hour. The tide reversal from ebb to flood is rapid; this is clearly related to the steepness of the tidal front. Tide reversal from flood to ebb coincides with the start of tidal fall; temporal and spatial water level gradients at tidal fall are smaller than during tidal rise and tide reversal from flood to ebb is therefore slower. The water level curve lags behind the velocity curve by approximately two hours (phase difference between peak flood (ebb) velocity and HW (LW)); this is close to the phase

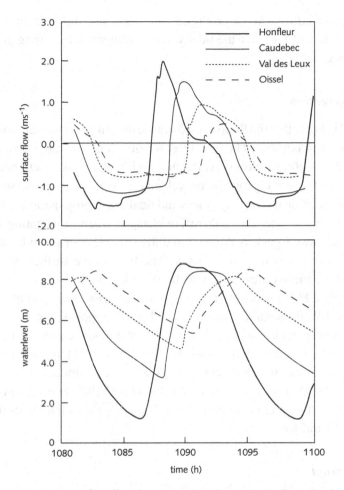

Fig. 4.21. Tidal curves for surface elevation and surface flow velocity from simultaneous measurements at several stations along the Seine estuary and tidal river during spring tide, redrawn from [55] with permission from Elsevier. The peak flood velocity is greater than the peak ebb velocity, in spite of the river discharge. Peak flood velocity coincides with greatest steepness of rising tide; the ebb tidal phase coincides with falling tide. Distance of tide stations from inlet: Honfleur 0 km, Caudebec 46 km, Val de Leux 90 km, Oissel 135 km.

lag of $T/8$ which might be expected for tidal propagation in a uniform channel dominated by friction, see Fig. 4.21. Peak flood velocity coincides with the greatest steepness of rising tide; the ebb tidal phase coincides with falling tide. Such a situation would be expected either for a standing wave, or for a frictionally propagating tide in a strongly convergent channel. At the head of the tidal river a standing wave is produced by reflection at the weir; in the seaward (estuarine) part of the river the tidal wave is influenced by strong geometrical convergence.

Tidal propagation

Figure 4.21 shows that the flood velocity at the mouth starts to decrease when the tide reaches a maximum. The tidal level remains high for some time and peaks a second time before it drops. At the first high water peak, frictional damping is strong and the wave propagation velocity is low. Later, at the second high water peak, frictional damping is less and tidal wave propagation is faster. The shape of the high water curve therefore changes when propagating landward; the first high water peak is damped and the steepness of the tidal wave front decreases before reaching high water. The tidal curve further upstream has a single short high water peak which coincides with slack water. The tidal curves imply a HW-propagation speed in the funnel-shaped part of the river of 13 m/s; the LW-propagation speed is much lower, approximately 7 m/s. For a frictionless propagating wave ($c^{\pm} = \sqrt{gD^{\pm}}$) we would expect respectively 12 and 9 m/s. The theoretical speed for frictionally dominated wave propagation in a strongly converging channel ($c^{\pm} = g(D^{\pm})^2/rL_b$) amounts to 17 and 6 m/s, taking $r = 0.003$. The best agreement is found with diffusion-type propagation in a weakly converging channel ($c^{\pm} = D^{\pm}\sqrt{2g\omega/r}$), which yields theoretical values of 13 and 8 m/s.

Sedimentation

The longitudinal distribution of suspended sediment is characterised by a turbidity maximum, see Fig. 4.5. The turbidity maximum is situated close to the salt intrusion limit; it moves up and down the estuary with the tide, in a way similar to the seawater intrusion front. Bed sediments in the Seine estuary and in the outer bay are predominantly muddy, with sandy sediments prevailing only along the channel thalweg. Large mud deposits are present in the outer bay and on the tidal flats bordering the channel at the north. Most of this fine material is of continental origin (clay minerals) and has been transported downriver during high river floods [122]. Strong mudflat erosion occurs during storm events, mainly due to locally generated waves. Observations indicate that the mass of eroded sediment is of the same order as the mud mass contained in the entire turbidity maximum. Under storm conditions highly turbid waters are dispersed over the estuary and the outer bay by lateral circulation. Sediment redeposition on the mudflats takes place when conditions return to normal after storm, but also just after peak river floods [273].

Sediment transport

Considerable sediment resuspension takes place also at spring tide; suspended sediment concentrations may reach up to $1 \, kg/m^3$ on average over the vertical [18]. Net annual sediment transport is determined by spring tidal conditions and by conditions of very high river discharge. In the estuarine part of the Seine, spring flood shear stresses are substantially higher than ebb shear stresses. Model simulations indicate that this is mainly due to tidal asymmetry, with a lesser contribution of salinity induced estuarine circulation [55]. Under conditions of low to average river discharge more sediment is suspended and transported by the flood current than by the ebb current. In addition, the settling period around HSW is larger than around LSW. The turbidity maximum is the result of this net upstream sediment transport. Seaward transport of fine sediment takes place at high river floods (discharge larger than $1000 \, m^3/s$) [273].

As mentioned, peak flood velocities exceed peak ebb velocities over almost the entire reach of the tidal river during periods of low river discharge. This explains the occurrence of sediment deposits of marine origin far upstream, in the docks of Rouen, 100 km from the sea inlet [182].

4.4.2. Western Scheldt

Geomorphology

The Western Scheldt, in the southern part of The Netherlands, is a meso/macrotidal estuary, opening to the Southern Bight of the North Sea, see Fig. 4.4. Sea water intrudes over a distance of about 80 kilometres; the tide propagates further upstream the Scheldt river as far as the weir at Gent, at 160 km from the sea inlet. The Western Scheldt estuary is situated in a large coastal plain, where other tidal basins also have developed, in particular the Eastern Scheldt basin, with a similar size as the Western Scheldt. During the Holocene transgression coastal barriers created a sheltered lagoonal environment in which large amounts of marine sand and clay were deposited [33]; fluvial inundation later turned the lagoon into a peat bog. Subsidence of these marsh areas due

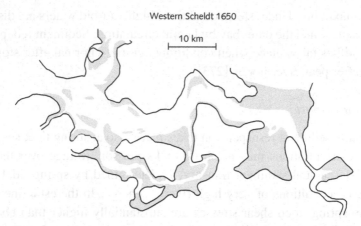

Fig. 4.22. Western Scheldt basin in the 17th century. Tidal flats are shaded. About 9×10^6 m^2 intertidal area is at present left of the former 3.4×10^7 m^2 [461].

to drainage and reclamation in Roman times (2000 BP) created conditions for renewed flooding by the sea. The current estuary was formed by storm surges in the early and late Middle Ages and deliberate dike breaching for military purposes; the greatest extent of the estuary was reached around AD 1600, see Fig. 4.22. Together with the basin area, the tidal prism also greatly increased; tidal currents became so strong that most of the peat layers at the surface were removed. Deep channels scoured the underlying sandy floor and only at some places erosion-resistant peat layers remained. In the 17th century the Western Scheldt became the main navigation route to Antwerp; before, most ships used to sail via the Eastern Scheldt. At later times the Western Scheldt basin area was greatly reduced, partly by sedimentation and partly by land reclamation. The latest reclamation works took place in the fifties of the past century, establishing the present basin configuration. About 9×10^6 m^2 intertidal area is left of the former 3.4×10^7 m^2 [461]. The total basin area has been reduced by more than 40%. However, the tidal prism has decreased only by 13%, mainly due to a substantial increase in the tidal amplitude. During the past century the navigation channel to Antwerp was deepened several times and at present it is maintained by continuous dredging at a minimum depth of 14 m below mean low water level.

The overall geometry of the estuary is convergent; the width is 5–6 km at the mouth and decreases to approximately 1 km at the Belgian–Dutch border (km 80). Characteristic morphologic parameters along the basin are shown in

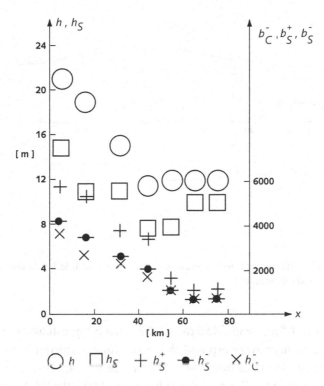

Fig. 4.23. Bathymetric parameters along the Western Scheldt basin. The average channel depth h decreases at the inlet but remains almost constant in the landward part of the basin; the average propagation depth h_S does not show strong along-channel variation. The channel width at low water b_C^-, the basin width at high water b_S^+ and the basin width at low water b_S^- all decrease from inlet to estuary head.

Fig. 4.23. The convergence length L_b is about 50 km, which is of the same order as the inverse tidal wavenumber $1/k = L/2\pi$. Intertidal flats occupy approximately 1/3 of the basin area. Most intertidal flats are sandy without vegetation; fine sediments are retained in salt marshes (schorren) along the outer banks of the estuary. The channel system meanders and braids between the outer estuarine banks, which are protected by dikes. Distinct flood- and ebb-dominant channels have developed almost everywhere in the estuary.

Tide and river discharge

The tide is semidiurnal and almost symmetric, with a small daily inequality and a moderate spring-neap tide variation. At the inlet the tidal range is about

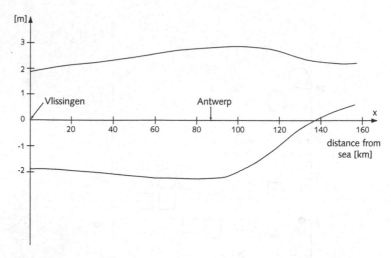

Fig. 4.24. Longitudinal distribution of maximum and minimum tide levels along the Western Scheldt estuary and the Scheldt river.

4 m (spring tide 4.5 m, neap tide 3 m); in the converging estuary, the tidal range increases to 5 m near Antwerp and decreases further upstream in the almost uniform river section to about 1.5 m at Gent, see Fig. 4.24. At the time of the greatest extension of the Western Scheldt, around 1600, the tidal amplitude was lower; it reached a maximum of 4 m at Antwerp. The discharge of the Scheldt river is small — on the order of 100 m³/s. The limit of seawater intrusion is situated near Antwerp or near to the Belgian–Dutch border at high river discharge. The estuary is well mixed; density induced estuarine circulation is small, typically less than 0.1 m/s.

Tidal amplification

First we investigate the influence of channel convergence on the increase of the tidal amplitude in the estuarine section (Fig. 4.24). For converging nonreflecting basins the tidal damping coefficient μ is given by (B.86)

$$\mu \approx \omega^2 L_b [c^{-2} - c_0^{-2}], \quad c_0^2 = g h_S. \tag{4.49}$$

Substituting $h_S = 10$ m for the average propagation depth (Table 4.3) and estimating the average wave speed $c = 10$ m/s from Fig. 4.26, we find $\mu \approx 0$. Even with a uncertainty of 10% in the values of h_S and c the amplification of the tidal

Table 4.3. Bathymetric parameters for the Western Scheldt (WS) and for the Eastern Scheldt (ES).

Parameter	WS high water [m]	WS low water [m]	ES high water [m]	ES low water [m]
b_S	3500	2500	5250	3250
b_C	2200	1900	2750	2500
D	17.0	13.0	16.5	13.5
D_S	10.5	10.0	8.6	10.4

range from 4 m at the mouth to 5 m at the head of the estuary ($\approx$ km 80) cannot be explained by channel convergence. This suggests that the amplification is produced by partial reflection at the head of the estuary. This suggestion is supported by the rapid HW propagation near the head of the estuary, visible in Fig. 4.26.

Tidal asymmetry

Typical tidal curves for water level and discharge are shown in Fig. 4.25 at the inlet (station A) and some 50 km upstream (station B). Tidal asymmetry is already apparent at the inlet; tidal rise is quite sharp in the hours preceding high water. This produces a peak-flood discharge which is some 30% larger than the peak ebb discharge. The times of high water (HW) and low water (LW) precede the times of flow reversal after flood (HSW = high slack water) and after ebb (LSW = low slack water) by about one hour, both at stations A and B. Flow reversal is not simultaneous over the cross-section; between bottom and surface the time difference does not exceed 10 minutes (at low slack water), but between the channel banks and the centre of the main channel the slack water time delay can be one hour. Because tidal velocity and tidal elevation are less than $\pi/2$ radians out of phase, a net seaward velocity compensates for Stokes drift. Due to this Stokes drift compensation the average peak flood and ebb velocities have comparable magnitude, but in the cross-section substantial differences occur between flood and ebb-dominated channels. The phase difference between tidal elevation and tidal discharge is greater than for frictionally dominated tidal propagation; this is attributed partly to the convergence of the estuarine geometry and partly to tidal wave reflection.

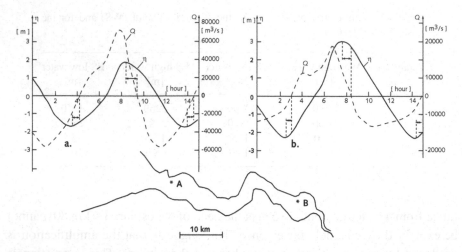

Fig. 4.25. Representative curves for tidal elevation and discharge in the Western Scheldt (Survey Department Rijkswaterstaat Zeeland): (a) station A (inlet), (b) station B (50 km upstream). Peak-flood discharge exceeds peak-ebb discharge; the time delay between HW/LW and corresponding slack tide is about one hour throughout the basin.

Tidal propagation

The tide-propagation speed has greatly increased during the past centuries. In the 17th century the HW and LW travel time from Vlissingen to Antwerp was around 5 hours. In 1900 the travel time had reduced to about 2.5 hours and at present it is around 2 hours. This strong decrease coincides with a similarly strong increase of channel depth and a decrease of intertidal area. High water propagates faster through the basin than low water, see Fig. 4.26; in the seaward part of the basin (the first 80 km), the average HW propagation speed $c^+ = 11.5$ m/s and the average LW propagation speed $c^- = 9$ m/s. This part of the basin has a multi-channel system and large tidal flats. The difference between HW and LW propagation speeds has remained almost the same during the past century. The observed propagation speeds are lower than one might expect for a standing wave, and are more representative of a progressive wave. However, the expression $c^\pm = (g D_S^\pm)^{1/2}$ for a frictionless progressive wave yields the same value $c \approx 10$ m/s for both HW and LW propagation (see Table 4.3). The propagation speed corresponding to friction-dominated tidal flow in strongly convergent basins, $c^\pm = g D^\pm D_S^\pm / r^\pm L_b$, yields values close to the observed ones ($c^+ = 11.7$ m/s, $c^- = 8.5$ m/s), taking $r^\pm = r = 0.003$ for the friction coefficient and using for $D^\pm$, $D_S^\pm$ the values of Table 4.3.

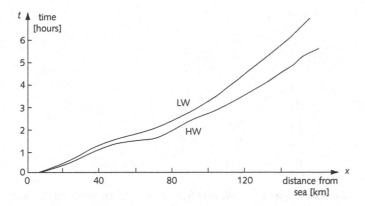

Fig. 4.26. Propagation of HW and LW tide along the Western Scheldt estuary and Scheldt river; HW propagates faster than LW.

In the landward part of the estuary (km 60 to 100) the tidal propagation speed is lower than in the seaward part: 9.5 m/s for HW and 7.4 m/s for LW. The morphology of this part of the estuary is more characteristic for a river tidal inlet: A single channel with converging width and a turbidity maximum. The LW propagation speed is fairly well predicted by the model for frictionally dominated flow in a strongly converging channel, but the predicted HW propagation speed is too high. The progressive wave formula $c^{\pm} = (gD_S^{\pm})^{1/2}$ and the formula for friction-dominated flow in a weakly converging basin, $c^{\pm} = (2g\omega D^{\pm} D_S^{\pm}/r^{\pm})^{1/2}$, both overestimate the wave speed for HW and LW, see Table 4.4.

Sediment transport

Major morphologic changes have taken place since the extension of the Western Scheldt basin in the 14th to 16th centuries. A large amount of sediment has been removed from the tidal channels and high sedimentation has occurred on the shallow subtidal and intertidal zones, where the accretion rate is estimated at about 3 cm/year [461]. In the past century channel deepening has continued; dredging is probably the major cause. In the 1930–1990 period the sub-tidal basin volume increased by some 7×10^7 m³. At the same time the storage volume of the Western Scheldt has decreased by 6×10^7 m³ through sedimentation and by 6×10^7 m³ through land reclamation; this decrease is partly compensated by the increase of the tidal amplitude (3.5×10^7 m³) and sea-level rise

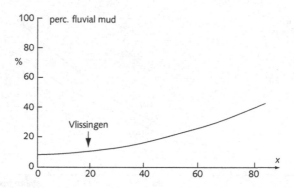

Fig. 4.27. In the seaward part of the Western Scheldt fine organic sediment deposits are mainly of marine origin. Data are based on carbon isotope analysis, after [389, 470].

(10^7 m^3). The vegetated marsh areas have accreted in height, but decreased in area, due to erosion. Dredging is concentrated at the junctions between flood- and ebb-dominated channels, where shoals are present, see Fig. 4.44. Most of the dredged material is deposited in secondary flood channels and along the channel banks in the estuary. The sand balance of the estuary is close to equilibrium; net accretion of the estuary is between one and two million m^3 per year, which is of the same order as the net import which would be required for keeping pace with sea-level rise. Most of the sediment transported in suspension is fine material (median grain diameter less than 100μm) of marine origin; in the estuarine section concentrations are less that 0.1 kg/m^3. Figure 4.27 shows the percentage of fine marine sediment in the basin as a function of distance to the inlet. The turbidity maximum is found in the transition zone between the marine and the fluvial sections, situated between the Belgian–Dutch border and Temse (20 km upstream of Antwerp). Here the average suspended sediment concentrations range between 0.1 and 0.2 kg/m^3 and most of the fine sediment is of fluvial origin [436]. Sediment is retained mainly in marsh areas; the retention process is strongly stimulated by marsh vegetation.

4.4.3. *Rotterdam Waterway and Rhine Tidal River*

Geomorphology

The River Rhine flows through several branches to the Southern Bight of the North Sea, see Fig. 2.16. At present, the outflow is directed most of the time through a single branch, the Rotterdam Waterway. The seaward part is an artificial canal of 15 km length dredged at the end of the 19th century; it has been deepened and widened several times during the 20th century. The present channel width and cross-section of 600 m and 6000 m^2 are almost constant up to Rotterdam, at 30 km of the inlet. Between 30 and 50 km from the inlet, the width and the cross-section decrease almost linearly to 300 m and 2000 m^2. A second major branch, the Haringvliet, was dammed in 1970. When river runoff is above average (2200 m^3/s), the excess Rhine water is discharged to the sea through sluices in the Haringvliet dam. The tide enters the Rhine estuary through the Rotterdam Waterway and propagates upriver into the two main Rhine branches, the Waal and the Lek-Nederrijn. Frictional damping limits the tidal intrusion length to about 80 km from the sea. The Lek-Nederrijn branches to the Rotterdam Waterway at 50 km from the inlet; the width and depth of the tidal river are almost constant. The mean depth is 5–7 m at average river runoff. Both the estuary and the tidal river can be roughly schematised to uniform tidal channels. The surface of the intertidal area is small compared to the channel surface. Forelands, which are present in the upper reaches of the tidal river, are flooded at high river runoff. The channel bed consists of medium sized sand.

Tide and river discharge

The average tidal range at the inlet is 1.75 m (spring tide 1.9 m, neap tide 1.5 m). The tidal range decreases upstream to almost zero at the Hagestein-weir in the Lek-Nederrijn, at 75 km from the sea. The average river discharge through the Rotterdam Waterway is about 1750 m^3/s, with an annual minimum and maximum of about 500 m^3/s and 4000 m^3/s, respectively. The average discharge in the Lek river is about 500 m^3/s. The limit of seawater intrusion is generally located near Rotterdam, at 20 km from the inlet. Suspended sediment concentrations in the river and in the estuary are rather low, typically on the order of 0.05 kg/m^3. Close to the intrusion limit the suspended sediment distribution has a maximum of about 0.1 kg/m^3. In the Rotterdam Waterway, maximum flood velocities are around 1.5 m/s and maximum ebb velocities around 2 m/s. A well developed estuarine circulation is responsible for substantial asymmetry of the vertical velocity distribution between flood and ebb; during flood the velocity is highest halfway between the surface and the bottom, while during ebb the highest velocity occurs at the surface. Near-bottom velocities are small during most of the ebb period, see Fig. 4.28.

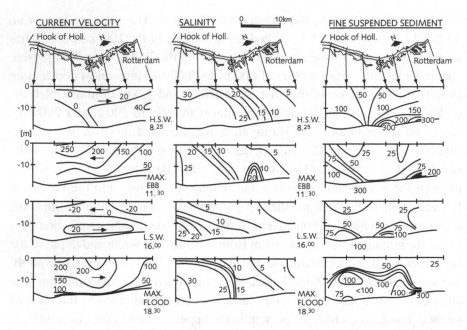

Fig. 4.28. Longitudinal distribution of velocity [cm/s], salinity [ppt] and suspended sediment [mg/l] in the Rotterdam Waterway during a tidal cycle (Survey Department Rijkswaterstaat, 9 September 1967). Estuarine circulation is well developed: Seaward directed velocity dominates near the surface and landward directed velocity dominates near the bottom. Salinity stratification is stronger during ebb than during flood. Suspended sediment concentrations are highest at maximum flood, but rather low in comparison with many other river dominated estuaries. A local turbidity maximum is visible at maximum flood and ebb near the seawater-intrusion limit.

Tidal asymmetry

The tide is strongly damped in the tidal river, as shown in Fig. 4.29 for average tidal and fluvial conditions. Tidal asymmetry develops during upstream tidal propagation; the period of tidal rise becomes shorter while the period of falling tide increases. The tidal discharge also decreases upstream; tidal velocities dominate in the estuarine zone but further landward river flow becomes dominant. The tide level curves lag behind the tidal discharge curves; the average phase lag is approximately one-and-a-half hour. This phase lag indicates that tidal propagation is dominated by friction; it has been shown earlier that a similar phase lag occurs for frictionally dominated tidal propagation in a uniform channel. The upstream increase of average water level is similar to the

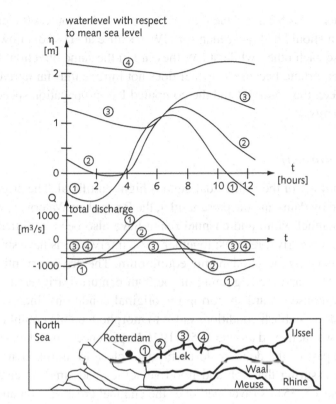

Fig. 4.29. Observed tidal curves for water elevation and discharge along the river Lek, redrawn after [115].

river-bed slope; flow acceleration produced by the water level slope is balanced by momentum transfer to the river bed [115].

Tidal propagation

High water propagates faster than low water in the tidal river Lek; the HW propagation speed relative to average river flow $c^+ \approx 7\,\text{m/s}$ and the LW propagation speed relative to average river flow $c^- \approx 4.5\,\text{m/s}$. For a frictionless propagating wave we have $c^\pm = \sqrt{gD^\pm}$; for HW ($D^+ = 7\,\text{m}$) we find $c^+ = 8.4\,\text{m/s}$ and for LW ($D^- = 5.5$ m) we find $c^- = 7.4\,\text{m/s}$. The observed propagation speeds are much lower. Better agreement is found with the propagation speed for friction-dominated flow, $c^\pm = D^\pm\sqrt{2g\omega/r^\pm}$. If we set $r^\pm = r = 0.003\,\text{m/s}$ we find

$c^+ = 6.7$ m/s, $c^- = 5.3$ m/s. One may argue that the friction coefficient for LW propagation should be higher than for HW, because at HW tidal flow and river flow oppose each other, while at LW they are in the same direction. Stratification is unimportant, because seawater does not intrude thus far upriver. Agreement between the observed and the computed LW propagation speed requires $r^- = 0.004$ m/s.

Sediment transport

The morphology of the Rhine tidal delta is highly artificial. The original inlets were closed by dams and the present inlet, the Rotterdam Waterway, was newly created. Channel width and channel depth have also been modified in many parts of the lower river reaches; together these interventions have strongly disturbed the existing morphodynamic equilibrium. The first interventions, at the end of the 19th and the beginning of the 20th centuries, triggered a morphodynamic response towards restoring the original conditions. Intense dredging was needed to establish conditions close to morphodynamic equilibrium; such a situation was achieved around 1950 [192]. Experience showed that the tidally dominated part of the delta could be brought close to morphodynamic equilibrium by assuring that ebb and flood velocities had similar magnitudes of approximately 0.55 m/s (averaged over the channel cross-section and over the tidal period) [116]. This condition is equivalent to maximum cross-sectionally averaged spring tide velocities for ebb and flood on the order of 1 m/s. It is not clear, however, whether this condition is also consistent with long-term morphodynamic equilibrium in the river-dominated reach of the tidal delta, as morphodynamic adaptation of the tidal river is a slow process, due to the modest fluvial sediment input of the Rhine River.

4.4.4. *Eastern Scheldt Tidal Basin*

Geomorphology

The Eastern Scheldt is a typical barrier tidal inlet, just north of the Western Scheldt inlet. It has formed under similar conditions, with similar sedimentary structure and composition. The Eastern Scheldt has a greater width than the Western Scheldt and accomodates two large flood deltas. The present length of the basin is 45 km; the landward portion of the two flood deltas was dammed in

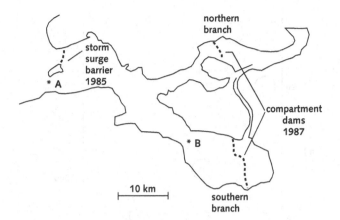

Fig. 4.30. Map of the Eastern Scheldt, indicating the storm surge barrier at the inlet completed in 1985. The tidal prism was further reduced by the construction of compartmenting dams in 1987.

the eighties of the past century. In the 8 km wide inlet a storm surge barrier was built, allowing tidal intrusion into the basin; the barrier is closed only during extreme storm surges (see Fig. 4.30). The Eastern Scheldt has a meandering multichannel system which was stabilised within embankments in the past centuries; the channels are separated by large sandy intertidal shoals, see Fig. 4.4. The Eastern Scheldt has no converging geometry; the characteristic channel and tidal flat dimensions are shown in Fig. 4.31, for the southern delta-branch. The average channel depth is similar to the Western Scheldt; in the seaward portion the Eastern Scheldt is deeper and in the landward portion it is shallower. The intertidal flat area in the Eastern Scheldt is larger than in the Western Scheldt; it occupies about half the basin area. The tidal flats are sandy; the fine sediment ($< 50~\mu$m) content is highest in the salt marshes near the basin boundaries, but remains generally low. The channel bed consists of medium sand in the mouth and fine to medium sand in the landward part of the basin.

Tide and river discharge

The tide curve at the inlet is similar to that at the Western Scheldt inlet, but the tidal range of about 3 m is smaller (spring tide 3.3 m, neap tide 2.3 m). The tidal range increases landward and reaches 4 m at the head of the basin. Fresh water inflow is small; before 1985 the inflow was kept constant at 50 m³/s and

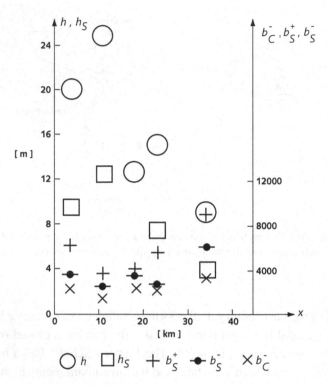

Fig. 4.31. Bathymetric parameters along the Eastern Scheldt basin. The average channel depth h and the average propagation depth h_S both decrease landward. The channel width at low water b_C^- is roughly constant; the basin width at high water b_S^+ and the basin width at low water b_S^- slightly increase towards the basin head.

at present the fresh water inflow is reduced to about $10 \, \text{m}^3/\text{s}$. In the major part of the basin the salinity is almost equal to the seawater salinity.

Tidal asymmetry

Figure 4.32 shows tidal elevation and velocity curves at the inlet and 30 km landward, near the head of the basin, for the period before 1985. The tidal curves indicate that the durations of tidal rise and tidal fall are similar. In the landward part of the basin the peak-flood and peak-ebb discharges have similar magnitude as well. Tidal velocity and elevation are almost $\pi/2$ radians out of phase. The geometry of the Eastern Scheldt produces tidal wave reflection at the landward boundary; one may therefore expect that the tidal wave in the landward part of the basin will be a standing wave. At the inlet the situation is

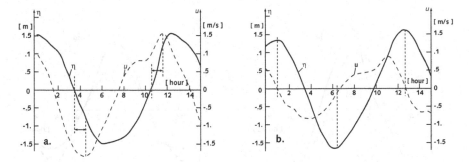

Fig. 4.32. Representative curves for tidal elevation and velocity in the Eastern Scheldt before 1985, at stations A (inlet) and B (30 km upstream), see Fig. 4.30. At the mouth there is a time delay of almost one hour between HW/LW and the corresponding slack water, characteristic of a partially standing tidal wave. Near the head of the basin there is no time delay, characteristic of a standing tidal wave. (Measurements by the Survey Department Rijkswaterstaat.)

different, as shown in the same figure. Here the velocity-elevation phase shift is approximately one hour less than the quarter tidal period. This implies that Stokes drift plays a role; on the average, the peak-ebb velocity at the inlet is higher than the peak-flood velocity. Ebb dominance at the inlet agrees with observations of net sediment export and a growing volume of the outer delta in the period before partial closure [103].

Tidal propagation

Simultaneous tide curves at the inlet and near the head of the basin are shown in Fig. 4.33 for the situation before 1985 and for the situation after partial closure. In the former situation high water propagated a little faster through the estuary than low water; the propagation time was 30'–60' for HW and 40'–60' for LW. The higher uncertainty for the HW times is related to the flat shape of the tidal curve around HW. The HW propagation time through the basin is approximately equal to the time delay at the inlet between high water and high slack water and the LW propagation time is approximately equal to the time delay at the inlet between LW and low slack water; this is in agreement with the theoretically expected propagation of a damped, reflected wave in a uniform basin, as shown in Appendix B.6. The following expression is derived for the HW-propagation time $t_{HW}(x)$ and the LW-propagation time $t_{LW}(x)$ through

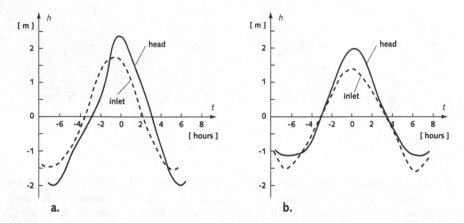

Fig. 4.33. Simultaneous tidal curves for average spring tide at the inlet and at the head of the Eastern Scheldt basin, for the situation before 1985 (a) and for the present situation, after partial closure (b). The propagation time decreased from 30'–60' in the former situation to about 10' in the present situation. This is mainly due to the strong decrease of momentum dissipation in the basin.

the basin (B.64):

$$t_{HW}(x) - t_{HW}(0) \approx \frac{r}{2g} \frac{1}{D^+ D_S^+} \left[l^2 - (x - l)^2 \right],$$

$$t_{LW}(x) - t_{LW}(0) \approx \frac{r}{2g} \frac{1}{D^- D_S^-} \left[l^2 - (x - l)^2 \right]. \tag{4.50}$$

Substituting the characteristic geometrical values for the Eastern Scheldt (see Table 4.3) and setting $r = 0.003$ for the friction coefficient, we find a propagation time of 37' at the head of the basin $x = l$, both for HW and LW.

In the present situation, after partial closure and partial damming of the flood deltas, the tidal velocities in the basin have decreased by more than 30%. This has greatly reduced energy dissipation and tidal wave damping. The tidal wave now has a standing wave character almost throughout the basin. Figure 4.33 shows that the phase difference between the tide-level variations at the inlet and at the head has become very small, on the order of 10'–15'. The corresponding reduction of the linearised friction coefficient can be estimated from expression (B.64), yielding $r = 0.001$ instead of $r = 0.003$ in the former situation. This suggests that the decrease of the friction coefficient is substantially stronger than the decrease of the tidal velocities. A possible explanation is the absence of ripple formation and associated form drag, because of insufficient current strength and accumulation of fine sediment on the channel bed.

Sediment transport before partial closure

Sediment was exported from the Eastern Scheldt in the period of enhanced tidal prism between 1965 and 1985. Measurements indicate a net seaward transport both for sand and for fine sediment ($< 50 \, \mu$m) on a yearly basis [437]. This is consistent with the very low concentrations of fine suspended sediment (typically less than $0.05 \, \text{kg/m}^3$) and the low mud content of tidal flats. Tidal basins generally act as a sink for fine sediment; the former Eastern Scheldt is an exception. It has been suggested that slack water asymmetry is a major cause of the export of fine sediment [119]. The ebb load of fine suspended sediment is higher than the flood load, probably because the period of tide reversal from flood to ebb is longer than the period of tide reversal from ebb to flood (see Fig. 4.32), resulting in greater sediment deposition at LSW than at HSW.

Sediment transport after partial closure

As mentioned, partial closure by the storm surge barrier and reduction of the flood delta areas in 1985 have decreased the tidal prism of the Eastern Scheldt by 30%. The strength of tidal currents in the seaward part of the basin was reduced to 70% of the original values and even less near the head of the basin. Concentrations of suspended sediment in the basin are still lower than in the past. Most of the fine sediment carried by flood flow into the basin is deposited in the channels. Medium and coarse sediments carried by the flood flow are captured in deep scour holes at the seaward side of the storm surge barrier; these holes prevent import of marine sand into the basin.

Tidal flat sedimentation and erosion

Compartmenting dams constructed in the Rhine–Meuse–Scheldt delta in the sixties have separated the Eastern Scheldt from other tidal inlets and from river inflow. These dams increased the tidal prism of the Eastern Scheldt by 10%; the condition of enhanced tidal prism lasted for almost 20 years. Successive changes of the tidal prism have had a significant influence on channel-shoal relief in the basin. In the period of enhanced tidal prism the topographic relief increased; channels deepened and the height and extent of tidal flats increased, see Fig. 4.9. The opposite process is taking place in the present situation of reduced tidal prism; tidal flats are eroding and channels are silting up, see

Fig. 4.10. These observations illustrate the sensitivity of basin morphology and tidal asymmetry to changes in current strength.

4.4.5. *Comparison*

Comparison of tidal wave propagation in the four tidal inlets leads to the following conclusions:

- The tidal amplitude increases landward from the inlet. An exception is the Rotterdam Waterway, where the tidal wave is not reflected and where the cross-sectional width is not convergent. The increase of the tidal amplitude in the other inlets is consistent with convergence of the basin width and/or with reflection of the tidal wave at the landward boundary. The tidal amplitude decreases in the upstream part of the tidal river where the channel width remains almost constant.
- Maximum tidal velocities are typically on the order of 1 m/s in tidal basins characterised by tidal sand flats and meandering flood and ebb-dominated channels. Maximum tidal velocities exceed 1 m/s in inlets with strong estuarine circulation (Rotterdam Waterway) or a well developed turbidity maximum (Seine).
- The phase difference between tidal flow and tidal elevation is smaller than $\pi/2$ radians; in the strongly converging Seine estuary the average time difference Δ_S between high/low water and corresponding slack water is only about half an hour and in the Western Scheldt Δ_S is about one hour. In the Eastern Scheldt the slack water phase lag Δ_S decreases to zero at the landward boundary, where the tidal wave is reflected; at the inlet the phase lag is about one hour. In the Rhine tidal inlet the slack water phase lag Δ_S is close to $\pi/4$, indicative of tidal wave diffusion.
- Landward from the inlet, tidal rise becomes steeper and tidal fall gentler. An exception is the Eastern Scheldt (former situation), where HW and LW propagate at approximately the same speed; the Eastern Scheldt is also the basin with the largest tidal flat area. In the Seine and the Western Scheldt, tidal rise is already faster than tidal fall at the inlet. At the Western Scheldt inlet the time difference Δ_{FR} between tidal fall and tidal rise is about half an hour; at the Seine inlet Δ_{FR} is more than 3 hours.
- The tidal wave speed is substantially lower than in the frictionless case. The observed HW and LW speeds are close to the predicted speed for the case

Table 4.4. Comparison of observed HW and LW propagation speeds c^+, c^- (accuracy $\approx \pm$ 2 m/s) with different idealised propagation models: Frictionless propagation, friction-dominated propagation for a weakly convergent basin and friction-dominated propagation for a strongly convergent basin. For the river tidal basins the propagation depth D_S is assumed to be equal to the channel depth D. The Scheldt refers to the landward part of the Western Scheldt basin and the Scheldt tidal river (km 60 to km 100). The observed HW and LW wave propagation speeds include the tidal and river current speed at HW and LW; at HW this current speed is generally upstream and at LW it is downstream, opposing wave propagation. After correction for the current speed the HW propagation velocity will be slightly smaller and the LW propagation speed will be some 0.5–1 m/s higher. In spite of several discrepancies, the friction-dominated propagation for a weakly convergent basin yields the closest correspondence to the observed propagation speeds.

Tidal inlet	Observed		$\sqrt{gD}$		$D\sqrt{2g\omega/r}$		gD^2/rL_b		Ref
	c^+	c^-	c^+	c^-	c^+	c^-	c^+	c^-	
Scheldt	9.5	7.4	11.7	9.9	13.4	9.6	12.8	6.5	
Elbe	7.5	6.0	10.8	8.9	11.5	7.7	11.8	5.2	[104]
Weser	9.5	4.3	9.8	7.8	9.3	6.0	23.9	9.8	[104]
Gironde	8.5	5.0	10.6	8.0	11.0	6.2	9.6	3.1	[84]
Seine	13.0	7.0	11.7	8.9	13.4	7.7	17.1	5.6	[55]
Hooghly	13.0	6.1	8.8	6.1	7.6	3.9	8.3	1.9	[308]
Rhine/Lek	7.0	4.0	8.4	7.4	6.7	5.3			

of friction-dominated tidal propagation. For other inlets too there is best agreement with the predictions of friction-dominated propagation in weakly converging basins, see Table 4.4. However, none of the idealised tidal models produces reliable estimates of HW and LW propagation velocities. The basin geometry is too severely simplified; part of the mismatch may be explained by changes of the shape of the tidal curve by higher order nonlinear effects. Data quality is also questionable, because observations have generally been made for other purposes.

- The difference between high water and low water propagation speeds is consistent with the assumption of separate propagation modes for high and low water depending on the high water geometry and low water geometry, respectively.

- Flood dominance is consistent with the decrease of tidal rise duration. In the Eastern Scheldt (former situation), ebb dominance seems to be mainly related to Stokes drift. Ebb dominance in the Rotterdam Waterway is related to river flow.

- Net import of sediment is consistent with flood dominance and net export with ebb dominance.

4.5. Equilibrium of River Tidal Inlets

What is an equilibrium?

A river tidal inlet is in morphodynamic equilibrium if the net (tide-averaged) sediment flux is the same everywhere along the tidal river and equal to the sediment flux upstream of the tidal limit. Can such a condition ever be satisfied? And what are the implications for the tidal wave in the river and for the river morphology? However, we first need to define what we consider as an equilibrium. An equilibrium at the tidal timescale is not necessarily an equilibrium at the seasonal timescale or at the timescale of decades or centuries. Also, a local equilibrium is not necessarily the same as an equilibrium at basin scale. Because of these complications, we will limit our discussion to necessary conditions for the existence of an equilibrium and leave open the question whether these conditions are also sufficient. Equilibrium requires that the different processes causing a gradient in the net sediment flux balance each other. The tide is one of the factors causing a gradient in the net sediment flux, as the contribution of tidal flow to the sediment flux vanishes upstream. We will therefore focus on the question under what conditions the gradient in the tide-induced net sediment flux vanishes everywhere along the river.

4.5.1. *Residual Sediment Transport*

Sediment load related to flow strength

To investigate this question several assumptions have to be made. The most important assumptions are in regard to the sediment load carried by the flow. We will assume that this load is related to the instantaneous depth-averaged flow strength; we disregard gradients in sediment composition and sediment availability along the river. A fourth power relationship between sediment load and flow strength is assumed (3.44), although a better representation of sediment transport might be obtained by a relationship with a lower velocity power, but including explicitly a threshold shear stress for sediment motion (3.38). The fourth power law mimics to some extent the existence of a threshold for sediment motion; this simpler formulation yields more transparent expressions and avoids the introduction of additional parameters. The results obtained are qualitatively similar to those produced with more complex sediment transport formulations. A more critical assumption is the neglect of temporal lag effects between flow strength and sediment load, but it is defendable in view of the duration of the tidal period. Another debatable assumption is the neglect of tidal variations in the shape of the velocity profile for stratified estuaries.

Accepting these simplifications the tide-averaged sediment flux is given by:

$$\langle q \rangle \propto \int_{b_C} \langle u^5 \rangle \, dy, \tag{4.51}$$

where the brackets stand for tide-averaging,

$$\langle q \rangle = \frac{1}{T} \int_0^T q(t) \, dt.$$

Here q is the total, cross-sectionally integrated sediment flux.

Tide-induced sediment transport components

We will express (4.51) as a function of the total discharge $Q = Db_C u = Q_R + Q_T$, where Q_R is the river discharge, $Q_T = b_C(h + \eta)u_T$ the tidal discharge ($\langle Q_T \rangle = 0$), b_C is the channel width and $D = h + \eta$ is the instantaneous average channel depth. We assume that there are no tributaries, so Q_R is constant along the estuary. Upstream of the tidal limit $Q_T = 0$; the fluvial sediment input is given by

$$\langle q \rangle \propto -Q_R^5 / h_R^5 b_R^4, \tag{4.52}$$

where h_R, b_R are the upstream river width and depth. Downstream of the tidal limit we have:

$$\langle q \rangle \propto \frac{1}{b_C^4} \left\langle \left(\frac{-Q_R + Q_T}{h + \eta} \right)^5 \right\rangle \approx \frac{1}{h^5 b_C^4} \left\langle (-Q_R + Q_T)^5 \left(1 - \frac{\eta}{h} \right)^5 \right\rangle. \tag{4.53}$$

For deriving these expressions it has been assumed that b_C varies only slightly over a tidal period and that $|\eta| \ll h$. The high velocity power implies that the result depends strongly on the difference between flood and ebb velocity. This difference is partly due to river discharge Q_R and partly to tidal asymmetry.

Tidal asymmetry is represented by the hypothetical tidal curve:

$$Q_T = Q_m \cos \omega t + Q_a \cos 2\omega t, \quad Q_m = \frac{Q_T^+ + Q_T^-}{2}, \quad Q_a = \frac{Q_T^+ - Q_T^-}{2},$$
$$\tag{4.54}$$

where Q_T^+ and Q_T^- are the peak-flood and peak-ebb discharges and $Q_m = hb_C U$. The expression (4.53) will be evaluated for the downstream river reach where $Q_R \ll Q_m$. This inequality holds for the major part of tidal rivers with a strongly upstream converging channel cross-section. We will also assume $|Q_a| \ll Q_m$ and only retain first order terms relative to the small quantities $Q_R/Q_m, Q_a/Q_m, \eta/h$. This yields:

$$\langle q \rangle \propto \frac{1}{h^5 b_C^4} \left(-5 Q_R \langle Q_T^4 \rangle + \langle Q_T^5 \rangle - 5 \left\langle \frac{Q_T^5 \eta}{h} \right\rangle \right). \tag{4.55}$$

Tidal asymmetry versus Stokes component

The terms in the right-hand side of (4.55) represent different transport processes that will be discussed below.

- The first r.h.s. term can be approximated by

$$-\frac{5}{h^5 b_C^4} Q_R \langle Q_T^4 \rangle = -\frac{15}{8h} Q_R U^4, \tag{4.56}$$

  where U is the tidal velocity amplitude. It represents the contribution of tidal flow to the fluvial sediment flux; downstream sediment transport is strongly enhanced by the combined effect of ebb flow and river discharge. We call this the term river-induced asymmetry flux. It would cause a downstream increase of the sediment flux in the absence of tide-induced asymmetry.
- The second term

$$\frac{\langle Q_T^5 \rangle}{h^5 b_C^4} = \frac{5}{4h} U^4 Q_a \tag{4.57}$$

represents the contribution of tidal asymmetry; we call this term the tide-induced asymmetry flux. It is directed upstream if flood duration is shorter than ebb duration. The difference $2Q_a$ between the peak-flood discharge Q_T^+ and the peak-ebb discharge Q_T^- can be related to the HW and LW propagation velocities by (4.42). If we use the expression (4.25) for the propagation velocity we find

$$\frac{\langle Q_T^5 \rangle}{h^5 b_C^4} = \frac{5 b_C U^5}{2T} \left(\Delta_{FR}^{inlet} + \frac{2x}{c^-} - \frac{2x}{c^+} \right)$$

$$= \frac{5 b_C U^5}{2T} \left(\Delta_{FR}^{inlet} + \frac{4ax}{h^2} \sqrt{\frac{r}{2g\omega}} \right), \qquad (4.58)$$

where Δ_{FR}^{inlet} is the difference between the durations of tidal fall and tidal rise at the inlet ($x = 0$). This contribution to the tide-induced asymmetry flux can be important because in many estuaries tidal asymmetry is already present at the inlet. In funnel-shaped estuaries the inlet width b_C can be very large; the contribution (4.58) then dominates the first term $-5 Q_R \langle Q_T^4 \rangle / h^5 b_C^4$. This would imply a net upstream sediment flux at the inlet; however, this upstream flux may be offset by the last term of (4.55).

- The last term, $-5 \langle Q_T^5 \eta \rangle / h^6 b_C^4$, is a seaward directed transport component compensating for Stokes drift. This term strongly depends on the phase lag φ between the times of HW/LW and corresponding slack water. Tidal elevation and phase lag φ are given by

$$\eta(t) = a \sin(\omega t + \phi), \quad \varphi = 2\pi \Delta_S / T. \qquad (4.59)$$

Neglecting Q_a compared to Q_m we find for the Stokes-drift contribution:

$$-\frac{5 \langle Q_T^5 \eta \rangle}{h^6 b_C^4} \approx -\frac{25a}{16h} b_C U^5 \sin \varphi. \qquad (4.60)$$

In many field situations the phase lag φ is small enough to approximate $\sin \varphi \approx \varphi$, but also large enough for competition between this term and the previous term (4.58). Because of their linear width dependence both terms are largest at the inlet ($x = 0$) and decrease upstream; they should approximate balance for morphologic equilibrium. This is equivalent to

$$\Delta_{FR}^{inlet} = \frac{5\pi a}{4h} \Delta_S. \qquad (4.61)$$

The contributions of the tide-induced asymmetry component (l.h.s.) and the Stokes component (r.h.s.) are compared in Table 4.5 for a few tidal inlets, based on literature data. It is not easy to draw a conclusion; the data were not recorded for this purpose and it is not clear whether they yield representative values, especially for the slack water time lag Δ_S. There are important differences

Table 4.5. Observed values of Δ_{FR} (difference between durations of falling and rising tide) and Δ_S (average slack water phase lag relative to corresponding HW/LW) at the entrance of several tidal inlets. The uncertainty in the value of Δ_{FR} is at least half an hour, as it strongly depends on the shape of the tidal curve, which is different for each tidal cycle. Uncertainty is involved also in the definition of the periods of tidal fall and rise, for instance in the case of double maxima or minima in the tidal curve. The uncertainty in Δ_S is also large ($\pm$ one hour), in particular at stratified estuaries, because of different times of tide reversal over the cross-section. The value of the relative tidal amplitude a/h pertains to the converging estuarine part of the inlet (not necessarily equal to the average value of a/h in Table 4.2). Comparison of the last and the first column gives an indication of the seaward sediment flux due to covariance of water level and velocity (Stokes contribution) relative to the landward sediment flux due to tidal asymmetry. In most cases the Stokes contribution to the sediment flux is of the same order as the contribution of tidal asymmetry at the entrance; for the Seine and the Rotterdam Waterway the tidal asymmetry is larger and for the Columbia River the Stokes contribution is larger.

Tidal inlet	Δ_{FR} hours	Δ_S hours	a/h	$5\pi a \Delta_S/4h$ hours	Ref
Western Scheldt	0.4	1.0	0.1	0.4	[379]
Thames	0.4	0.5	0.25	0.5	[308]
Severn	0.0	0.0	0.15	0.0	[448]
Seine	3.0	0.7	0.32	0.9	[55]
Columbia	0.0	1.2	0.13	0.6	[191]
Ord	0.0	0.0	0.25	0.0	[497]
Hooghly	0.5	0.6	0.25	0.6	[308]
Rotterdam Waterway	1.0	1.5	0.085	0.5	[379]
Eastern Scheldt	0.33	0.8	0.1	0.3	[379]
Vlie Inlet	1.0	1.0	0.24	0.9	[379]

among the inlets; the tidal asymmetry at the inlet is absent at some inlets (Severn, Ord) and strong for others (Seine). The slack water lag ranges between 0 (Severn, Ord) to more than one hour (Columbia River). The general tendency seems to be an approximate balance between the contributions related to tidal asymmetry and to Stokes drift.

River-induced asymmetry versus tide-induced asymmetry

The tidal velocity amplitude U is generally high throughout the funnel-shaped part of the inlet. Therefore the river-induced asymmetry flux $-5Q_R \langle Q_T^4 \rangle$ and the tide-induced asymmetry flux $\langle Q_T^5 \rangle$ make significant contributions to the total sediment transport in the funnel; the other contributions decrease upstream from the inlet, except the river-induced transport Q_R^5. The river-induced asymmetry flux and the tide-induced asymmetry flux are both substantially stronger than the river sediment flux landward of the tidal limit. Uniformity of the sediment flux then requires that these two terms be approximately in balance. According to (4.56) and (4.57), this balance is equivalent to

$$Q_R = 2Q_a/3 = (Q_T^+ - Q_T^-)/3. \tag{4.62}$$

Equilibrium is attained if the difference between the maximum discharge during flood, $Q^+ = -Q_R + Q_T^+$, and the maximum discharge during ebb, $Q^- = Q_R + Q_T^-$, is equal to the river discharge Q_R. Dredging experience in the tidally dominated lower reach of the Rhine delta indicates that equilibrium is attained when Q^+ and Q^- are approximately equal. This is not very different from (4.62), considering that the tidal discharge is much larger than the river discharge. The tidal discharge asymmetry Q_a can be related to basin topography according to (4.58). The equilibrium relation is then expressed as:

$$Q_R = xb_C \frac{16aU}{3hT} \sqrt{\frac{r}{2g\omega}}. \tag{4.63}$$

In the idealised representation of river tidal inlets the width b_C is an exponentially decreasing function, $b_C \propto \exp(-x/L_b)$, and the depth h is assumed constant. In a strongly converging, friction-dominated basin the tidal amplitude a and the tidal velocity amplitude U are to be constant as well. Under these assumptions, the balance (4.63) can only be satisfied locally, in the region where the variation xb_C is approximately constant. This implies that the balance can be realised just locally around $x = L_b$. In that region we have

$$Q_R = L_b b_C(L_b) \frac{16aU}{3hT} \sqrt{\frac{r}{2g\omega}}. \tag{4.64}$$

We have shown earlier that L_b is well approximated for many inlets by the relation (4.9). Substitution in (4.64) yields

$$b_C^{equil} = \frac{3\pi e}{8} \sqrt{\frac{2g\omega}{r}} \frac{Q_R}{U^2} \approx 3.2 \frac{Q_R}{U^2}. \tag{4.65}$$

In this expression $b_C^{equil} = eb_C(L_b)$ is the inlet width ($e = 2.72$) and we have assumed $r = c_D U_c \approx 0.003$.

4.5.2. Equilibrium Inlet Morphology

Why does the equilibrium width depend on river discharge?

The expression (4.65) implies that the inlet width mainly depends on river discharge. However, the velocity associated with river discharge is generally an order of magnitude smaller than tidal velocity. There is convincing and widely accepted evidence that the area of the inlet cross-section depends primarily on the tidal prism, see Sec. 4.2.6. This raises the question: Why does the inlet width depend on river discharge? The answer to this question is inherent to the considerations leading to the result (4.65). In the first place, the influence of river flow on downstream sediment transport is amplified by the tide, as it creates an asymmetry between flood and ebb. At some distance from the inlet this river-induced ebb-flood asymmetry produces the largest contribution to downstream sediment transport. The only balance is provided by ebb-flood asymmetry related to tidal wave asymmetry, caused by different propagation speeds of HW and LW. Both opposing contributions strongly depend on tidal characteristics; the tidal dependence is partly cancelled, however, when both contributions are balanced, leaving a primary dependence on river discharge. This happens upstream of the inlet, where the inverse width $1/b_C$ increases almost proportionally to the distance x. At the inlet, other mutually opposing transport contributions play a dominant role, in particular the contribution related to Stokes drift and offshore tidal asymmetry. The balance of these contributions places no constraints on the inlet width, however. The condition which fixes the inlet width is set upstream; it determines the width downstream because of the approximately exponential convergence rate. According to Fig. 4.8 this convergence rate mainly depends on tidal characteristics and is almost independent of river discharge. These considerations together explain why the tidal inlet width is related to river discharge, as indicated by the result (4.65).

The inlet width as equilibrium indicator for existing tidal basins

Does the predicted equilibrium inlet width have practical relevance for real river tidal inlets? Can it tell, for instance, how close existing tidal inlets are to morphologic equilibrium? To answer this question we compare the predicted equilibrium width with the measured width of several river tidal inlets. The measured inlet widths are given in Table 4.6, based on data referenced in

Table 4.6. Inlet location and inlet width.

Number	Tidal inlet	Inlet location	b_C (m)
WS	Western Scheldt	Bath	1000
EL	Elbe	Cuxhaven	5000
WE	Weser	Bremerhaven — 10 km	5000
TH	Thames	Mucking	800
HB	Humber	Grimsby	2500
TA	Tamar	Weir — 20 km	1000
SV	Severn	Barry	20000
GI	Gironde	Pointe de Grave + 20 km	6000
SE	Seine	Honfleur	1500
HU	Hudson	Yonkers	900
PT	Potomac	Chesapeake	5000
DE	Delaware	Bay inlet	15000
SA	Satilla	river mouth	1700
CL	Columbia River	Sand Island	2500
OR	Ord	Pantin Island	3000
HO	Hooghly	Saugar	10000
FL	Fly	Tirere	22000

Table 4.2. The location of the river inlet is specified, because often the outer part of the inlet system does not match the characteristics of a river tidal inlet. In some cases the inlet is formed by hard structures (Columbia, Gironde, Hudson, Ord, Severn, Tamar) and in other cases the outer part of the inlet corresponds to a barrier tidal basin (Elbe, Humber, Weser, Western Scheldt, Thames). The outer part of these river tidal inlets has been excluded. It is assumed that tidal velocity asymmetry is not strongly developed at the inlet or that it is offset due to Stokes drift. The predicted equilibrium width b_C^{equil} is estimated for the river inlets of Table 4.6, using the data of Table 4.2. The data refer to spring-tidal conditions; spring-tidal currents are substantially larger than neap-tidal currents and most sediment transport takes place in the period around spring tide. The corresponding value of the cross-sectionally averaged velocity U has been estimated from published data; it should be mentioned that reliable data, representative of average spring-tidal conditions in each estuary, are scarce in the literature. The average river discharge value is used everywhere for consistency. Conditions of average discharge are probably not representative of long term sediment transport; a higher discharge value would be more appropriate. The most representative value depends on discharge statistics.

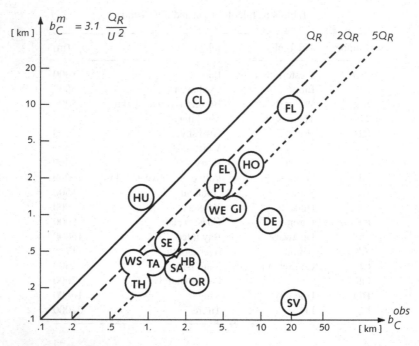

Fig. 4.34. Observed inlet width of river tidal basins compared to the width derived from the morphologic equilibrium relationship (4.65), for different representative river discharges Q_R. The basin names are abbreviated as indicated in Table 4.6. For many inlets the computed equilibrium width is consistent with the observed width provided the river discharge representative for equilibrium sediment transport is taken between two and five times the mean river discharge Q_R (corresponding to the dotted lines).

Equilibrium or not?

In Fig. 4.34 the computed equilibrium inlet widths are compared to the observed widths. Based on this comparison several remarks can be made:

- For most basins the observed inlet width is larger than the computed equilibrium width corresponding to average discharge. However, the widths of the
Hudson and the Columbia are substantially smaller; these two estuaries will
pass fluvial sediment to the sea even if the river discharge is below average.
In general, upstream sediment transport due to tide-induced asymmetry is
larger than downstream sediment transport due to river-induced asymmetry,
except for the above two estuaries.
- For many inlets the computed equilibrium width matches approximately the
observed width, if the representative river discharge is taken between two

and five times the average value (corresponding to the dotted lines); five times the average discharge is a typical magnitude of the maximum annual discharge of rain rivers in temperate climate zones. For these inlets, the inlet width is consistent with the condition of morphodynamic equilibrium, if it is assumed that equilibrium is determined by river discharges higher than the average. In the case of long-term morphologic equilibrium, fluvial and marine sediments will be retained in the estuary only up to the amount required to keep pace with sea-level rise. Consistency with the equilibrium condition does not imply that all these basins are actually in equilibrium with sea-level rise; it may happen that sediment supply from both fluvial and marine sources is too small. An example is the Potomac, where the model predicts only a weak tidal asymmetry and where fluvial sediment input is small [322].

- According to the model (4.65) a few inlets are far from morphologic equilibrium; their width is much larger than the equilibrium width. Marine sediment import dominates over river-related seaward transport, even at high river discharge. This is the case for the Severn and the Delaware and to a lesser degree also for the Ord and the outer part of the Humber estuary. The Severn and Delaware receive small inputs of fluvial and marine sediments [322, 483]; these inputs are insufficient for shaping an equilibrium inlet morphology. Studies of the sedimentation history of the Severn also point to the absence of morphodynamic equilibrium [194].

- Several inlets export fluvial sediment only when the river discharge is much higher than the average discharge. This is the case for the Gironde, Humber and Ord; these estuaries have a high concentration of fine suspended sediment and a well developed turbidity maximum. Part of the fluvial sediment is probably retained in these estuaries. The Gironde, for instance, is close to equilibrium [69], yet a substantial fraction of the fluvial sediment remains trapped in the estuary [5]. The Humber estuary is filled mainly with marine sediment [244], in spite of a high fluvial sediment input. This suggests that the Humber is still in an infilling stage. The Ord river has extreme runoff variability; almost the entire annual discharge occurs within a three-month period [497]. Peak river discharges are therefore an order of magnitude higher than the average discharge and the same holds for the peak sediment fluxes.

4.5.3. *Other Phenomena Influencing Sediment Transport*

Some words of caution are needed, however. The model on which these conclusions are based is an extreme simplification and ignores several important processes.

Estuarine circulation

The influence of estuarine circulation has been disregarded thus far. Estuarine circulation is produced by the horizontal density gradient which is associated with sea water intrusion, see Sec. 4.7.2. At high river discharges, representative of morphologic equilibrium, sea water can be almost completely flushed out of the inlet. In that case estuarine circulation is suppressed; examples include the Seine [18], the Weser [176], the Elbe [395] and the Columbia [215]. In many other river inlets a well developed estuarine circulation is present, even at high river discharge. The residual velocities related to this circulation may exceed the velocity related to river runoff, especially in the deeper channel sections. This casts a doubt on the reliability of the model for estuaries with strong estuarine circulation. At least two aspects of estuarine circulation need to be considered: (1) Tidally averaged upstream flow along the bottom together with shear stresses that are substantially higher during flood than during ebb. This is the case, for instance, in the Hudson [51], the Gironde [5] and the Tamar [452]. (2) Significant density stratification, especially during ebb tide. This density stratification decreases the bottom friction experienced by LW tidal propagation and therefore reduces tidal wave asymmetry. This may explain why for several estuaries the difference between observed HW and LW propagation speeds is smaller than the difference predicted on the basis of identical friction coefficients for HW and LW propagation, see Table 4.4. The processes (1) and (2) have opposite effects on the net sediment flux; if these effects are comparable the dominance of bottom shear stress at flood relative to ebb is similar with and without density-induced processes. This is consistent with the finding that the gross features of tidal propagation in estuaries can be rather well reproduced by models in which density effects are ignored [360]. In this case the net upstream sediment fluxes with and without density-induced processes have comparable magnitude. Numerical model computations indicate that for the Seine estuary, for instance, the turbidity maximum is fairly well reproduced by tidal asymmetry, even if estuarine circulation is ignored [55].

Fluid mud

In river tidal inlets with a well developed turbidity maximum a fluid mud layer may form at the bottom. An example is the Gironde [5], where a strong turbidity maximum extends over a zone between 30 and 80 km from the inlet, see Fig. 4.5. In this zone a fluid mud layer is formed at neap tide and high river discharge, with densities up to $250 \, \text{kg/m}^3$. Most of this fluid mud is brought into suspension during spring tide. Fluid mud has also been observed in the Thames [332] and the Fly [195]. The presence of fluid mud at the bottom affects tidal propagation; the bed roughness is decreased and tidal propagation may be no longer friction-dominated. It is not clear if fluid mud also plays a role in periods of high river discharge and spring tide, during which most sediment transport takes place.

Suspension and erosion time lags

Suspension and erosion time lags have been ignored in the model. The delayed response of the suspended load to velocity variations induces a net transport of fine sediment which depends on tidal asymmetry in a different way than rapidly responding suspended load. In Sec. 4.7 this will be discussed in more detail; it will be shown that suspension and erosion lag effects are particularly important in tidal basins with large intertidal areas. This is less relevant for the river tidal inlets of Table 4.6.

Flood and ebb dominated channels

Flood and ebb discharges are in general not evenly distributed over the channel cross-section. In some cases the inlet consists of two or more channels which convey different fractions of the tidal flow during ebb and flood; this may be related, for instance, to channel meandering or to the influence of earth's rotation. Examples are the Gironde [5] and the Fly [494], where the difference between flood and ebb discharge in each channel is greater than the river discharge. But single-channel inlets may also exhibit significant lateral differences in flood-ebb asymmetry, for instance, the Hudson [51] and Tamar [450]. In these estuaries fluvial sediment escapes to the sea via the ebb-dominated inlet channel (or via the ebb-dominated part of the inlet). This implies that the sediment budget of the estuary may be close to equilibrium, even if the river discharge is too small for neutralising flood dominance related to tidal wave asymmetry over the cross-section. In the Gironde, for instance, the dredged southern inlet channel mainly carries flood flow; flood dominance is enhanced in this channel by estuarine circulation. A large part of the ebb flow is conveyed by the shallower northern channel; during high river discharge the tide-averaged near-bottom flow in this channel is directed downstream, allowing fluvial sediment to escape the estuary [5].

Wind influence

Suspension and subsequent export of fine sediment from mud flats is strongly stimulated by wind-induced waves. An example is the Seine, with large mud-flats bordering the main channel near the mouth. The highest sediment export from the the estuary occurs during storm events, when the mudflats are eroded by wind waves and highly turbid waters are dispersed over the outer bay [273]. Many other observations exist of sediment export caused by wave-induced mudflat erosion, for instance in the Wadden Sea [105].

Non-uniform tidal asymmetry

Some of the estuaries are shallow (Tamar, Ord, Humber) and have a large amplitude-to-depth ratio. This invalidates several approximations made in the derivation of the model results. Moreover, in the case of a large amplitude-to-depth ratio, new phenomena may come into play. A few examples follow.

- The assumption that tidal velocity and tidal elevation are out of phase by $\pi/2$ radians does not hold over the entire channel cross-section. Uncles and coworkers have observed ebb-dominant flow over the intertidal areas in the otherwise strongly flood-dominant Tamar estuary [451]. They explain this phenomenon by the inertial delay of flow reversal in the deeper parts of the channel relative to the shallow parts. Therefore the ebb-flow is concentrated at the shallow channel banks during a few hours after high water.
- In the Ord estuary, Wright *et al.* have observed strong ebb flow in the deepest part of the main channel [497]. Although the peak flood-discharge in the Ord is much higher than the peak-ebb discharge, bedforms in the deepest channel parts have their steepest slope oriented in ebb direction. They explain this phenomenon by draining of tidal flats at the end of the ebb period and

subsequent concentration of ebb flow in the deepest part of the channel, when water levels are so low that higher lateral parts of the channel bed are dry. The authors suggest that flood-dominant sediment transport in the shallower parts of the cross-section may be partly offset by ebb-dominant sediment transport in the deepest channel parts. If this is true, the Ord estuary might be closer to morphodynamic equilibrium than suggested by the simple one-dimensional model. A similar phenomenon may also occur in the Humber. In this estuary a downstream displacement of sandy bed sediments is observed in periods of high river flow [353].

- In estuaries with a large amplitude-to-depth ratio the contribution of increased ebb velocities due to Stokes drift is more important than in estuaries with a smaller amplitude-to-depth ratio. In the Tamar, model computations indicate that there is a small downstream residual flow near the bottom in the shallow upper part of the estuary, which can be explained in this way [452].

Model as a tool for diagnosis

The above discussion illustrates that simple flow models may give a misleading picture of the net sediment transport in river tidal inlets. One may wonder what is the use of idealised models if there are so many limitations? It is clear that the simple model (4.63) is not adequate for predictive purposes. The strength of idealised models is their transparency; therefore they should be considered and used as a tool for analysis of concepts, that helps define and resolve questions about phenomena observed in the field. The models themselves do not contain much knowledge, yet they can be very useful for developing better understanding of estuarine morphodynamics.

4.6. Equilibrium of Barrier Tidal Inlets

4.6.1. *Equilibrium Morphology*

Equilibrium relationship

In barrier tidal inlets, sediment transport is mainly related to the tide. These basins are not former river valleys and and generally they receive no significant river inflow. In shallow areas, sediment is stirred by wind waves and some transport may occur due to wind driven currents. Tidal currents are, by far, the most important sediment transport agent. The strength of the tidal currents is modulated by the neap-spring tide cycle and by storms, but the main periodicity is the tidal period.

As for river tidal inlets, morphodynamic equilibrium of barrier tidal inlets requires uniformity of the long term average sediment flux. Neglecting river discharge and fluvial sediment input, the equilibrium condition reads (see (4.55))

$$\langle q \rangle \propto \frac{1}{h^5 b_C^4} \left(\langle Q_T^5 \rangle - 5 \left\langle \frac{Q_T^5 \eta}{h} \right\rangle \right) = 0, \qquad (4.66)$$

with the following notations: sediment flux q, tidal discharge Q_T, tidal elevation η and tidal amplitude a. The channel depth h and channel width b_C are representative basin-averaged values. Using (4.54) and (4.59) this becomes

$$\langle q \rangle \propto \quad Q^+ - Q^- - \frac{5a}{8h}(Q^+ + Q^-) \sin \varphi = 0. \qquad (4.67)$$

The asymmetry between peak-flood (Q^+) and peak-ebb (Q^-) discharges is related, according to (4.41), to the difference in duration between ebb and flood Δ_{EF}. We distinguish between ebb-flood asymmetry caused by a difference between fall and rise of the offshore tide, Δ_{FR}^{inlet} and ebb-flood asymmetry which develops within the basin, Δ_{EF}^{basin}. Using the definition of the phase lag φ (4.59), we find from (4.67) that equilibrium requires

$$\Delta_{FR}^{inlet} + \Delta_{EF}^{basin} = (5\pi a/4h)\Delta_S. \qquad (4.68)$$

Tidal asymmetry versus Stokes contribution

Equation (4.68) expresses the balance between landward sediment transport due to tidal asymmetry (l.h.s.) and seaward transport related to Stokes drift (r.h.s.). In Table 4.5 the tidal asymmetry at the entrance (first l.h.s.-term) is compared with the Stokes contribution (r.h.s.-term), for a few tidal basins along the Dutch coast (Western Scheldt, Eastern Scheldt, Vlie Inlet); the order of magnitude of both terms is similar. The importance of the Stokes contribution for these basins can be estimated from Figs. 4.25, 4.32 and 4.35, showing a phase delay of approximately 1 hour. Theoretical expressions for Δ_{EF}^{basin} and Δ_S are derived in the appendix (B.66), for uniform or weakly converging tidal

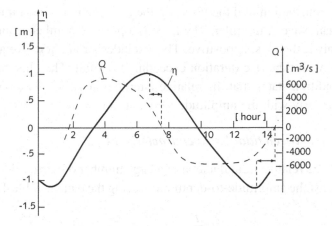

Fig. 4.35. Simultaneous tide curves of water elevation η and tidal discharge Q at the Vlie Inlet (Wadden Sea), near the entrance. The average slack water phase lag relative HW/LW is approximately one hour, indicative of a partially standing tidal wave and of an important Stokes drift contribution. (Survey Department Rijkswaterstaat.)

basins (h, b_C, b_S approximately constant). These expressions read

$$\Delta_{EF}^{basin} = \frac{rl^2}{g}\left(\frac{1}{D^- D_g^-} - \frac{1}{D^+ D_g^+}\right),$$

$$\Delta_S = \frac{rl^2}{4g}\left(\frac{1}{D^- D_S^-} + \frac{1}{D^+ D_S^+}\right), \tag{4.69}$$

where l is the length of the basin, $r \approx (8/3\pi)c_D U$ is the linear friction coefficient, $D^\pm$ the channel depth at HW and LW and $D_S^\pm$ is the propagation depth at HW and LW. We are looking for simple expressions which can be compared with published data, accepting the qualitative character of this comparison; therefore we will not distinguish between channel width at LW and HW and consider $b_C^+ \approx b_C^- \approx b_C$, and further we assume that the tidal variation of depth η and surface width b_S are small compared to the tide-averaged values, $a \ll h$, $b_S^+ - b_S^- \ll b_S$. For the equilibrium condition (4.68) we thus find the expression

$$\frac{h}{a}\frac{b_S^+ - b_S^-}{b_S^+ + b_S^-} = 2 - \frac{5\pi}{16} + \frac{c^2}{lU}\Delta_{FR}^{inlet}, \tag{4.70}$$

where c is the frictional tidal propagation speed (4.25) and where we have replaced lab_S/hbc by U/ω according to (4.9). The first term on the r.h.s. is the contribution of tidal asymmetry generated inside the basin, the second term is the Stokes contribution and the last term the contribution of asymmetry of the offshore tidal wave at the inlet. The l.h.s. is a positive number; equilibrium is possible only if the r.h.s. is positive. This excludes strong negative asymmetry at the inlet, with tidal rise duration exceeding tidal fall. The l.h.s. of the equilibrium condition states that the relative intertidal area $(b_S^+ - b_S^-)/(b_S^+ + b_S^-)$ increases linearly with the amplitude-to-depth ratio a/h.

Relationship between tidal basin characteristics

In Fig. 4.36 the relative tidal flat area of a large number of barrier tidal basins is plotted against the amplitude-to-depth ratio, using the data in Table 4.2. Coastal

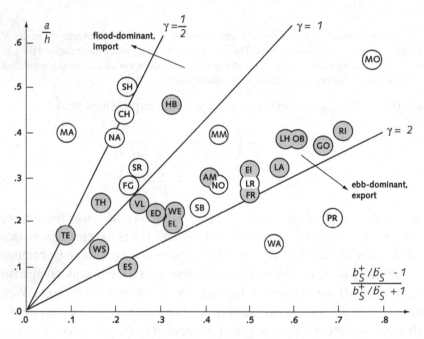

Fig. 4.36. Morphodynamic equilibrium of barrier tidal basins depends primarily on the ratio γ of the relative tidal flat area $(b_S^+ - b_S^-)/(b_S^+ + b_S^-)$ and the relative tidal amplitude a/h. No tidal asymmetry occurs at narrow microtidal shelves; in this case tidal basin equilibrium corresponds to $\gamma \approx 1$. At large meso/macrotidal shelves offshore tidal asymmetry may offset the Stokes contribution; in this case tidal basin equilibrium corresponds to $\gamma = 2$. Tidal basins at large meso/macrotidal shelves (in particular the European northwestern shelf) are represented by grey dots; most of these basins are close to $\gamma = 2$. Tidal basins situated at narrow microtidal shelves (in particular the US Atlantic shelf) are represented by white dots; the average γ-value of these basins is close to $\gamma = 1$. Net sand import may be required for keeping pace with sea-level rise; this may partly explain smaller γ-values. Tidal basin abbreviations are given in Table 4.2.

lagoons without well defined tidal channels and tidal flats have been excluded; these lagoons are far from equilibrium as sediment supply is insufficient to keep pace with sea-level rise. Many of such lagoons are present along the narrow-shelf US Atlantic and Gulf coasts [325, 357]. Figure 4.36 shows that most of the reported basins are compatible with an increasing relative tidal flat area as a function of the amplitude-to-depth-ratio. All these basins lie within the range

$$\frac{b_S^+ - b_S^-}{b_S^+ + b_S^-} = \gamma \frac{a}{h}, \quad 0.5 < \gamma < 2. \tag{4.71}$$

Wide-shelf basins versus narrow-shelf basins

The range $1 < \gamma < 2$ corresponds to partial compensation of the Stokes contribution by offshore asymmetry (full compensation: $\gamma = 2$ and no compensation: $\gamma = 1$); most of the tidal basins along the northwest European shelf fall in this range. The northwest European shelf is characterised by meso to macrotidal ranges and shallow depth (in particular along the Dutch coast and the Wadden Sea coast), with significant distortion of the offshore tidal wave. The value of γ decreases when the offshore wave asymmetry becomes smaller. Along the US Atlantic coast offshore tidal distortion is substantially smaller than on the northwest European shelf [281]; this is related to the smaller amplitude-to-depth ratios of the tide along the US Atlantic coast, compared to the European north-western shelf. At most inlets along the southern US Atlantic coast offshore tidal asymmetry offers hardly any compensation for the Stokes contribution. Most of the inlets of the US Atlantic coast fall in the range $0.5 < \gamma < 1$. The absence of offshore tidal asymmetry could be one of the reasons why the import of marine sediment in many tidal lagoons along the US Atlantic coast is insufficient to keep pace with sea-level rise, contrary to, for instance, the tidal lagoons along the Wadden Sea. Tidal asymmetry offshore of the UK east coast is also rather small; the Thames and the Humber inlets are both in the range $0.5 < \gamma < 1$.

There is rather strong scatter of the observed inlet geometries relative to the equilibrium relationship (4.71). However, it should be noted that several tidal inlets do not meet the conditions underlying the derivation of this relationship. For instance, for many inlets the ratios a/h and $(b_S^+ - b_S^-)/(b_S^+ + b_S^-)$ are not much smaller than 1. In the derivation of (4.71) it has been assumed that the fluvial sediment input can be ignored. For the Price (PR) and Wachapreague (WA), two small inlets along the US Atlantic coast, this condition does not hold. These inlets have experienced infill in excess of sea-level rise [325] by non-marine sources and consist of tidal creeks bordered by large marshes and mudflats. They are ebb-dominated and deliver sediment to the sea [158]. Another small microtidal inlet along the US Atlantic coast, the Manasquan (MA), is flood-dominated and also outside the equilibrium range. This inlet has a strongly constricted mouth and a converging basin, more characteristic of a river tidal inlet than a barrier tidal inlet.

With the exception of Price, Wachapreague and Eastern Scheldt, all tidal basins with offshore tidal asymmetry lie above the line $\gamma = 2$, and most tidal basins without offshore tidal asymmetry lie above the line $\gamma = 1$. This suggests that these basins have a tendency towards flood dominance. This tendency may be explained by the sediment import needed for keeping pace with sea-level rise.

Morphodynamic equilibrium

Figure 4.36 suggests that tidal basins tend to a morphodynamic equilibrium by developing a morphology which counteracts flood dominance related to off-shore tidal asymmetry and ebb dominance related to Stokes drift. In shallow basins the excess propagation speed of HW relative to LW is greater than for deep basins. This excess HW propagation speed increases ebb duration relative to flood duration, producing stronger flood currents than ebb currents. The HW propagation speed is slowed down by the presence of intertidal areas, by diversion of flood flow to these areas. In shallow basins larger intertidal areas are needed to counterbalance the excess HW propagation speed, than in deep basins. This explains the equilibrium requirement of increasing intertidal area with increasing amplitude-to-depth ratio, see Fig. 4.37. This is the case when strong offshore tidal asymmetry (large γ) offsets ebb dominance related to Stokes drift. In the absence of offshore tidal asymmetry (small γ), ebb dominance related to Stokes drift need to be compensated by internally generated flood dominance; in this case large intertidal areas will not develop.

Offshore tidal asymmetry and long-term basin evolution

The role of offshore tidal asymmetry for the long-term evolution of tidal basins was already mentioned for the tidal lagoons along the US Atlantic coast; it was suggested that the absence of offshore tidal asymmetry (equal duration of

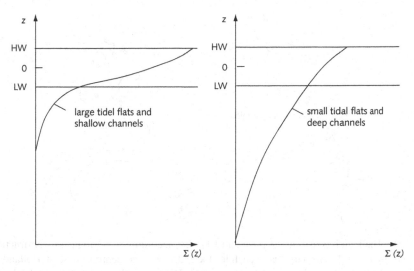

Fig. 4.37. Hypsometric curves for two different tidal basin geometries which are both consistent with morphologic equilibrium. $\Sigma(z)$ is the basin surface area as a function of the distance z from a reference surface level.

tidal rise and tidal fall) may explain the relative weakness of tidally induced sand import [357]. The flood delta of these lagoons is small compared to the tidal inlets along the European Atlantic and North Sea shelves, and sand import in many US Atlantic lagoons is insufficient for keeping pace with sea-level rise.

The opposite evolution has taken place along the Dutch coast between 5000 BP and 3000 BP. When marine transgression reached the present Dutch coastline, some 6500 BP, the sea was able to intrude the low lying plain of the rivers Rhine, Meuse and Scheldt. Large tidal inlets developed in the southern, central and northern parts of The Netherlands, scouring deep channels in the Pleistocene floor, see Fig. 4.38. The tidal inlets in the southern and northern parts of the Netherlands still exist, but those in the central part have disappeared, in particular the large Bergen Inlet. What made the difference between the evolution of these tidal inlets? Figure 4.39 shows the distortion of the tidal wave along the Dutch coast; tidal asymmetry is greatest at the coast of central Holland, where tidal rise is almost five hours shorter than tidal fall. The inlets on the central Dutch coast therefore experienced much stronger flood currents than ebb currents [460], causing a rapid infill of the basins and the development of large intertidal areas. The rectifying effect of these intertidal areas on

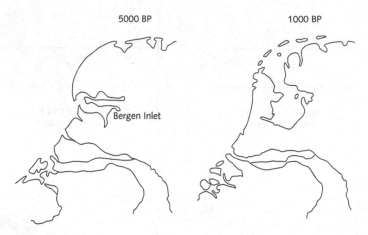

Fig. 4.38. The Dutch coastline 5000 BP and 1000 BP, according to geological reconstructions, redrawn after [505]. The large Bergen Inlet located along the central coast of Holland has completely filled in and the coastline was closed. Further to the north, the Texel inlet and Zuyderzee basin have developed during the past 1000 years.

HW and LW propagation was probably insufficient to offset flood dominance imposed by the offshore tide. Around 3500 BP the central Dutch coastline was closed.

Evolution of Eastern Scheldt and Texel Inlet

In Fig. 4.36 Texel Inlet lies on the line $\gamma = 1/2$ and the Eastern Scheldt on the line $\gamma = 2$. This difference cannot be explained by offshore tidal asymmetry; one would expect their places in the diagram to be reversed, because offshore asymmetry is stronger at Texel than at the Eastern Scheldt. The Eastern Scheldt responded to an increase of the tidal prism; tidal flats have extended and channels have deepened. The Eastern Scheldt was in an exporting mode, before its partial closure in 1985. This explains its place in the diagram, slightly below the line $\gamma = 2$. The Texel Inlet responded to the loss of its major flood basin, the Zuyderzee, which was closed in 1930, see Fig. 4.40 (top). In the situation before closure, the tide in the Texel Inlet was a damped propagating wave. The phase difference between tidal elevation and velocity at the mouth was well below $\pi/2$ radians; there could be equilibrium because asymmetry-induced import was offset by Stokes drift compensating ebb dominance, see Fig. 4.40 (bottom). After closure of the Zuyderzee the nature of the tide changed into an

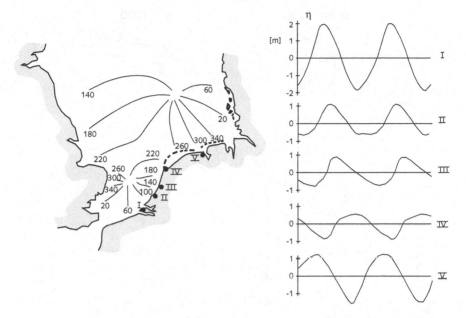

Fig. 4.39. While propagating along the Dutch coast from south to north the tidal wave becomes increasingly distorted. Asymmetry between tidal rise and fall is most pronounced along the coast of central and northern Holland (between III and IV). This asymmetry is mainly due to the relative shallowness of the inner shelf in this stretch of the coast, see Appendix C.2. Further to the north and along the Wadden Sea the tidal wave is part of the amphodromic system of the central North Sea, which develops less asymmetry than the amphidromic system of the Southern Bight.

almost standing tidal wave, with only minor Stokes drift. The basin has been importing sediment since, but the intertidal area is still too small to completely offset the flood-dominance at the inlet due to the offshore tide. This explains the position of Texel Inlet above the line $\gamma = 1$. Figure 4.36 illustrates the opposite evolution of these basins in response to human interventions.

Local differences in flood and ebb dominance

In the simple one-dimensional representation of tidal inlets we have ignored cross-sectional differences in flow strength. Such differences can be quite substantial though. This is illustrated in Fig. 4.41, for the inlet channel of the Eastern Scheldt's southern flood delta. The northern part of the inlet channel is ebb-dominated, while the southern part is flood-dominated. Hence, even if

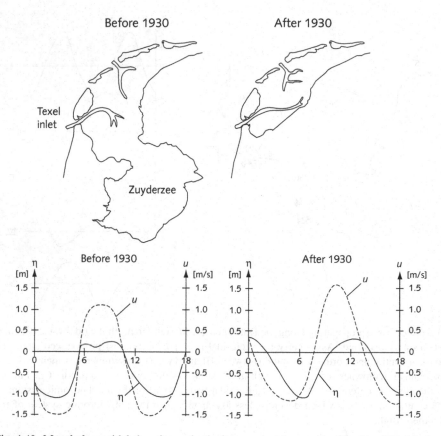

Fig. 4.40. Morphology, tidal elevation and velocity curves at the mouth of the Texel Inlet, before and after closure of the Zuyderzee. The tidal elevation curve is measured, the velocity is computed with a numerical two-dimensional tidal model. Before closure in 1930, the Texel Inlet was the main inlet channel of the Zuyderzee. Tidal energy was almost completely dissipated in the shallow Zuyderzee; the nature of the tide at Texel was a progressive damped wave, with almost no phase difference between tidal elevation and velocity. Texel Inlet was ebb-dominated due to Stokes drift and due to longer flood duration than ebb duration. After closure of the Zuyderzee the tide was reflected at the closure dam. A phase shift of more than one hour between tidal elevation and velocity decreased the Stokes drift; the flood duration became shorter than the ebb duration and the tide became flood-dominated.

the inlet channel is ebb-flood neutral as a whole, a net import or export of sediment will take place. Local flood or ebb dominance are often stronger than the cross-sectionally integrated flood or ebb dominance due to tidal asymmetry. Therefore the sediment balance of an inlet cannot be attributed solely to tidal asymmetry. If, for instance, the deepest parts of the channel are ebb-dominated

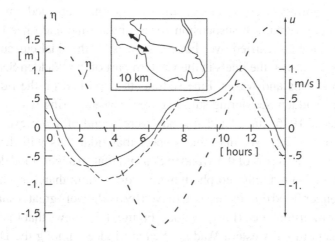

Fig. 4.41. Residual circulation in the inlet channel to the southern flood delta of the Eastern Scheldt. The solid line is the depth-averaged velocity in a vertical near the southern channel bank and the dotted line the depth-averaged velocity in a vertical near the northern channel bank; both are determined from simultaneous measurements. The southern part of the channel is flood dominated, while the northern part is ebb dominated. (Survey Department Rijkswaterstaat.)

and the shallower parts are flood-dominated, flood dominance will prevail over a larger cross-sectional width than ebb dominance; this may produce a net flood-dominated sediment transport. The meandering character of tidal channels sets a limit to the length of flood or ebb-dominated channels. At the end of a flood or ebb channel sediment is deposited due to flow divergence; in this way shoals develop at the transition between flood and ebb channels. Therefore, import or export of sediment by local flood dominance or ebb dominance generally affects only a limited portion of the basin. Tidal asymmetry may also change along the basin; however, the asymmetry characteristics of the tidal wave will generally persist over longer distances than the individual flood and ebb channels. For that reason tidal asymmetry plays a major role in the overall sediment import or export of tidal basins.

Waves and storms

Sediment load is strongly increased under storm conditions with high waves. This may cause additional sediment import in sheltered basins, because of a heavily loaded incoming flood current. By contrast, in open basins sediment export will be enhanced, because wave stirring is most effective on the intertidal areas. Fine sediments are more sensitive to storms than medium or coarse sediments; they remain in suspension for a longer time and once brought in suspension, they are carried over large distances by the tidal current. Among all Dutch tidal basins, the inlets in the western part of the Wadden Sea are most exposed to storms and waves. Fine sediments deposited on the tidal flats in the western Wadden Sea during summer are removed during winter. After the storm surge of 1976, fine Wadden sands were found offshore even at 20 km from the coast [492]. We may thus expect that tidal flats will develop less easily in the western part of the Wadden Sea and that stronger flood dominance is required for maintaining morphodynamic equilibrium than for other similar basins. Stronger flood dominance is achieved by a relatively greater shallowness and smaller intertidal area (low γ-value). Figure 4.36 shows that the Texel and the Vlie Inlets in the western Wadden Sea are indeed among the Dutch tidal basins with the lowest γ-value. A low γ may be indicative of storm and wave influence on morphodynamic equilibrium.

4.6.2. *Morphodynamic Stability*

Management questions

In the foregoing we have established the condition for morphologic equilibrium imposed by tide-induced sand transport. Many existing tidal basins have morphologic characteristics which are consistent with this equilibrium condition. But does the equilibrium condition also ensure a stable equilibrium? In the following we will consider this question, which is very relevant from a management perspective. How does a tidal basin respond to interventions, such as channel dredging or tidal flat reclamation? Will it keep pace with sea-level rise, or not? In order to answer these questions we will slightly perturb a tidal basin at equilibrium and look what happens.

Stability analysis

For given tidal boundary conditions, we may express the cross-sectionally averaged net sediment flux at the inlet, $\langle q \rangle$, as a function of basin morphology: $\langle q \rangle = f(\xi_1, \xi_2, \ldots, \xi_n)$. Here $\xi_1, \xi_2, \ldots, \xi_n$ represent parameters describing basin morphology, for instance, channel depth $\xi_1 \equiv h$, low-water basin width $\xi_2 \equiv b_S^-$, high-water basin width $\xi_3 \equiv b_S^+$, channel width $\xi_4 \equiv b_C$ etc. In case of equilibrium the net sediment flux is uniform along the basin; if the basin is closed at the landward boundary and if it receives no fluvial sediment, the net sediment flux equals zero. Then we intervene by slightly changing some of the morphologic parameters. This intervention influences tidal asymmetry and therefore causes a departure of the net sediment flux from its zero-value, corresponding either to a net import or to a net export of sediment. Import will reduce the basin volume V and export will increase the basin volume. This increase or decrease is expressed by the relation

$$V_t = -\langle q \rangle. \qquad (4.72)$$

The basin volume V depends on time through the basin parameters $\xi_1(t), \ldots, \xi_n(t)$. Two situations may occur: (1) the change of basin volume V tends to restore the initial zero sediment flux or (2) the change of V enhances the departure of the net sediment flux from its zero-value. In the first case the initial morphologic equilibrium is said to be stable; some time after a small morphologic perturbation a new equilibrium is established, which is close to the initial equilibrium. In the second case the morphologic equilibrium is unstable; after a slight perturbation the morphology moves further away from the initial equilibrium state.

Depth instability

We consider the case that the initial perturbation is limited to one parameter, the channel depth $h \equiv \xi_1$. The departure of ξ_1 from the equilibrium state is designated ξ_1' and we assume $\xi_1' \ll \xi_1$. In that case $V_t \approx V_1 \xi_{1t}'$ and $\langle q \rangle \approx q_1 \xi_1'$, where V_1 and q_1 are the ξ_1-derivatives at equilibrium, $V_1 \equiv V_{\xi_1}$ and $q_1 \equiv \langle q \rangle_{\xi_1}$. The relation (4.72) now reads

$$\xi_{1t}' = -(q_1/V_1)\xi_1'. \qquad (4.73)$$

The basin volume V increases with increasing depth, thus $V_1 > 0$. With increasing depth, tidal asymmetry changes such that the duration of tidal rise increases compared to tidal fall; the net sediment flux therefore becomes negative, $q_1 < 0$. From (4.73) it follows that the rate of change of ξ_1' is positive; the depth moves away from its equilibrium value. Basin morphology is thus unstable against depth perturbation.

Dynamic coupling of depth with tidal flat area

Such an instability conflicts with the existence of tidal basins. Apparently we have overlooked something. Indeed, the parameters $\xi_1, \xi_2, \ldots, \xi_n$ are not independent; we have to take into account the dynamic coupling between these parameters. This dynamic coupling has been discussed in Sec. 4.2.6. An increase of channel depth means a decrease of the tidal velocity amplitude U below the critical value for maintaining the tidal flats. The tidal flats will therefore erode; the width and height of the tidal flats will decrease and the eroded material will settle in the channel. By this process the channel depth will decrease and return to its initial value. The amplitude of the tidal velocity will also be restored at its critical value. However, the tidal flat area is decreased. This decrease of tidal flat area influences tidal asymmetry such that the duration of rising tide is decreased relative to falling tide. The net sediment flux therefore changes to import; the channel depth is further decreased and the amplitude of the tidal currents will exceed the critical value. The channel bed is eroded, channel depth is increased and the eroded material is used for restoring the tidal flats. The basin goes through a cyclic process of increasing and decreasing channel depth and corresponding decreases and increases of the tidal flat area.

A simple model of dynamic depth-width coupling

The model is based on Eq. (4.72), in which we introduce the dynamic coupling of channel depth and tidal flat area. This coupling is expressed by the equation

$$V_2 \xi'_{2t} = q_T. \tag{4.74}$$

Here $\xi_2 \equiv b_S^-$ is the basin width at low water; the departure ξ'_2 of this width from equilibrium increases with increasing channel depth, mainly due to dominance of wave-induced erosion over tide-induced accretion. The amount of eroded tidal flat sediment is given by $V_2 \xi'_2$, where $V_2 \equiv V_{\xi_2} > 0$ is the ξ_2-derivative of the basin volume at equilibrium, see Fig. 4.42. The transverse sediment flux from tidal flat to channel is designated by q_T; it is a function of the average maximum tidal velocity U. This transverse flux is zero on average if U is equal to the critical velocity U_c for tidal flat stability (tide-induced accretion compensates for wave-induced erosion). A small departure from equilibrium $U' = U - U_c$ produces a transverse sediment flux $q_T = q_{TU} U'$, with negative

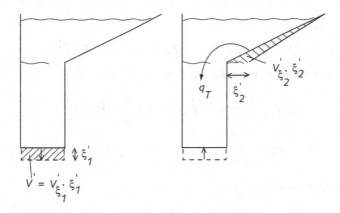

Fig. 4.42. A small channel depth increase ξ_1' produces a small increase in basin volume, $V' = V_{\xi_1}\xi_1'$. The increase of channel depth is compensated by a lateral sand input q_T from tidal flat erosion. Tidal flat erosion increases the basin width at low water, $\xi_2 \equiv b_S^-$; the amount of eroded sand equals $V_{\xi_2}\xi_2'$.

q_{TU} (U-derivative of q_T at equilibrium). The increase of depth ξ_1' produces a decrease of the maximum tidal velocity U; this decrease is to first order given by $U' = U_{\xi_1}\xi_1'$, with $U_{\xi_1} < 0$. We therefore may write $q_T \approx q_{TU}U_{\xi_1}\xi_1' = q_{T1}\xi_1'$, with $q_{T1} = q_{TU}U_{\xi_1} > 0$. Equation (4.74) thus expresses the decrease of tidal flat area at low water as a function of excess channel depth,

$$V_2\xi_{2t}' = q_{T1}\xi_1'. \tag{4.75}$$

The above argument implies that the time variation of the basin volume in Eq. (4.72) depends not only on ξ_1 but also on ξ_2 and the same holds for the departure of the sediment flux from equilibrium. Equation (4.72) can thus be written as

$$V_t = V_1\xi_{1t}' + V_2\xi_{2t}' = V_1\xi_{1t}' + q_{T1}\xi_1' = -\langle q \rangle = -q_1\xi_1' - q_2\xi_2', \tag{4.76}$$

where q_2 is the ξ_2-derivative of $\langle q \rangle$ at equilibrium. It is assumed that the longitudinal flux $\langle q \rangle$ refers to sediment import to (or export from) the channel; sediment import to (or export from) the tidal flats takes place through transverse transport according to (4.74), without affecting the basin volume. The parameter ξ_2' can be eliminated from (4.75) and (4.76) by an additional time derivation; the result is a linear second order differential equation:

$$\xi_{1tt}' + (\sigma_1 + \sigma_T)\xi_{1t}' + (\sigma_2\sigma_T)\xi_1' = 0, \tag{4.77}$$

where

$$\sigma_1 = q_1/V_1, \quad \sigma_2 = q_2/V_2, \quad \sigma_T = q_{T1}/V_1. \tag{4.78}$$

The solution of this equation is given by

$$\xi_1'(t) = \xi_1'(0)e^{\sigma_i t}\left(\cos \sigma_r t + \frac{\sigma_i}{\sigma_r}\sin \sigma_r t\right),$$

$$\sigma_i = -\frac{1}{2}(\sigma_1 + \sigma_T), \quad \sigma_r = \sqrt{\sigma_2\sigma_T - \sigma_i^2}. \tag{4.79}$$

Timescale estimates

We now will determine rough estimates for the time scales σ_i^{-1} and σ_r^{-1}. First we express the tidally averaged sediment flux $\langle q \rangle$ as a function of basin geometry by using the earlier derived expressions (4.57), (4.42), and (4.69),

$$\langle q \rangle = \alpha b_C \langle U^5 \rangle \approx \frac{5}{4}\alpha b_C U_c^5 (Q^+ - Q^-)/Q_m,$$

$$(Q^+ - Q^-)/Q_m = 2\Delta_{EF}/T \approx \frac{2rl^2}{gTb_ch^3}[2a(b_S^+ + b_S^-)$$

$$- h(b_S^+ - b_S^-)], \tag{4.80}$$

with $\langle q \rangle = 0$ at equilibrium. This yields

$$q_1 \equiv \partial q/\partial h \approx -\alpha U_c^5 \frac{5rl^2}{2gTh^3}(b_S^+ - b_S^-),$$

$$q_2 \equiv \partial q/\partial b_S^- \approx \alpha U_c^5 \frac{5rl^2}{2gTh^2}. \tag{4.81}$$

An estimate of the transverse sediment flux q_T is derived from the tidal flat erosion/accretion figures for the Eastern Scheldt [103], which indicate that a decrease of the maximum tidal velocity with $pU \approx 0.25U$ produces a tidal flat erosion of 2–3 cm per year, i.e., the tidal flats are lowering at a rate of $w_e \approx 8.10^{-10}$ m/s. The tidal flat surface is approximated by $l(b_S^+ - b_S^-)$, where l represents the basin length. The transverse sediment flux for a unit decrease of maximum tidal velocity is then given by $dq_T/dU \approx -w_e l(b_S^+ - b_S^-)/pU_c$. We also have $dU/dh = d(Q_m/b_ch) = -U_c/h$. This yields

$$q_{T1} = (dq_T/dU)(dU/dh) \approx w_e l(b_S^+ - b_S^-)/ph. \tag{4.82}$$

The change of tidal channel volume per unit change of depth, $V_1 \equiv \partial V/\partial h$, is estimated at $V_1 \approx lb_C$. The change of tidal flat volume per unit change of

LW width can be approximated by $V_2 \equiv \partial V^{flat}/\partial b_S^- \approx al$, according to the schematisation of Fig. 4.42.

Collecting the results, estimates can be derived for the time scales σ_1^{-1}, σ_2^{-1} and σ_T^{-1}. For a typical tidal basin (length $l = 20\,\text{km}$, intertidal area of the same order as the channel area, $b_S^+ - b_S^- \approx b_C$, $U_c = 1\,\text{m/s}$, friction coefficient $r = 0.003\,\text{m/s}$ and sediment transport coefficient $\alpha = 10^{-4}$) we find $-\sigma_1^{-1} \approx 120$ years, $\sigma_2^{-1} \approx 25$ years and $\sigma_T^{-1} \approx 50$ years. For the time scales σ_i^{-1} and σ_r^{-1} we find $-\sigma_i^{-1} \approx 100$ years and $\sigma_r^{-1} \approx 35$ years.

Cyclic morphodynamic behaviour

The solution (4.79) shows that a depth perturbation brings the basin into a cyclic response mode. The amplitude of this cycle of channel depth increase-decrease and tidal flat decrease-increase, has a negative growth rate σ_i and thus decreases over time, for the basin parameters considered in the numerical example. The tidal basin morphology is stable against depth perturbation, but the cycle period is long (on the order of 200 years) and the damping rate is small. The result depends in particular on the basin length l, as the inverse time scale $-\sigma_1$ increases with l. For very long basins the growth rate σ_i may therefore become positive.

The simplicity of the mechanism by which cyclic morphodynamic behaviour is initiated is particularly interesting. There are many causes of cyclic tidal inlet behaviour related, for instance, to competition between different channels, see Sec. 2.5. Cyclic morphodynamic behaviour is best documented for the channel-shoal system at the inlet of tidal basins [232]; it is generally believed that such behaviour exists throughout tidal basins. Fluctuations in tidal basin adaptation to interventions have been observed for Texel Inlet, with a time scale on the order of 50 years [135]. However, the uncertainty in the data is large, and fluctuations may also be ascribed to external conditions, such as storm events.

4.6.3. *Tidal Basin Response to Change*

Response to channel dredging

The depth perturbation discussed in the previous section mimics in a certain way an incidental, basin-wide dredging operation. We therefore might consider the foregoing as a model for the basin response to a large scale dredging operation. This response is depicted schematically in Fig. 4.43. However, this picture is too

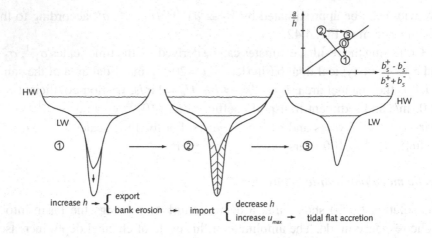

Fig. 4.43. Different stages of the morphodynamic response to channel dredging.

simple; the dredging strategy (places of removal and disposal) plays a role too. Dredging is normally performed in the main tidal channel, which carries most of the ebb flow. The channel is not deepened everywhere; dredging is concentrated in particular on the shoals present at the intersection of flood and ebb dominated channel sections, see Fig. 4.44. In general, the morphologic response to channel dredging consists of: (1) increased sedimentation in the intertidal zone and (2) increase in tidal amplitude. Such a response has been observed after heavy dredging, for instance, in the Western Scheldt [240], Ribble [463] and Mersey estuaries [308]. The increase in tidal amplitude may compensate the loss of storage volume in the basin due to sedimentation [461]. A greater part of the tidal flow will be concentrated in the main channel, at the expense of secondary channels; flow reduction and rapid infilling of these secondary channels has been observed, for instance, in the Ribble estuary [463]. Most of the sediment is delivered by intertidal sandbanks around these channels; these sandbanks experience erosion.

Flow concentration in the main channel is enhanced by the construction of training walls. Accelerated accretion of shallow subtidal and intertidal zones is observed as an immediate consequence of training walls, in the Mersey, the Ribble, the Lune estuary [308] and the Seine [18]. In the Western Scheldt, infilling of secondary flood channels has been further stimulated by disposal of dredged material. Therefore, the deepening of the main channel need not

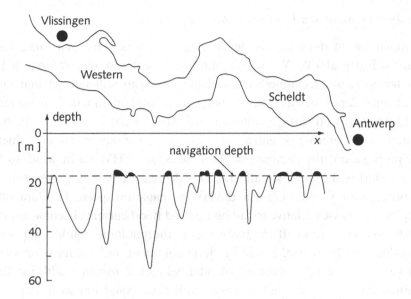

Fig. 4.44. Depth profile of the Western Scheldt basin along the channel axis. Dredging in the Western Scheldt navigation channel is concentrated at the shoals at the junctions of flood and ebb-dominated channels. The dredged material is not removed from the basin, but mainly disposed in secondary flood channels and along the channel banks.

necessarily imply a decrease in the flow velocity; such a decrease has not been observed in the Western Scheldt [240]. This may explain why the observed response to channel dredging (increased tidal flat sedimentation) is opposite to the response described in the previous section. The dredging activities and the resulting change in the flow pattern increase sediment mobility, especially that of the fine fraction. In the Ribble estuary, the outer delta has acted as major sediment source [463]. Channel deepening and reduction of the basin width at low water both contribute to increasing the LW-propagation speed relative to the HW-propagation speed; the ebb duration will shorten relative to the flood duration and the basin may turn from flood dominant into ebb dominant. In the Western Scheldt and the Ribble estuary the evolution towards ebb dominance is offset by a decrease of HW basin width; this decrease is due to accretion of intertidal areas above HW level and reclamation of intertidal areas. In the Western Scheldt the duration of rising tide relative to falling tide has not changed significantly during the past century, in spite of considerable channel deepening.

Response to mean sea-level rise (embanked basin)

A sudden rise of the mean sea level reduces the intertidal area, since fewer shoals will dry at LW. We will assume that the basin surface area at HW does not increase; reclaimed land outside the basin will not get inundated. The channel depth increases somewhat, but the tidal prism will increase more, due to the increased storage volume of the basin. Tidal velocities in the main channels will therefore be enhanced. Among the parameters which influence tidal propagation, the decrease of the ratio b_S^+/b_S^- (HW basin width to LW basin width) is the most significant. This decrease causes a decrease of the LW-propagation speed relative to the HW-propagation speed. The duration of rising tide decreases relative to falling tide and flood currents become stronger relative to ebb currents. If the basin was in morphologic equilibrium before the sudden sea-level rise, it will be flood dominant and sediment importing afterwards. The general increase of tidal velocities, together with the flood dominance of tidal sediment transport, will cause sand import in the basin. This sand import will be primarily directed to the tidal flats, which may restore their height relative to sea level. According to this scenario (see Fig. 4.45) the basin will adjust to mean sea-level rise. Imported sand is initially supplied by the outer delta, which therefore becomes smaller.

Field observations over the past decades show that the outer deltas of the Wadden basins are eroding, together with the adjacent coastal stretches, see Fig. 4.46; this is thought to be mainly due to mean sea-level rise and associated sand import into the Wadden basins. At the Texel Inlet and the Frisian Inlet, sand import is also affected by the closure of the Zuyderzee (1930) and Lauwerszee (1970) [293]. In the foregoing it has been shown that tidal asymmetry and

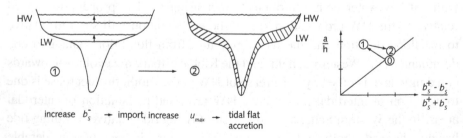

Fig. 4.45. Different stages of the morphodynamic response to sea-level rise.

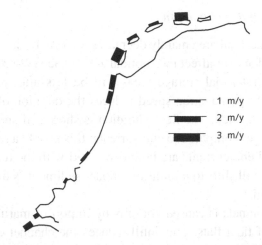

Fig. 4.46. Locations along the Dutch coast with greatest ongoing coastal erosion; most of these locations are near tidal inlets. The total volume of eroded material in the coastal zone is of the same order as the average tidal basin infill required to keep pace with sea-level rise. An equivalent sand volume, approximately 6 million cube on an annual basis, is supplied to the Dutch coast by artificial nourishment of the shoreface and the beach.

sediment import caused by an increase of LW-basin width is small; we may thus expect a time delay between sea-level rise and basin response. Average sedimentation rates after closure of the Zuyderzee and Lauwerszee are on the order of 1 cm/year [293]. This suggests that the Wadden Sea basins will not be able to keep pace with a sea-level rise of 1 cm/year or more.

Response to mean sea-level rise (non-embanked basin)

In the case that mean sea-level rise does increase the HW basin surface, the basin response will be different. If the increase of HW basin width exceeds the increase of LW basin width, the basin will become ebb-dominated. The situation is similar to the polder inundations which have occurred in the past in the Rhine-Meuse-Scheldt delta after major storm surges. The tidal prism increases and channels are scoured to accommodate the increased tidal flow. The inundated land turns into a new flood delta with channels and tidal flats. A substantial part of the eroded sediment will be exported and deposited in the ebb tidal delta; the volume of the ebb tidal delta will increase.

Response to tidal flat reclamation

Reclamation of intertidal area mainly decreases the HW basin width, assuming that reclamation does not affect the subtidal zone. The HW-propagation delay caused by filling intertidal storage areas will be less after reclamation; the increase of the HW-propagation speed reduces the duration of the rising tide relative to the falling tide. The flood duration is shortened and flood currents will become stronger compared to ebb currents; this is only a relative increase, because flood and ebb currents are both decreased with the reduction of tidal prism. The basin will shift to an infilling mode; sediment is first deposited in the channel system.

Infill of the channels is caused not only by import of marine sediment but also by erosion of tidal flats. This infill reduces the channel cross-section; it increases the tidal velocities and finally restores the capability of tidal currents to counteract tidal flat erosion. At this stage the basin is still flood dominant, because erosion has caused a reduction of intertidal area in addition to the decrease caused by land reclamation. The ongoing import of marine sediment will now serve to rebuild tidal flats. This process goes on, reducing the HW-propagation speed until is has become similar to the LW-propagation speed. At that point the basin achieves a new morphologic equilibrium. The process is schematically depicted in Fig. 4.47. The channel cross-section in the new equilibrium situation is smaller than in the former situation. The basin has

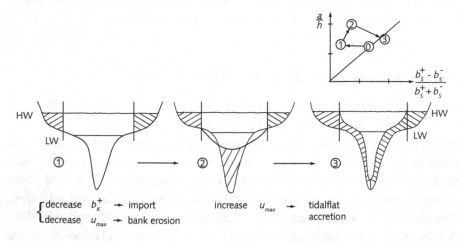

Fig. 4.47. Different stages of the morphodynamic response to tidal flat reclamation.

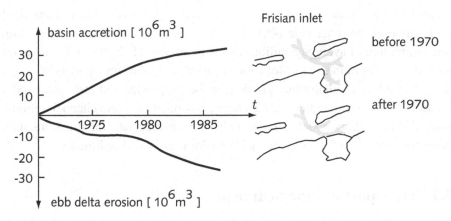

Fig. 4.48. Sedimentation in the Frisian Inlet and simultaneous erosion of the ebb-tidal delta after closure of the Lauwerszee flood delta.

imported sediment, which has mainly been supplied by the ebb-tidal delta; the volume of the ebb-tidal delta has decreased. These phenomena have been observed after closure of the Lauwerszee, which in the past was part of the flood delta of the Frisian Inlet, see Fig. 4.48.

The above description is general and qualitative; the detailed process of morphologic adaptation is far more complex. Some parts of the original channel system are more strongly affected by tidal flat reclamation than others. Therefore the original channel system will be reshaped; this process involves significant sediment displacements. The adaptation time scale of the basin therefore not only depends on sediment transport through tidal asymmetry, but also on other processes, such as the reconfiguration of the main channel system.

Other models for basin adaptation to change

The basic assumptions underlying the foregoing morphodynamic tidal basin model are (1) the dominant role of tidal asymmetry for sand import or sand export and (2) the existence of an equilibrium between tidal flat accretion due to tidal currents and tidal flat erosion due to wave action. In the literature several other models have been proposed for describing the adaptation of a tidal basin to sea-level rise or to tidal flat reclamation. In these models other assumptions are used, based on empirical relationships. One type of models is based on equilibrium relationships between channel cross-section and tidal prism (see 4.2) on the one side and between high-water basin area and intertidal basin area on the other side [464, 108]. Another type of models is based on the assumption of average equilibrium sediment load [112, 466, 170], together with other assumptions. In [466] for instance, sediment transport is assumed to be related to a longitudinal gradient in the suspended sediment load and a relationship is assumed between average suspended load and channel volume. Such diffusion type models are probably more appropriate for situations with fine sediments than for situations with medium or coarse sediments.

4.7. Transport of Fine Sediment

What are fine sediments?

Fine cohesive sediments consist mainly of clay and detritus particles with a diameter on the order of 1–10 microns, which hardly settle as individual particles. In the water column they tend to form flocs with a much higher settling velocity (typically on the order of 0.1–1 mm/s), which may break up when being resuspended. The characteristics of fine cohesive sediments were discussed in Sec. 3.2.3. Non-cohesive silt with a grain diameter below 63 μm is also considered part of the fine sediment fraction because of the low settling velocities on the order of 1 mm/s.

Major transport processes

Many tidal basins are deposition areas for fine sediment. This is primarily due to their sheltered location. Wave action prevents deposition of fine sediments on the shoreface (nearshore zone in front of the beach), except in coastal zones with massive mud supply (e.g. the mud coasts of Guiana and Surinam). Several processes have an important influence on the transport of fine sediment, in particular: Tidal dispersion, estuarine circulation and tidal asymmetry. These three processes will be discussed in this section, but we will limit ourselves to the most essential notions. There are two reasons: (1) extensive literature is available about the subject, especially on tidal dispersion and estuarine circulation [150, 337] and (2) morphologic feedback to transport of fine sediment is less developed than for medium or coarse sediment, except in cases of very high suspended concentration. The response of fine sediment deposits to changes in flow patterns is different from sandy deposits. Fine sediment deposits are in general either highly mobile or highly consolidated; this behaviour opposes the generation of bedforms. At very high suspended concentration, sediment transport may take place through density-induced turbidity currents or mud avalanching. Such conditions will not be considered in the following.

4.7.1. Tidal Dispersion

Similarity with dissolved substances

Fine sediments are easily kept in suspension, even at low current velocity. Once suspended, they move with the tidal current during most of the tidal cycle, or even during the entire cycle if they are suspended high enough in the water column. In the Dutch tidal basins typically half of the suspended sediment remains in suspension at slack water, see for instance, Figs. 4.55 and 4.56. The displacement of these particles is similar to the displacement of dissolved substances. This does not hold for particles which are deposited during part of the tidal cycle. Settling and resuspension may produce a strong and consistent net displacement of fine sediment; these transport mechanism will be discussed in the following sections. The tidal dispersion mechanism refers to residual transport related to non-uniformity of the spatial flow distribution. It tends to smooth the distribution of suspended sediment by dispersing particles away from high concentration zones. For the description of this dispersion process we will disregard settling and resuspension.

Scale dependency of dispersion processes

Dispersion is the general designation of transport processes which smooth gradients in the concentration of dissolved or suspended matter. These transport processes correspond to the mixing of water masses, i.e., to the relative displacement of water parcels without net displacement of water volume. The dispersion coefficient is a measure of the efficiency of mixing processes; it can be expressed as the ratio of a squared length scale X^2 and a time scale T. The length scale X is a measure of the distance over which water parcels are displaced relative to each and the time scale T is a measure of the time involved. Dispersion processes occur at many different spatial and temporal scales; the definition of concentration is scale dependent accordingly. The term diffusion is normally used for any dispersion process that takes place at sub-model scales.

Brownian motion is the dispersion process at the smallest temporal and spatial scale (molecular scale). Turbulence is the most powerful mechanism for dispersal of dissolved or suspended matter at the scale of water depth. Morphologic time scales are generally quite long; therefore we focus on dispersion processes at temporal and spatial scales which are much longer than the scales of Brownian motion or turbulence. We are mainly interested in transport along the tidal basin in longitudinal direction; the corresponding large scale mixing processes are known as longitudinal dispersion.

Random walk

Smoothing of concentration peaks will not occur if water parcels move all together in a coordinated way. If, for instance, all individual water parcels return to their initial position after a tidal cycle, there will be no dispersion; the same is true if all water parcels are displaced over an identical distance. On the other hand, if initially close water parcels travel over very different distances in a tidal cycle (see for an example Fig. 4.49), an initial concentration peak will be smoothed. If these individual parcel displacements after a tidal cycle are randomly distributed, there is practically no chance that a concentration peak is created from an initially smooth concentration distribution. This illustrates that dispersion can be described as a random walk process at the time scale of a tidal cycle or several tidal cycles. The distance over which water parcels move away from each other depends on the strength of their respective net

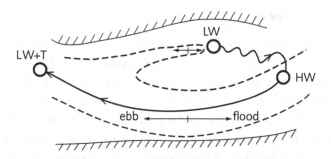

Fig. 4.49. Schematic representation of dispersion by the combined action of lateral tidal velocity gradients and cross-sectional mixing. A water parcel at low water in a shallow secondary channel moves landward with low flood velocity. Transverse currents and horizontal eddies take the water parcel towards the main channel, where it arrives at around HW. Subsequently it travels with the strong ebb current in the main channel far seaward from its initial position at low water.

tidal displacements and the time scale T_A at which these net displacements become statistically uncorrelated. The time scale T_A is defined as the correlation time scale of water parcels in the same cross-section and we call X_A^2 the average quadratic net displacement of these water parcels away from the original cross-section over the period T_A. We assume that the semidiurnal tide has a cyclic character and that the net discharge is zero; the sum of all individual displacements thus equals zero. Random walk theory then states that the average quadratic displacement X^2 of water parcels after a time $t \gg T_A$ increases proportionally with time t.

Dispersion equation

An equivalent statement is that the tidally and cross-sectionally averaged concentration $c_0(x, t) \equiv \langle \overline{c} \rangle$ satisfies the dispersion equation

$$c_{0t} = D_L c_{0xx}, \tag{4.83}$$

with a dispersion coefficient D_L given by

$$D_L = X_A^2 / 2T_A. \tag{4.84}$$

In Eq. (4.83) the dispersion coefficient has been assumed to be independent of x. In the presence of a residual discharge Q_R and taking into account other sediment transport processes, the dispersion equation reads

$$c_{0t} - u_R c_{0x} + \langle q_x \rangle / A_0 = D_L c_{0xx} + \langle Er - De \rangle / h. \tag{4.85}$$

In this equation the river discharge velocity $u_R = Q_R / A_0$ is also assumed to be independent of x; $\langle q \rangle$ is the net tidal sediment flux due to tidal asymmetry and estuarine circulation, A_0 is the tidally averaged cross-sectional area and Er, De are the local erosion and deposition rates.

Dispersion coefficient

It is important to note that the dispersion coefficient (4.84) depends on the flow distribution during the tidal cycle. However, D_L is independent of the concentration distribution $c_0(x, t)$, if it is assumed that the cross-sectional mixing time scale T_A is much smaller than the time scale at which sediment particles travel in the longitudinal direction through the basin [118]. In that case D_L can be derived experimentally from the observed tidally averaged salinity distribution $\langle S(x, t) \rangle$ in a tidal basin. If the river discharge Q_R is constant during a time long enough for the establishment of an equilibrium salinity distribution $S_{eq}(x)$, then the tidally averaged salt flux through any estuarine cross-section equals zero,

$$u_R S_{eq} + D_L S_{eq_x} = 0. \qquad (4.86)$$

The cross-sectionally averaged salinity distribution S_{eq} is often a smoothly varying function of x. In that case the sea water intrusion length L_S can be approximated by $L_S \approx S_{eq}/|S_{eq_x}|$, giving

$$D_L \approx u_R L_S. \qquad (4.87)$$

Typical values of the dispersion coefficient D_L are in the range 100–300 m^2/s, as shown in Fig. 4.50 for the Eastern Scheldt and the Ems–Dollard tidal basins. Even higher values up to 1000 m^2/s have been found in tidal basins with a very large width-to-depth ratio (≥ 1000) and complex geometry (meandering and braiding channels, tidal flats), like the Wadden Sea basins [509] and Chesapeake Bay [17]. In such basins tidal dispersion counteracts the formation of a strong turbidity maximum. Strong turbidity maxima are seldom observed in wide tidal basins with complex geometry.

Dispersion processes

Processes responsible for tidal dispersion are related to the spatial flow structure. Spatial inhomogeneity is generated by vertical shear, lateral depth variations, dead zones and residual circulation. The greatest contribution to dispersion is provided by cross-sectional fluid exchange between zones of high and low tidal velocity at the timescale of the tidal period [150, 118, 512]. An example is shown in Fig. 4.49. In shallow well mixed estuaries vertical mixing takes place over a timescale much shorter than the tidal period; differences in flow velocity over the water column therefore do not strongly contribute to longitudinal dispersion in the absence of stratification. Lateral mixing over the entire basin width often takes place at a timescale substantially greater than the tidal period, but local mixing over lateral tidal flow gradients takes place at the tidal timescale and therefore causes strong tidal dispersion.

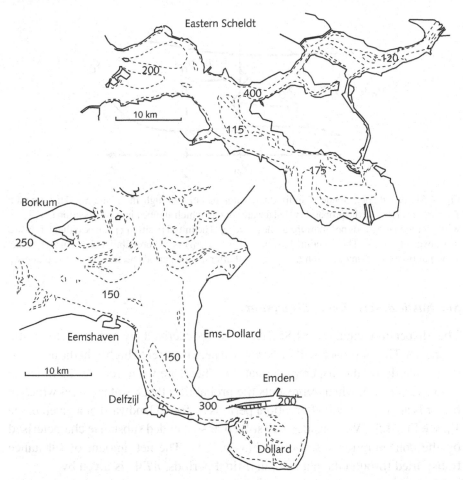

Fig. 4.50. Experimentally determined values of the longitudinal dispersion coefficient [m²/s] in the Eastern Scheldt and the Ems–Dollard tidal basins.

Residual circulation contributes most to tidal dispersion if the timescale for mixing across the circulation is very long [117]. Vertical circulation is most effective in the presence of density stratification; otherwise lateral circulation yields a greater contribution to longitudinal dispersion. Analytical estimates of dispersion coefficients are at best qualitative, due to the complexity of the flow structure in most tidal basins. The best estimates of dispersion coefficients are derived either from observations or from numerical models [376].

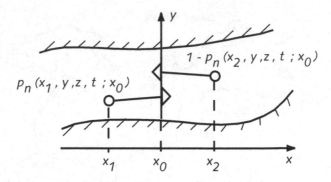

Fig. 4.51. Net displacement of individual water parcels through the plane x_0. The function $p_n(x_1, y, z, t; x_0)$ is the probability that a water parcel which at time t is at any location (x_1, y, z), will after n tidal periods be landward of the plane x_0. The initial position x_1 can be either landward or seaward from x_0. The function $1 - p_n(x_2, y, z, t; x_0)$ is the probability that a water parcel which at time t is at any location (x_2, y, z) will after n tidal periods be seaward of the plane x_0.

Stochastic description of dispersion

The dispersion equation (4.85) can also be derived by using a stochastic approach. This approach will be described here in order to highlight the assumptions underlying the dispersion equation. Therefore we introduce the function $p_n(x, y, z, t; x_0)$, which represents the probability that a water parcel which at time t is at (x, y, z) will after n tidal periods be found landward of a given plane x_0, see Fig. 4.51. We consider a dissolved or suspended substance characterised by the concentration distribution $c(x, y, z, t)$. The net amount of substance transported through the plane x_0 in n tidal periods, $nT\Phi$, is given by

$$
nT\Phi(x_0) = \int_0^{x_0} dx \int\int_A p_n\, c\, dydz - \int_{x_0}^{l} dx \int\int_A (1 - p_n)\, c\, dydz
$$

$$
= \int_0^{x_0} AP_n\, c_0 dx - \int_{x_0}^{l} A(1 - P_n) c_0 dx
$$

$$
+ \int_0^{l} dx \int\int_A p_n\, c'\, dydz. \tag{4.88}
$$

In this equation the following notations are used: l is the basin length, A is the cross-sectional area at x_0 at time t, P_n is the cross-sectional average of p_n and $c'(x, y, z, t) = c(x, y, z, t) - c_0(x, t)$. We assume that the longitudinal variation of the concentration is represented by a linear function:

$$c_0(x, t) = c_0(x_0, t) + (x - x_0) c_{0x}(x_0, t) + \cdots ,$$

Then after substitution we find

$$\Phi(x) = Q_R c_0(x, t) - D_L A c_{0x} + \cdots + \Phi', \tag{4.89}$$

with

$$Q_R = \frac{1}{nT} \left[\int_0^x A P_n dx' - \int_x^l A(1 - P_n) dx' \right], \tag{4.90}$$

$$D_L(x) = \frac{1}{2nT} \int_0^l (x' - x)^2 \frac{\partial P_n(x'; x)}{\partial x'} dx' \tag{4.91}$$

and

$$\Phi' = \frac{1}{nT} \int_0^l dx' \int \int_A P_n(x', y, z, t; x) \, c'(x', y, z, t) \, dydz. \tag{4.92}$$

The expression for the dispersion coefficient D_L corresponds to the average quadratic distance travelled by water parcels arriving at the cross-section x after n tidal periods; this is equivalent to the result of the random walk model (4.84). The dispersion equation now reads

$$c_0(x, t + nT) - c_0(x, t) + \frac{nT}{A} \Phi_x = 0. \tag{4.93}$$

This is equivalent to (4.85), except for the last term Φ' in Eq. (4.89). The integral (4.92) expresses the correlation between the initial location of a fluid parcel in the cross-section and the net displacement after n tidal periods. For dissolved substances this term decreases to zero when nT becomes much larger than the cross-sectional mixing time scale; in that case fluid parcels no longer remember their initial location in the cross-section.

Suspended matter is not dispersed by estuarine circulation

Suspended matter has a systematic tendency to sink to the lower part of the vertical. The residual displacement of suspended particles therefore remains correlated with estuarine circulation; the integral (4.92) does not vanish for large n. This implies that suspended matter can accumulate at the landward limit of estuarine circulation, while dissolved substances cannot. For suspended matter, estuarine circulation cannot be treated as one of the tidal dispersion mechanisms, but needs to be considered separately, in relation to the process of particle settling.

4.7.2. *Estuarine Circulation*

Definition

Seawater intrusion is responsible for an inhomogeneous density distribution in river tidal basins. In seawater the pressure increases faster with depth than in fresh water; at the inlet ($x = 0$) the increase of pressure with depth is thus greater than at the seawater intrusion limit ($x = L_s$), see Fig. 4.52. This implies that near the bottom a longitudinal landward pressure gradient exists in addition to the pressure gradient related to surface slope. This density-induced pressure gradient generates, in comparison to the homogeneous situation, an additional landward directed flow component near the bottom. For mass balance reasons, there is a compensating seaward flow component near the surface. The density-induced landward flow near the bottom and the compensating seaward flow near the surface are together called 'estuarine circulation', see Fig. 4.53. Estuarine circulation interacts with tidal flow and river runoff. Therefore it is not a steady circulation superimposed on the homogeneous flow pattern, but a tidally modulated circulation. Its influence on sediment transport is greatest during flood, when tidal flow and estuarine circulation have the same direction near the bottom.

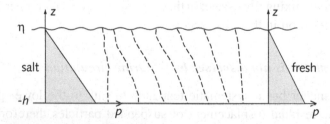

Fig. 4.52. Increase of pressure with depth. Due to the higher density of seawater compared to fresh water, the pressure increases more strongly with depth in seawater than in fresh water. This produces a landward directed pressure gradient, which is the highest near the bottom.

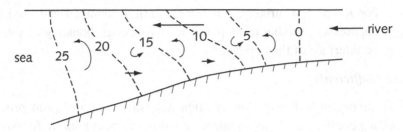

Fig. 4.53. Schematic representation of tidal mean salinity distribution and estuarine circulation in a partially mixed estuary. The dotted lines are isohalines and the figures indicate salinity in ppt.

Stratification

Seawater inflow is concentrated in the lower part of the water column, while fresh water outflow is concentrated in the upper part of the water column. The greater density of seawater hinders turbulent fluid uplift and counteracts mixing with less dense surface water. Energy for turbulent mixing in the water column is mainly provided by tidal motion and to a lesser degree by river runoff and the estuarine circulation itself. Vertical mixing in the river plume seaward of the inlet is often due to energy input from wind driven surface currents. If the tidal energy input is low, mixing between the underlying seawater and the upper fresh water will be slow. A decrease of mixing energy leads to stratification, with a sharp interface between salt and fresh water. This is a self-stabilising process, because stratification will damp turbulent motion, while damping of turbulent motion increases stratification.

Richardson number

Stratification is characterised by the Richardson number Ri. It represents the ratio of local potential energy, $g \Delta\rho(\Delta z)^2/12$, due to stratification $\Delta\rho = \rho_z \Delta z$, in a small portion of the water column of height Δz, and twice the maximum local kinetic energy for mixing, $\rho(\Delta u)^2 \Delta z/12$, due to velocity shear $u_z = \Delta u/\Delta z$,

$$Ri = g\rho_z/\rho u_z^2. \qquad (4.94)$$

Due to the self-stabilising stratification effect the Ri-number provides a sharp criterion for distinguishing between stratified and non-stratified fluids. The transition between the two conditions is at $Ri = 1/4$. For $Ri < 1/4$ mixing between denser and less dense fluid layers takes place through turbulent mixing. For $Ri > 1/4$ turbulence is almost entirely damped and mixing is strongly suppressed; mixing takes place mainly through breaking of internal wave at the interface of the two layers.

Mixing coefficients

Vertical mixing in turbulent flow is often described as a diffusion process, with mixing coefficients N for momentum (eddy viscosity) and K for passive dissolved substances (diffusivity). For homogeneous flow N and K mainly depend on shear velocity and depth. Density-induced damping of turbulent mixing strongly decreases the mixing coefficients; this decrease can be related to the Richardson number. The following expressions are derived from observations [318],

$$N \approx \frac{N_0}{(1 + 10Ri)^{\frac{1}{2}}}, \quad K \approx \frac{K_0}{(1 + \frac{10}{3}Ri)^{\frac{3}{2}}}, \tag{4.95}$$

where N_0, K_0 are the mixing coefficients for homogeneous flow. Even for weak stratification, $Ri \approx 1/4$, mixing is substantially decreased. For strong stratification, $Ri > 1/4$, vertical mixing is not adequately described by diffusion.

During part of the flood tide, density differences may act in the opposite way and increase vertical mixing. This occurs in particular in shallow macrotidal estuaries, when seawater is carried by strong near-surface flood currents on top of partially mixed estuarine water. This destabilises the water column and enhances vertical mixing by convective overturning. This process contributes substantially to the removal of stratification during flood [360].

Salt wedge

In the case of weak tidal velocities and low river flow little mixing energy is available. The estuary is strongly stratified and seawater inflow along the bottom is hardly diluted by mixing with fresh water runoff. Seawater may intrude quite far inland; the intrusion length is determined by friction at the salt-fresh interface and by the bottom slope [393, 337]. In this case an almost stagnant salt wedge develops without substantial sediment transport. Examples are the Rhone River and the Kattegat between the North Sea and the Baltic Sea. Salt-wedge type tidal intrusion may also occur in the case of strong tide and high river discharge, as shown in Fig. 4.54 for the Fraser River. At the end of the ebb period the salt wedge is entirely flushed out of the estuary.

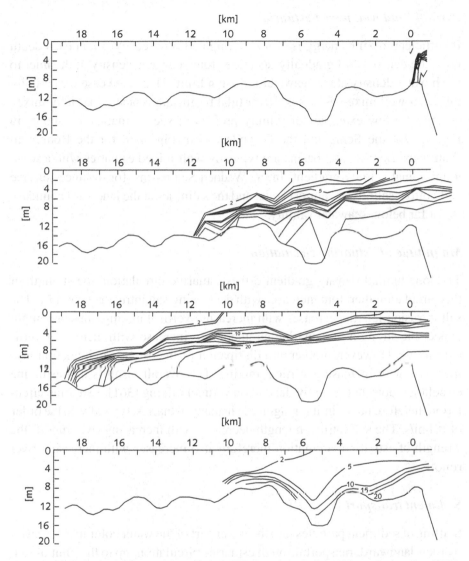

Fig. 4.54. Salinity in the Fraser River during different phases of the tide, redrawn after [171]. Top graph: End of ebb; the salt wedge has been flushed out of the estuary, leaving a sharp salinity front at the mouth. Second graph: Mid-flood; salt wedge is advancing up the estuary at a speed of 0.7 m/s. Third graph: End of flood; maximum intrusion condition with a temporarily arrested wedge. Fourth graph: Mid-ebb; the salt wedge has collapsed, and high salinity water is confined to a thin layer at the bottom.

Partially and well mixed estuaries

If sufficient mixing energy is available (high eddy viscosity N), at each depth level the density will gradually decrease from seawater density at the inlet to fresh water density at the seawater intrusion limit. This is the case for partially mixed or well mixed estuaries. River tidal basins are typically partially mixed estuaries; a few examples of salinity profiles in such estuaries are shown in Fig. 4.5 for the Seine and the Gironde and in Fig. 4.28 for the Rotterdam Waterway. Barrier tidal basins are typically well mixed estuaries with a weak longitudinal salinity gradient. In the Wadden Sea basins, for instance, surface and bottom salinity are almost equal and the salinities at the landward boundary is not far below seawater salinity.

Magnitude of estuarine circulation

The longitudinal density gradient drives estuarine circulation; the strength of this circulation therefore increases with decreasing salt intrusion length L_s. The salt intrusion length decreases with increasing vertical mixing; thus one might expect the strength of estuarine circulation to increase with increasing vertical mixing. However, momentum dissipation of the estuarine circulation also increases with increasing vertical mixing. The overall result is that estuarine circulation does not strongly depend on vertical mixing [361]. Estuarine circulation therefore has a similar magnitude in many estuaries, typically on the order of 0.1 m/s. The salt intrusion length decreases with increasing river runoff; the strength of estuarine circulation therefore also increases with increasing river runoff.

Sediment transport

Settling of sediment particles into the lower part of the water column is followed by their landward transport through estuarine circulation, up to the limit of seawater intrusion. This upstream sediment transport contributes to the formation of a turbidity maximum near the seawater intrusion limit. The presence of estuarine circulation invalidates the model for tidal river equilibrium of Sec. 4.5, at least for average discharge conditions. This is consistent with the comparison of observed basin widths with predicted equilibrium widths, see Fig. 4.34. This comparison shows that in most cases, downstream sediment transport exceeds upstream transport only for river discharges much larger than the average. At very high river discharges the influence of estuarine circulation becomes minor, for two reasons. The first reason is seaward displacement of the seawater intrusion limit; in many cases seawater is flushed out completely at the end of the ebb period. The second reason is the strong mixing energy provided by high river flow; through this mixing fine sediment particles are moved up higher in the water column where the strong surface current transports them to the sea.

Numerical models

Estuarine circulation does not strongly influence sediment transport at high river flow. However, for river discharges around the average, estuarine circulation contributes to the formation of a turbidity maximum, close to the seawater intrusion limit. Simulations with numerical models indicate that such a turbidity maximum also appears in the absence of estuarine circulation, due to tidal asymmetry. This stresses the importance of tidal asymmetry, not only for net transport of sand, but also for net transport of fine sediment.

4.7.3. Tidal Asymmetry

Suspension delay

Tidal transport of fine sediment differs from tidal transport of medium or coarse particles in at least one major aspect: The delayed response of fine sediment suspension to tidal variation. This response delay is significant relative to the tidal timescale $\omega^{-1} = T/2\pi$. There are several causes:

- the time needed for fine particles to settle to the bottom when the current is weakening, see Fig. 4.55; this holds in particular for particles which have been suspended high in the water column,
- the time needed for deposited particles to be resuspended when the bottom shear stress increases above the critical threshold for erosion,
- the time needed for particles to diffuse over the water column.

Due to the response delay the suspension retains substantial 'memory' of its past trajectory. A water column moving over a mud patch will keep a significant amount of mud in suspension a long distance away from the patch. Data of fine suspended sediment sampled at a fixed location in a tidal basin cannot, in general, be interpreted in terms of local erosion and deposition.

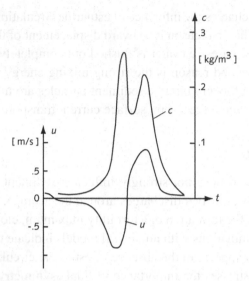

Fig. 4.55. Current velocity and fine suspended sediment ($\leq 63\ \mu$m) concentration in a tidal channel close to the head of the Ameland Inlet basin, redrawn after [355]. The settling time is too long for deposition during the period of low slack water (short interval between maximum ebb and flood flow), in spite of the small channel depth (less than 2 m).

Moving reference frame

A way to handle this difficulty is by considering a water column moving with the tidal current. This dampens the influence of concentration fluctuations advected from other locations, but does not rule out advection completely; the water column moves at different speeds near the bottom and near the surface. We use this moving-frame approach, as it makes the influence of tidal asymmetry more transparent than the fixed-frame approach. Tidal variation of the suspended sediment concentration measured from a vessel floating with the mean tidal current is shown in Fig. 4.56. The peak concentration at maximum flood is mainly due to suspension of fine sand. The variation of the suspended sediment concentration exhibits the expected behaviour: The degree of settling is related to the duration of the slack water period (currents below the critical strength for erosion) and the overall variation is delayed relative to the tidal variation of current strength by about 1–2 hours.

Net tidal sediment flux

For investigating net tidal fluxes we will assume that the moving reference frame returns to its initial position after a tidal period. This assumption is

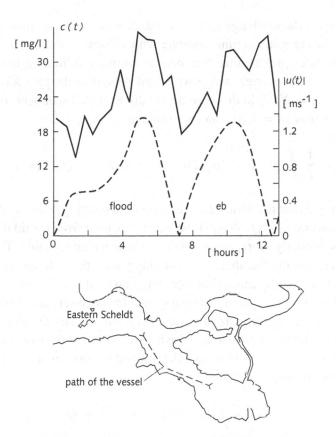

Fig. 4.56. Current velocity and depth-averaged suspended sediment concentration measured from a vessel that followed a buoy floating with the mean tidal current [119]. The experiment was carried out in the Eastern Scheldt in 1984. Most of the suspended material is fine sediment (< 63 μm), except at maximum flood. The flood peak velocity is somewhat higher than the ebb peak velocity. The slack water asymmetry is the opposite of that in the Wadden Sea: The period of LSW is longer than the period of HSW. The net fine sediment flux in the Eastern Scheldt is on average directed seaward [437]. The tidal variation of the suspended sediment concentration exhibits an overall time lag relative to the tidal current of 1–2 hours. (Survey Department Rijkswaterstaat.)

reasonable only if tidal discharge is more than two orders of magnitude larger than river runoff and if estuarine circulation is small; for barrier tidal basins these assumptions generally hold. We ignore horizontal residual circulation; its influence needs to be considered separately. Residual circulation contributes mainly to the net sediment flux through dispersion; this contribution can be estimated by means of dispersion coefficients, as discussed earlier. The velocity of the moving reference frame is taken as $u = Q/A_C$, where Q is the instantaneous tidal discharge and A_C is the channel cross-sectional area. $X(t)$ is the path of the water column moving with velocity $dX/dt = u$. If we consider cyclic tides and assume that there is no phase difference between tidal channel flow and water exchange with tidal flats, then the path $X(t)$ is cyclic too, $X(t + T) = X(t)$. In that case the tidally averaged sediment flux through the cross-section $x_0 = X(0)$ can be written as

$$\langle q \rangle = \frac{1}{T} \int_0^T dt \int_{x_0}^{X(t)} [b_C(De - Er) - (A_{C_t} + Q_x) c] \, dx, \qquad (4.96)$$

where b_C is the channel width, De the deposition (settling) rate, Er the suspension (erosion) rate, $-(Q_x + A_{C_t})$ the water flux to or from the tidal flats and c the cross-sectionally averaged suspended sediment concentration. This integral simply expresses the fact that sediment only passes through the moving plane if it is deposited on the channel bed or stored on tidal flats before arrival of the moving plane $X(t)$; sediment that is suspended from the channel bed or inflows from the tidal flats before arrival of the plane will not pass the plane and must be subtracted. Now we assume that the time variation of the suspended load in the cross-section, $(A_C c)_t$, is mainly determined by erosion-sedimentation and by sediment exchange with tidal flats,

$$(A_C c)_t \approx b_C(Er - De) + (A_{C_t} + Q_x)c. \qquad (4.97)$$

We have neglected gradients in along channel advection $(Qc)_x$; this is not a safe assumption locally, but averaged over the area crossed by the moving plane it seems reasonable. In that case the integral (4.96) can be rewritten, after partial integration, as

$$\langle q \rangle = \frac{1}{T} \int_0^T Q(X(t), t) c(X(t), t) dt. \qquad (4.98)$$

The net sediment flux is expressed now as an integral of the instantaneous flux through the moving plane. Terms related to the cross-sectional variation of flow velocity and suspended sediment concentration are left out of consideration. The expression (4.98) shows that the net import or export of fine sediment can be determined by measuring the suspended concentration in a frame moving along the channel with the average tidal flow.

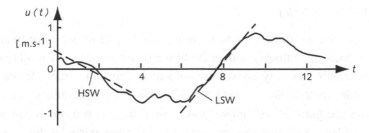

Fig. 4.57. Tidal variation of the current velocity u measured in the inner part of the Ameland Inlet. The velocity variation is much faster around low slack water than around high slack water. (Survey Department Rijkswaterstaat.)

Velocity asymmetry in a moving frame

Postma, in his famous analysis of sedimentation in the Wadden Sea, first pointed to the existence velocity asymmetry in tidal basins and its effect on net import of fine sediment [355]. Postma considered primarily a difference between duration of high slack water and low slack water and not a difference between flood peak and ebb peak velocities. He noted that tidal currents will be stronger around LSW than around HSW if the channel cross-section at LSW is much smaller than at HSW, see Sec. 4.3.3. Such an asymmetry is visible in recordings of the tidal current in the Wadden Sea at a fixed location near the landward basin boundary, as shown in Fig. 4.57. A second source of velocity asymmetry is due to the moving frame; even if there is no tidal velocity asymmetry at a fixed location, the velocity variation can be asymmetric in a moving frame. Around HSW many sediment particles move over tidal flats, where the current velocity is low; around LSW sediment particles are moving through the channels, where velocities are much higher. The two sources of velocity asymmetry add up; hence, for a suspended particle moving with the current, the period of low current velocities is much shorter around LSW than around HSW. Therefore more particles will settle to the seabed at HSW than at LSW. This implies that at the beginning of the flood period most particles which have been suspended during ebb are still in suspension, while at the beginning of the ebb period most particles are deposited. As it takes some time for sediment particles to get resuspended, the suspended sediment concentration at the beginning of flood flow is substantially higher than at the beginning of ebb flow. Even if the strength of flood and ebb currents is equal, the flood flow, on average, will carry more sediment than the ebb flow and a net import of sediment will result, see Fig. 4.59.

Tidal width asymmetry

When following a sediment particle on its tidal trajectory we have to consider both the velocity asymmetry and the bathymetric asymmetry. Around low water the particle will probably move somewhere in the main inlet channel, with considerable water depth, even at LSW. There is a high probability that towards high water the particle will move onto a tidal flat, with a much lower water depth, even at HSW. For this reason too, sediment particles have a higher probability of reaching the bottom at HSW than at LSW. This phenomenon becomes visible in the expression (4.96) if we consider the tidal flat to be part of the channel; in that case the tidal flat exchange term vanishes and the channel width b_C is replaced by the total basin width b_S. In the case of large tidal flats this width is much larger at HSW than at LSW; hence the contribution of sedimentation and erosion at HSW to net landward transport is much stronger than the contribution of sedimentation and erosion at LSW to net seaward transport. The net landward displacement of a sediment particle due to these different mechanisms is depicted in Fig. 4.58.

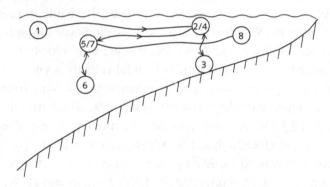

Fig. 4.58. Net tidal displacement of a sediment particle due to settling-erosion lag. The particle starts at 1 near the inlet at the onset of flood. Due to settling lag it is transported in suspension over almost the entire flood excursion distance. At high slack water the particle has arrived in shallow water, where it settles on the bottom at 3. When the ebb flow starts, the current is at first too weak to dislodge the particle from the bottom. Later it is suspended in the ebb current, but the remaining ebb excursion is shorter than the previous flood excursion. Towards low slack water the particle starts settling at 6, but the duration of the LSW period is too short and the distance to the bottom too long to reach the bottom. The particle is picked up by the flood flow immediately after slack water and travels landward again over the entire flood excursion. Towards HSW it starts settling, at a location 8 landward of the previous settling location 3.

Wave influence

Wind waves have an important influence on tidal flat sedimentation. Fine sediment will not accumulate on exposed tidal flats, except in periods of calm weather. In the Wadden Sea, for instance, mud deposits are found on many tidal flats during summer, but on an annual basis mud accumulates only on the most sheltered tidal flats. Fine sediment deposited on the tidal flats in calm periods is resuspended during storms and carried seaward by the ebb flow. The erosion-sedimentation asymmetry is reversed; erosion-sedimentation lag in the channel at low water dominates over erosion-sedimentation lag on the tidal flat at high water. Under storm conditions much sediment remains in suspension; dispersive transport may then play an important role in equalising sediment concentrations within the tidal basin and exchanging sediment with the near coastal zone.

A simple model

The integral expression (4.98) may serve as starting point for a semi-quantitative analysis of suspension-lag effects on net tidal sediment transport. Following Groen [180], we assume that the settling and erosion time lags are comparable and given by T_s; expressed in radians we call the time lag $\phi_s = \omega T_s$. This time lag is defined as the time scale at which the suspended sediment concentration adapts to an equilibrium concentration c_{eq}. In fact, a distinction should be made between high slack water and low slack water. In the moving frame, more sediment remains in suspension during LSW than during HSW, because of the greater average water depth at LSW compared to HSW. The adaptation time scale should therefore be taken longer at HSW than at LSW. However, this effect is ignored in the model for simplicity. The settling-erosion lag relation then reads, in a frame moving with the average tidal flow,

$$c_t = \frac{1}{T_s}(c_{eq} - c),\qquad(4.99)$$

or equivalently,

$$c(t) = c_0 + \frac{e^{-t/T_s}}{T_s}\int_0^t e^{t'/T_s}(c_{eq}(t') - c_0)dt',\qquad(4.100)$$

where the initial concentration $c_0 \equiv c(0)$ has to be chosen such that the exponential terms in the integral vanish.

We will assume that the equilibrium concentration c_{eq} depends quadratically on the tidal flow velocity u, with constant proportionality factor α, $c_{eq} = \alpha u^2$. We will also assume that the velocity u in a moving frame exhibits slack water asymmetry, such that the duration of high slack water is longer than the duration of low slack water,

$$u(t) = U(\sin \omega t + \epsilon \sin 2\omega t), \tag{4.101}$$

where the relative strength of the second harmonic tidal component is much smaller than the main component, $\epsilon^2 \ll 1$. The velocity u, the equilibrium concentration and the delayed suspended sediment concentration, according to (4.100), are shown in Fig. 4.59 for $\epsilon = 1/3$. Substitution in (4.98) for constant cross-section A_C yields, to first order in ϵ,

$$\langle q \rangle = \epsilon \alpha U^3 \frac{\pi \phi_s^3}{2(1 + \phi_s^2)(1 + 4\phi_s^2)}. \tag{4.102}$$

For a suspension time lag of 2 hours ($\phi_s \approx 1$) and asymmetry parameter $\epsilon = 1/3$, the net sediment import is 25% of the total flood import, according to the

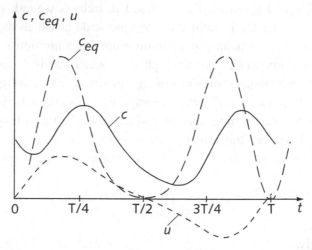

Fig. 4.59. Current velocity u (4.101) with asymmetry parameter $\epsilon = 1/3$, equilibrium suspended sediment concentration $c_{eq} = \alpha u^2$, and delayed suspended sediment concentration according to the model (4.99). The high slack water period at $t = T/2$ is much longer than the low slack water period. The erosion-sedimentation delay timescale T_s has been taken as 2 hours. This delay, together with the slack water asymmetry, results in a lower suspended sediment concentration during ebb than flood.

Table 4.7. Conditions for net transport of fine suspended sediment in barrier tidal basins.

	Basin conditions	Offshore conditions	Morphology
Import	duration HSW > duration LSW	offshore HW long compared to LW, tidal range large compared to channel depth	intertidal area small compared to basin area, channel depth small compared to tidal range
	max flood current > max ebb current	offshore tidal rise faster than fall tidal range large compared to channel depth	intertidal area small compared to basin area, channel depth small compared to tidal range
	more settling at HSW than at LSW in moving frame		intertidal area large
	no significant waves	calm weather	sheltered tidal flats, vegetation
Export	duration HSW < duration LSW	offshore HW short compared to LW, tidal range small compared to channel depth	intertidal area large compared to basin area, channel depth large compared to tidal range
	max flood current < max ebb current	offshore tidal rise faster than fall, tidal range small compared to channel depth	intertidal area large compared to basin area, channel depth large compared to tidal range
	more settling at LSW than at HSW in moving frame		intertidal area small
	high waves	storm	exposed tidal flats, bare

Table 4.8. Conditions for net sand transport in barrier tidal basins.

	Basin conditions	Offshore conditions	Morphology
Import	max flood current > max ebb current	offshore tidal rise faster than fall, tidal range large compared to channel depth	intertidal area small compared to basin area, channel depth small compared to tidal range
Export	max flood current < max ebb current	offshore tidal rise not faster than fall, tidal range small compared to channel depth	intertidal area large compared to basin area, channel depth large compared to tidal range

simple model. Many effects have been ignored; the simple model is not adequate for reliable estimates but provides qualitative insight into the influence of some parameters on net tidal import of fine sediment.

Comparison of conditions for net sediment transport

We conclude this chapter with an overview of the conditions for net import or net export of fine sediment in barrier tidal basins; these conditions are summarised in Table 4.7. They can be compared to the conditions for net import or net export of sand indicated in Table 4.8. Some conditions are similar, others are different. Fine sediment accumulates in the landward part of tidal basins where the channel depth is small and where tidal flats are large and sheltered. Sediment may be trapped here, even if export dominates in the seaward part of the basin. Export of fine sediment from the landward part of the basin only occurs during storms. Large intertidal flats (compared to the basin surface area) will favour export of medium to coarse sediment if the depth of tidal channels is large relative to the tidal range. The overall basin length also plays a role; in very long shallow basins the tidal wave will have a progressive character near the entrance because the reflected tidal wave is damped. This produces an important Stokes contribution and ebb-dominated transport of medium to coarse sediment near the basin entrance. In such situations a tidal basin exports sand and at the same time imports fine sediment.

Chapter 5

Wave-Topography Interaction

5.1. Abstract

Interaction of the sandy shoreface ...

The land-sea transition zone consists of various materials, such as rock, boulders, gravel, sand, silt and mud. Sand is the most common type of substrate of coasts which are exposed to moderate wave action. At sandy coasts the transition from sea to land is formed by a narrow coastal fringe with a rather steep bottom inclination, on the order of 1/100. This coastal fringe is referred to as shoreface. Further offshore, where water depth exceeds 10 m, the average bottom slopes are weaker, on the order of 1/1000. This zone is called the inner shelf. Locally, bottom slopes may be steeper, because of seabed structures such as dunes and ripples.

... with incoming breaking waves ...

The dynamics of the shoreface is largely determined by energy dissipation from wind waves propagating onto the coast. Part of these waves ('sea') is generated by local wind fields on the continental shelf or the nearby ocean; another part, called 'swell', is generated by remote storms on the ocean and has substantially greater characteristic wave period and wavelength. Before reaching the coast, sea and swell wave fields have often gathered wind energy over hundreds of kilometres. The breaking of waves on the shoreface converts this wind energy partly to heat and partly to forces acting on the seabed. These forces are generally stronger than forces from other water motions acting on the shoreface, such as tides and wind-driven or density-driven circulations. Sand coast dynamics is therefore mainly determined by wave action. Much sediment is suspended and set in motion, particularly in the surf zone, where waves are breaking.

... produces a variety of seabed patterns ...

This motion is not just a symmetric oscillation. Due to the nonlinear nature of wave motion in shallow water on the one hand, and the nonlinear nature of sediment transport on the other hand, sediment particles experience a net displacement. This net displacement varies in space and time, especially due to seabed topography. As a result of the interaction between gradients in sediment transport and gradients in topography, morphologic structures emerge on various scales, such as longshore and transverse bars, rip cells, cusps and ripples. These structures often have a temporary character. Their development may be 'forced', for instance, through resonance with incoming waves. They may also emerge 'spontaneously' as free instabilities. Coastal morphology is therefore continuously changing. Complexity and variability are a natural characteristic of coastal morphology. Waves are the dominant morphodynamic agent. However, sea level change, sediment supply and sediment type also play an essential role in coastal evolution at large time scales.

... which are studied in this chapter

This chapter aims to explain the origin and evolution of coastal morphology. The focus is on underlying processes more than on accurate methods for predicting coastal evolution. It is organised in four sections. In the first section the most essential notions and processes are outlined, setting the context for the different topics dealt with later. The section starts with a qualitative description of the major characteristics of coastal morphology, followed by an introduction to the role of wind waves in coastal morphodynamics and the role of sea-level rise. Some aspects of large-scale coastal behaviour are discussed with emphasis on the problem of long-term forecasting. The last subsection provides an introduction to feedback processes in the coastal system leading to instability and pattern formation. The second section deals with classical wave theory, which can be found also in many textbooks: Linear waves, wave asymmetry, radiation stress and wave-driven sediment transport (mathematical details can be found in the appendix). The third section considers coastal morphodynamics in the horizontal plane, beginning with large-scale coastline adaptation to wave forcing. In the following subsections different processes are discussed which play a role in the formation of shoreline patterns, such as shore-normal or shore-oblique bars and rip cells. A separate subsection is dedicated to the beach cusp phenomenon, which was described already in the introduction to Chapter 2. The fourth section considers coastal morphodynamics in the vertical shore-normal plane. Some models of the average coastal equilibrium profile are discussed; one of these models is further extended to provide a possible mechanism for the emergence of longshore bars (breaker bars) in the surf zone.

5.2. Morphodynamics of Sandy Coasts

This section is intended to familiarise the reader with some general aspects of sandy coast morphodynamics; a more quantitative discussion of the physical processes is postponed till later. First a qualitative description is given of typical morphologic features of sandy coasts and conditions are indicated under which these features are observed. An introduction is given to the major physical processes, which are related to shallow water wave dynamics. It is shown that sea-level rise plays an important role in the long-term large-scale evolution of sandy coasts. At shorter time scales the morphology of sandy coasts is highly variable. Much of this variability can be attributed to instabilities and symmetry breaking inherent to wave-topography interaction.

5.2.1. *Morphology of the Sea-Land Transition Zone*

Spatial scales

Sandy shorelines look fairly smooth when considered from a large-scale point of view, and this also holds for the bottom contour lines. This smoothness reflects the high mobility of the seabed which adapts to gradients in large-scale forcing by waves, wind and tides; these gradients are generally small. Large scale patterns, at the scale of 10 km and more, are often (but not always) caused by gradients in the substrate, by coastal inlets or deltas and by man-made structures. Changes in these large-scale patterns are slow, exceeding several decades or even centuries [425]. However, a closer look at the coastal morphology yields a very different impression, see Fig. 5.1. At smaller scales (one km and less) a great variety of patterns shows up, both in alongshore and in cross-shore direction. These patterns can change quite rapidly, over periods

Fig. 5.1. A closer look at the coast reveals a variety of structures... (Coast of Goeree, Holland).

of a day up to a few years. The smallest structures, bottom ripples, even evolve on the timescale of minutes.

The coastal profile

A cross-shore section of the coast is called a coastal profile. Coastal profiles typically have a small slope of the order of 1/1000 at the inner shelf zone, about one kilometre up to about ten or twenty kilometres offshore, with water depths typically ranging from about 10 to 30 metres. Landward of the inner shelf the seabed slope becomes much steeper, on the order of 1/100; this portion of the coastal profile is called the 'shoreface', see Fig. 5.2. In the most landward part of the shoreface, which includes part of the intra-tidal beach, waves are breaking and surfing onshore. We call $x = x_{br}$ the offshore distance at which waves start breaking (the breaker line). The position of the breaker line depends on the characteristics of the incoming waves, in particular the wave height H. The region landward of x_{br}, with water depths roughly between one and ten metres, is called 'surf zone'. In the surf zone we often find large, mostly shore-parallel structures, the longshore bars or breaker bars. The seabed slope in the

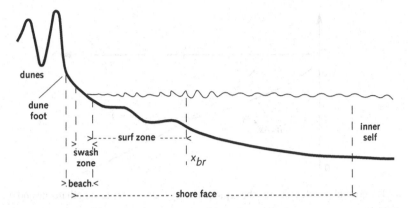

Fig. 5.2. On the shoreface different zones can be distinguished; each of these zones is shaped according to a particular type of morphodynamic feedback.

surf zone is highly variable because the location and pattern of these structures are irregular and variable over time. The average shape of the coastal profile corresponds to a landward increasing slope; many coastal profiles, in particular under microtidal conditions, can be approximated at some distance from the shoreline by a concave curve of the type [99, 101]

$$h(x) = Ax^{2/3}, \tag{5.1}$$

where x is the distance offshore from the shoreline, h the average water depth and A a site-specific coefficient, depending on local characteristics of the sediment, waves and tides. This expression can also be obtained from the assumption of uniform wave-energy dissipation per unit volume in the surf zone, as will be shown later [57, 99]. The 2/3 exponent in the power law (5.1) is not representative for all coastal profiles. Sometimes other exponents yield a better fit; laboratory experiments and field observations indicate that retreating coasts have a steeper profile, corresponding to a lower exponent of about 0.3 [107]. Laboratory experiments also show that the 2/3 exponent does not hold in the zone where waves start breaking [482]. Around the breaker line the shape of the coastal profile is more often convex than concave [498, 39]. Seaward of x_{br} the coastal profile is normally concave; there is some evidence that this part of the shoreface can be represented by a similar power law as for the surf zone [229, 32], see Fig. 5.3

$$h(x) = C(x - x_0)^{2/3}, \tag{5.2}$$

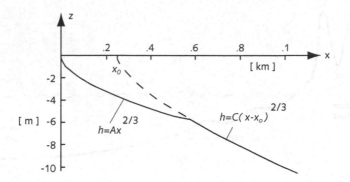

Fig. 5.3. The coastal profile can often be represented by a 2/3 power law, both inside and outside the surf zone.

where C is a coefficient and where x_0 can be related to the depth h_{br} at the breaker line x_{br}, by equating $h_{br} = h(x_{br})$.

Profile classification

The overall steepness of the coastal profile, i.e., the coefficient A, may strongly differ for different coasts. Field data indicate [101]:

$$A = 0.067 w_s^{0.44} [\mathrm{m}^{1/3}], \qquad (5.3)$$

where w_s is the average sediment fall velocity expressed in cm/s. Steep coastal profiles are generally associated with strong swell (long wave period) and coarse sediment (high fall velocity), while gentle coastal profiles are associated with dominance of short-crested waves and fine sediment [259]. Therefore coastal profiles are often characterised by the non-dimensional fall velocity [175, 98]

$$\Omega = H_{br}/(w_s T), \qquad (5.4)$$

where H_{br} is the wave height at which waves start breaking and T is the peak spectral period. Wright and Short [499] developed a beach typology based on field surveys in Australia covering a broad range of characteristic wave and sediment conditions; they showed that morphologic beach characteristics could be related to the non-dimensional fall velocity Ω. They distinguished a category of 'reflective' beaches, corresponding to $\Omega < 1$, a category of 'intermediate' beaches, corresponding to $1 < \Omega < 6$, and a category of 'dissipative' beaches, corresponding to $\Omega > 6$. The general validity of this classification suggests that coastal profiles tend to an equilibrium profile associated with the prevailing hydrodynamic forcing and sedimentary characteristics [57, 59]. The steepness of reflective beaches is generally greater than suggested by the 2/3 power law. This is not necessarily in contradiction with uniform wave-energy dissipation per unit volume, because the energy of reflected waves has to be subtracted from the energy of incident waves. The corresponding profile is given by an expression of the form [21, 32]

$$x = (h(x)/A)^{3/2} + B(h^2(x)/A)^{3/2}, \qquad (5.5)$$

where B is a coefficient and $x < x_{br}$.

Tidal beaches

Although shoreface dynamics is dominated by wave action, tidal influence cannot be neglected. The major tidal effect is an alternating onshore and off-shore shift of the breaker line and the surf zone; the tide therefore stretches the width of the surf zone. Around low water the surf zone is located further offshore, in a flatter portion of the coastal profile, where shoreface dynamics has a dissipative character. Around high water the surf zone is located onshore, in a steeper portion of the coastal profile, where shoreface dynamics has a more reflective character, especially if the tidal range is large [67, 406]. The tidal sweep effect tends to inhibit the development of long-shore bars in the surf zone [198, 276]. Several attempts have been made to generalise the coastal profile model of Dean (5.1) to the tidal case [301, 32], by considering the tidal range H_{tide} or the relative tidal range H_{tide}/h_{br} as an additional parameter. For megatidal conditions, corresponding to a tidal range on the order of 10 metres, the surf zone exhibits three distinct segments [276] with different slopes β: A steep, coarse-grained, reflective high tidal zone with slope $\beta \approx 0.1$; a mid-tidal zone with fine to medium sand and moderate slope $\beta \approx 0.02$ and a very dissipative, featureless low tidal zone with fine sands and a weak slope $\beta \approx 0.005$. The low tidal zone is characterised by shoaling wave conditions and rapid migration of the breaker line. For moderate wave conditions and tidal range between 2 and 5 metres, the intertidal beach has a moderate slope $\beta \approx 0.01$; it is characterised by low-amplitude ridge-runnel structures (height less than 1 m) with a cross-shore spacing of 50 to 150 m and a length of 100 to 500 m.

Ripple fields

Tidal beaches exhibit at low water a great diversity of patterns resulting from the mutual interaction between waves and and seabed. The aforementioned ridge-runnel structures often go together with large fields of bed ripples (height up to a few centimetres and wavelength up to a few tenths of centimetres); these ripple fields occur mostly in sheltered zones in runnels or behind swash bars. They are caused by low-energy water motions (velocity below 50 cm/s) induced either by waves (shore-parallel ripple crests) or by currents draining the runnels at low tide (cross-runnel ripple crests). Fields of wave-induced ripples are also found offshore, on the shoreface and the inner shelf. These ripple fields contain both short wave ripples and large wave ripples, the latter category with wavelengths of order 1 m and height of order 5 cm. The generation mechanism of wave ripples has been discussed in the Chapter on Current-Topography Interaction, Sec. 3.3.6.

The swash zone

At the end of the breaking process, the incident waves transform into bores that run up the beach. The up-running wave bores are called 'swash' and the zone where the bores propagate is called the swash zone. Sediment carried by swash flow is deposited on the beach and nourishes beach ridges (also called beach berms); these ridges migrate up the beach as swash bars, see Fig. 5.4. The return flow (backwash) takes beach sediments offshore. If the bed is coarse-grained and permeable and if wave periods are long, the swash flow will partly percolate to the beach aquifer; such conditions are favourable for accretion and for the generation of steep beach slopes. In that case the shoreline may exhibit cusp-like structures or bars pointing seaward. Beach cusps are frequently observed on reflective beaches; the cusps develop, often within a day, under conditions of calm weather with shore-normal incident waves; they decay at the same timescale under storm conditions with obliquely incident waves [303].

Longshore rhythmic structures

Beach cusps on reflective beaches and ridge-runnel structures on dissipative, intertidal beaches are examples of alongshore rhythmic patterns. A multitude of other morphologic patterns can be generated along sandy coasts, especially

Fig. 5.4. Sequence of beach ridges at the meso-tidal coast of Terschelling (The Netherlands).

under low to moderate wave conditions and small wave incidence angles (close to shore-normal) [143, 61]. The shoreline may exhibit, for instance, cuspate patterns with a larger spacing than beach cusps (giant cusps), in the order of 100–2000 m; they are observed along the US Atlantic barrier coast [113] and along the Dutch coast, see Fig. 5.22. These cuspate structures are associated with shore-normal bars [260] or with rip-current embayments and crescentic longshore bars [259]. A rhythmic pattern of shore-oblique bars may be generated if the waves approach the shore under a small angle [374]. Rhythmic patterns of shoreline cusps exist even at scales of tenths up to a few hundred kilometres, as illustrated by the spits in the Sea of Azov (Fig. 5.5) and the capes along the USA Atlantic coast. A very oblique average wave incidence angle is thought to be responsible for the generation of these megacusps [14].

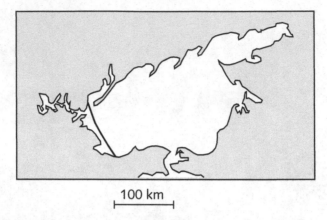

100 km

Fig. 5.5. The northern coast of the Sea of Azov is characterised by a sequence of large-scale sand spits. The development of these spits can be explained by growth of shoreline instabilities by positive feedback from longshore currents, driven by almost coast-parallel winds. See also Fig. 5.16.

Rip cells

Alongshore rhythmic patterns also occur in the sub-tidal zone. During storm periods the longshore bars in the surf zone are emphasised and straightened, while migrating seaward; during periods of low to moderate wave activity the longshore bars tend to break up into a sequence of crescentic bars. Under these conditions the wave-induced flow over the bars has a net shoreward component and the bars migrate shoreward. This shoreward flow is balanced by a net seaward flow which is channelled between the crescentic bars. The velocity of these rip currents may exceed 1 m/s. The alongshore spacing of the rip channels is on the order of one hundred to over one thousand metres. Most of the alongshore structures have an ephemeral character, except the mega-scale capes. They may be destroyed during a single storm, especially storms which generate strong longshore currents. The after-storm morphology is almost uniform alongshore.

Upper beach

The most shoreward part of the coastal profile is formed by the upper beach, also called supra-tidal or sub-aerial beach. It is covered by seawater only under exceptional conditions. It is nourished with sand from the lower (intra-tidal) beach, mainly by aeolian processes. The upper beach generally has a steeper equilibrium slope than the lower beach; down-slope gravity induced sand transport becomes significant only at a steep beach slope. The upper beach is not always sandy; gravel, rock and concrete structures are frequently found as well. Sandy upper beaches may develop into a dune area; sometimes the transition is gradual with low foredunes, but a more abrupt transition, marked by a steep dunefoot, may also occur. Sand is blown from the beach into the dunes; in this way considerable amounts of sand, from a few m^3/m up to a few tenths of m^3/m on a yearly basis, can escape landward. The average aeolian sand supply to the dunes along the Dutch coast is estimated at about $10\,m^3/my$ [462] and therefore contributes significantly to coastline retreat. The width of the upper beach determines to a great extent the sand supply to the dunes. But many other local beach features, including sediment composition, grain sorting (more than grain diameter) and vegetation, also play a significant role; therefore it is very difficult to make reliable predictions of aeolian sand transport across beaches [25].

5.2.2. *Wind Waves over a Shallow Seabed*

Waves travel without losing much energy ...

The transition zone between land and sea receives a high energy input from waves which deliver their energy when breaking on the coast. This energy has been accumulated in the wave field by wind forcing on the ocean and the continental shelf. Wave energy can travel over large distances compared to the wavelength, without much energy loss due to bottom dissipation. The short wave periods allow the development of only a very thin turbulent bottom boundary layer. Wave propagation may therefore be considered as an almost frictionless process. Most energy is lost by wave spilling at the water surface, which occurs when waves become too strongly peaked and loose their stability.

... but become asymmetric

Wave peaking is related to the nonlinear character of wave propagation, and is mainly due to the curvature of the wave surface (it does not occur in tidal waves, which have an almost flat wave surface). Wave-peaking also influences the wave-orbital water motion; the orbital motion in the wave propagation direction is stronger than the opposite motion, but also of shorter duration. This phenomenon is called 'wave asymmetry' or 'wave-skewness'. It has consequences for the net displacement of seabed sediments, which are more sensitive to the strong water motion in the wave propagation direction than to the slower opposite motion.

Single sine representation of an irregular wavefield

Waves arriving at the shore have different characteristics, depending on their origin. Waves coming from the ocean have the greatest wavelength; these swell waves have survived because of their gentle surface slope and related small energy dissipation. Waves generated by local wind fields have smaller wavelengths. They propagate shoreward from different directions; the wave field consists of short-crested random waves peaked in all directions. More or less linear wave crests become apparent close to the shore, due to the decrease of directional wave spreading by refraction. These sea waves are superimposed on the longer swell waves; together they form a complicated wave spectrum. Nevertheless, in simple models the incident wave field is often represented by a single (monochromatic) sine shaped wave, with the frequency of the most energetic part of the spectrum and a wave height equal to the root-mean-square wave height, H_{rms} [157, 418]. This sine wave reproduces the average energy of the wave field. Instead of H_{rms}, the significant wave heigh H_s is often used in practice; they are related by $H_s = \sqrt{2} H_{rms}$. A monochromatic wave representation can be used to understand certain statistical properties of wave interaction with the seabed, but it should be kept in mind that it is a very crude simplification of reality. A more realistic representation includes the modulation of the wave amplitude (wave groups) due to interference of different waves with similar but unequal periods within the wave spectrum.

Wave breaking

When a wave travels from offshore into shallow water the wave amplitude and the asymmetry increase. When the water depth gets smaller than about two to five times the wave amplitude the wave starts breaking [24]. Wave breaking does not depend only on the wave height-to-depth ratio, but also on the wave shape. More accurate breaking criteria involve the ratio of particle speed to wave speed at the crest or the particle acceleration at the crest [364].

Spilling or plunging

In the case of gentle shoreface slope and strongly peaked waves, breaking starts at the wave crest. This breaking process is called 'wave spill'; air is entrained into a roller, which is pushed forward at the wave front. The breaking process continues with a decreasing wave amplitude and a growing roller volume, until the wave takes the form of a bore propagating onto the beach, see Fig. 5.6. In the case of a steep shoreface slope and/or less peaked waves, wave breaking has a plunging character (Fig. 5.7), with production of strong vortex motions and sediment uplift from the seabed. After initial breaking, waves may reform and break a second time. The breaking process continues until formation of a wave bore, similar to that in the spilling breaker case.

Wave saturation

The zone between initial breaking and the beach is called the surf zone. Observations on gently sloping beaches show that the wave height-to-depth ratio H/D ratio in this zone becomes almost constant when the breaking wave is

Fig. 5.6. Spilling breakers on a gently sloping beach. Waves start breaking when the wave height becomes comparable to water depth, often at the outer longshore bar. Landward of the bar the wave-height-to-water-depth ratio is decreased. Waves may reform in the trough between the outer and inner bar and break a second time at the inner bar, close to the beach.

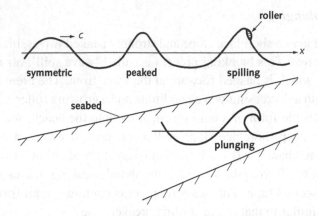

Fig. 5.7. Wave transformation on the shoreface. In shallow water the wavelength decreases and the wave surface slope becomes steeper. The wave then loses symmetry and the wave surface becomes peaked, with higher shoreward than seaward orbital velocities. When the wave height becomes comparable to water depth the wave starts breaking. On gently sloping beaches wave spilling is the most common breaking mode; on steep beaches plunging breakers will occur more frequently.

transforming into a bore [440]; the ratio is then close to 0.5 [210]. Constancy of the wave height-to-depth ratio in the surf zone is called 'wave-saturation':

$$H(x) = \gamma_{br} D(x), \qquad (5.6)$$

where $H = 2a$ is the wave-height, D is the total wave-averaged water depth and γ_{br} the breaker coefficient. Estimates of γ_{br} range from 0.4 to 1.3 [67].

Undertow

Wave breaking is a complicated process, which is still challenging mathematical modellers. Just before breaking, the water speed at the crest increases and approaches the wave speed, while a strong return current develops at the bottom. This return current, called undertow, opposes shoreward transport caused by wave asymmetry. In some cases this may result in a net seaward sediment transport. A return flow is present both in breaking and non-breaking waves; it balances the shoreward Stokes drift due to the covariance of water level and wave-orbital motion. In breaking waves the return flow is concentrated close to the bottom, while in non-breaking waves the maximum return flow occurs higher in the water column [371].

Breaker bars

Waves do not all break at the same depth, due to the irregularity of the wave field. Yet, wave breaking is often concentrated in a few surf zone strips, corresponding to shore-parallel submarine bars, the longshore bars or breaker bars. The breaking point is not located at the bar crest, but just onshore in the bar trough. Field observations and laboratory experiments indicate that wave breaking produces convergence of sediment transport at the bars [442, 137, 482], by stronger seabed stirring and stronger near-bottom return flow at the onshore side of the bar than at the offshore side. This feedback stimulates the development of these bars.

Radiation stress

Wave breaking in the surf zone entails not only a shoreward decrease of wave amplitude, but also a shoreward decrease of wave-orbital velocities. During shoreward orbital motion, more shoreward momentum enters the surf from offshore than is transferred from the surf zone to the beach. During seaward orbital motion more seaward momentum is transferred from the surf zone to offshore than enters the surf zone from the beach. The result is a net gain of shoreward momentum in the surf zone or, equivalently, a net loss of seaward momentum. This contributes, in the case of shore-normal wave incidence, to a net shoreward pressure gradient and in the case of shore-oblique wave incidence, to a net alongshore pressure gradient. This phenomenon is described in mathematical terms by means of 'radiation stresses'. The radiation stresses also incorporate the net effect of nonlinear contributions to the wave-induced pressure gradient, see Appendix D.3. Breaking-induced gradients in radiation stresses produce a shoreward increase of the average sealevel and a longshore current in the surf zone. This longshore current generates a substantial longshore sediment flux, in particular during storms, when wave breaking causes intense sediment suspension in the surf zone. Currents produced by radiation stresses play an important role in the morphology of the nearshore zone. But on the other hand, the radiation stresses themselves are very sensitive to seabed disturbances, which locally affect the intensity of wave breaking. Therefore the interaction between seabed morphology and radiation stresses has a rich potential of generating seabed instability and morphologic patterns. An example is the development of rip channels, which results from seabed instability inherent to the mutual interaction of seabed morphology and radiation stresses.

For monochromatic sinusoidal waves simple analytical expressions (D.20) can be derived for the radiation stresses. These analytical expressions will be used even in cases where (1) waves are not sinusoidal (i.e., in the breaker zone), (2) waves are not monochromatic (wave spectrum) and (3) the wave incidence angle is variable (wave directional spreading). Ignoring these effects leads in general to an overestimate of the radiation stresses [146]. However, the qualitative features of the radiation stresses are not strongly affected by these simplifications.

Wave refraction

In shallow water (wavelength substantially larger than water depth) the wave propagation velocity depends on depth; the smaller the depth, the slower waves propagate. A wave field approaching the coast under a non-zero angle will propagate at different velocities according to depth. In Fig. 5.8 we characterise the wave field by hypothetical crestlines, perpendicular to the propagation direction. The part of the crestline which is closest to the shore (shallowest water) propagates more slowly than the part of the crest further offshore. Hence,

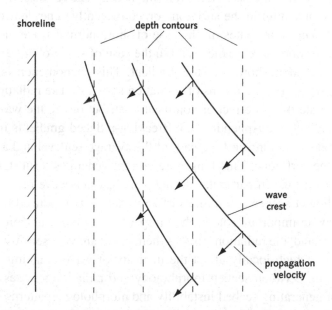

Fig. 5.8. The wave propagation speed increases with depth; an obliquely incident wavefront is therefore refracted towards the shoreline.

further away from the coast the crestline moves faster shoreward than closer to the coast. This implies that the crestline rotates towards a shore-parallel orientation, see Fig. 5.8. This phenomenon is called refraction and is described by Snell's law ('propagation velocity proportional to the sine of the angle between wave ray and bottom-slope direction') or by the condition that the field of wavenumber vectors $\vec{k}$ is irrotational, $\vec{k} = \vec{\nabla} f$.

Refraction will distribute wave energy non-uniformly over the surf zone if the seabed is not flat. As an example we assume a pattern of alternating shoals and troughs along the shore. The incident wave field will then be refracted towards the shoals. Wave breaking on the shoal produces a gradient in the radiation stress which drives a current toward the shoal. The sediment carried by this current may nourish the shoal and and amplify the trough-shoal morphology, see Fig. 5.9. This will only happen for low-energy waves, assuming that we may neglect the seaward transport caused by an increase of seabed stirring and near-bottom return flow.

Edge waves

Incident waves that do not lose all their energy when breaking, will be partially reflected on the coast. This may occur for swell waves on coasts with a steep bottom slope (reflective coasts). If these waves approach the coast under a non-zero angle they will become trapped at the coast by refraction. When moving away from the coast the crestline will rotate until the propagation is shoreward again. Such coastally trapped waves are called 'edge waves'. Wave reflection

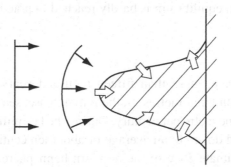

Fig. 5.9. Wave refraction concentrates wave breaking at a shoal and increases locally the radiation stresses. This drives sediment-laden flow onto the shoal and contributes to shoal growth.

on coasts with a weak bottom slope ('dissipative coasts') only occurs with infra-gravity waves (waves with a substantially larger wavelength than regular wind-generated waves). Incident and reflected edge waves may form a standing wave and generate a pattern of rhythmic bedforms structured to the nodes and antinodes [48]. At dissipative coasts the generation of edge waves can be related to the occurrence of wave groups [165]. Bound long waves associated with the wave groups partly become free waves after breaking of the short waves. However, the interaction of these free and bound long waves with bathymetry is not yet clear.

5.2.3. *Response to Sea-Level Rise*

Unbalance of large-scale longshore and cross-shore transport

Gradients in longshore and cross-shore sediment transport produce changes in coastal profile and coastline position. Longshore transport is nourished by sediment supply from rivers or by erosion of nearby sediment deposits; it is depleted by sediment withdrawal to adjacent back-barrier basins or by sedimentation in upstream coastal stretches. Cross-shore transport to or from the inner shelf depends on the balance between wave-induced shoreward transport and gravity induced down-slope transport. Cross-shore transport at the top of the coastal profile, the subareal beach, depends on the balance of aeolian inland transport and storm-induced dune erosion [328]. In an equilibrium situation the coastal profile will adjust in such a way that the gradients in longshore and cross-shore transport exactly balance in a time averaged sense. One of the major reasons that such an equilibrium is hardly reached in practice is sea-level rise.

Transgression

Since the last ice age, more than fifteen thousand years ago, the sea level has risen over one hundred metres. In the initial phase, sea-level rise was very rapid, up to about one metre per century. During the last millennia sea-level rise has strongly slowed down, to an average of about ten centimetres per century in the temperate climate zone of the northern hemisphere. Sea-level rise has produced great changes in the coastline of low lying coastal zones. Former coastal plains have been drowned and at most places the coastline has shifted landward over great distances. Large amounts of sediment have been carried

Fig. 5.10. New coastline (solid line) and former coastline (dashed line) after sea-level rise. (a) Coastal transgression in response to sea-level rise: Barrier sands are deposited on the submerged coastal zone. (b) Coastal retreat in response to sea-level rise: Sediment is eroded from the submerged coastal zone.

landward by wave and tide-induced shoreward sediment transport and have been deposited in the newly submerged coastal areas, see Fig. 5.10(a). This process, which is called 'transgression', may proceed in several ways, depending on the rate of sea-level rise, the availability of sediment (supply or removal) and the steepness of the bottom slope [53].

Coastal retreat

In the case of a steep coastal slope and limited sediment availability (fluvial sediment supply smaller than sediment withdrawal for infilling of back-barrier basins), the coastline is continuously retreating. Sediment delivered by erosion of newly submerged areas is used to compensate the deficit of sediment supply and to maintain the shoreface profile at equilibrium, see Fig. 5.10(b). Coastal retreat continues until the sediment-trapping back-barrier systems are filled [325].

Barrier formation

In the case of sufficient sediment supply and a seabed slope close to equilibrium, wave-driven shoreward transport will build up a coastal barrier [383], as illustrated in Fig. 5.11. If sea-level rise is not too fast, this barrier may keep pace with sea level and eventually act as sediment source for dune formation. Barriers are not necessarily attached to the shore but may enclose coastal lagoons. (The Ameland reef may be viewed as a small coastal lagoon, see Sec. 2.5.2.) Storm-driven sediment wash-over will cause landward migration of these barriers, a process which may eventually go on until shore attachment. If sea-level rise is

Fig. 5.11. In the case of sufficient sediment supply wave-driven shoreward transport builds a coastal barrier (coast of the island Terschelling in the Wadden Sea).

fast, it may happen that the barrier does not build up fast enough. In that case the barrier will be overstepped and a new barrier is formed further landward. The barrier remains become part of the submarine relief of the inner shelf. They will not necessarily become regressive seafloor relics; sometimes they are remodelled as submarine ridges in interaction with the prevailing current and wave conditions [433], see also Sec. 3.4.

5.2.4. *Large-Scale Coastal Behaviour*

Response to incident wave field

Large-scale coastal behaviour is governed by redistribution of sediment tending to minimise long-term averaged gradients in sediment transport. In this way the coastline is adjusted to the average large-scale forcing by waves, wind and tides. The general tendency is alignment of the shoreline in a direction perpendicular to wave incidence. The angle of wave incidence is determined not only by the direction of prevailing wind and fetch, but it is also influenced by the near-shore submarine depth contours, due to wave-refraction. Submarine depth contours reflect geological patterns, man-made structures or patterns imposed by tides and wind driven currents. Sediment transport is also influenced by sediment supply from nearby coastal inlets and deltas also plays a role. Therefore coastlines are not necessarily straight.

Prediction

Models for predicting large-scale coastal behaviour are still in a rudimentary stage. Large-scale long-term sediment transport is the sum of many small-scale, fluctuating processes for which no reliable parameterisations are available as yet. These processes often contribute with opposite signs, especially when the coastal system is not far from equilibrium. Inaccuracies in the parameterisation of these small-scale contributions add up to large errors in the sum and to erroneous predictions for the large-scale long-term evolution. Successful upscaling requires precise knowledge of the mutual feedback between small-scale constituent processes that determines their aggregate larger-scale behaviour [109]. A strategy for upscaling small-scale processes in view of long-term coastal prediction is the 'cascade approach' or 'scale-hierarchy approach' [87]. In this approach the coastal system is modelled according to a hierarchy of scale levels. At the highest scale level the coastal system is defined as a system of interacting morphologic units such as, for instance, shoreface, inner shelf and back-barrier basin. Each level of scale consists of interacting morphologic units defined at the next lower level of scale; morphologic evolution equations are formulated at the scale of these interacting morphologic units. This approach is still subject of ongoing research, focusing on the appropriate definition of morphologic units and their aggregate dynamics. Another complementary approach to large-scale and long-term coastal prediction is based on geological records obtained from bottom cores, interpreted with the help of dating techniques. The geological reconstructions obtained this way provide important evidence for calibrating or validating aggregate-scale models.

5.2.5. *Pattern Formation*

Inherent limitation to detailed predictions

There is no unique relationship between change in external conditions and coastal response [208]. The wave field and wave-driven currents interact with coastal morphology in a nonlinear way. From this interaction morphologic patterns arise on spatial and temporal scales which have no direct relationship with patterns in the original topography or in the offshore wave field. The first suggestions that these patterns might originate from instability of the local coastal morphology relative to hydrodynamic conditions appeared in the late 1960s [416]. At present it is broadly accepted that many patterns in the sedimentary coastal environment arise in this way. The instability of the system involves a great sensitivity to small perturbations, which cannot be captured in deterministic models. Very different patterns may develop starting from similar morphologies. This imposes an inherent limitation on detailed predictions of the small-scale coastal evolution at longer time scales. However, statistical properties of these patterns can often be derived from the basic physical properties of the morphodynamic interaction process.

Perturbation growth and symmetry breaking

The process of pattern formation due to wave-topography interaction is in several respects similar to current-topography interaction. Through this process the system evolves without external intervention from a state with little or no structure and high symmetry to a more structured state with less symmetry. Pattern formation is inherent to the dynamics of the system and it has an irreversible character as long as external forcing conditions remain the same. The trigger for this evolution can, in principle, be infinitesimally small if the initial symmetric (i.e., unstructured) state is unstable against perturbation. The development to a more structured and less symmetric state is therefore inevitable. This corresponds to what is observed in nature; if in the field no patterns are present or emerging, an initial perturbation will decay and will not start off pattern formation. Inversely, if patterns are present, they will automatically reappear after being eliminated by an external intervention. Pattern formation results from the nonlinear nature of the interaction between morphology and hydrodynamics and becomes manifest as soon as an infinitesimal perturbation is applied to an unstable symmetric initial state. Analysis of initial perturbation growth is therefore an adequate method for revealing the basic nature of pattern-generating feedback mechanisms; it will be used throughout this chapter.

Instability mechanisms

Wave-topography interaction produces seabed instability through several mechanisms. Relevant mechanisms have been mentioned before and are summarised below:

- Instability related to frictional delay of the momentum response in the wave boundary layer to seabed perturbations; this instability is responsible for the generation of wave-induced ripples, as described in Sec. 3.3.6.
- Instability related to the development of wave asymmetry and subsequent wave breaking on a sloping seabed. The gradients in sediment transport related to wave asymmetry and wave breaking result in feedback to seabed morphology and provide, for instance, a mechanism to generate longshore bars in the surf zone.
- Instability related to radiation stresses in the surf zone. Seabed topography produces gradients in radiation stresses, related to wave breaking and wave refraction; this topography is developed, in turn, through residual flow patterns produced by the radiation stresses. Positive feedback generates seabed patterns such as rip channels, crescentic bars and transverse bars.
- Instability related to swash flow on a sloping beach. Diversion of swash flow by cuspate beach structures produces a circulation pattern which provides a positive feedback to the generation of these structures.

What is the practical use of understanding pattern formation?

Wave-topography interaction may produce more instability mechanisms, which have not yet been fully explored. Predictive modelling of these processes is only possible to a limited extent, even if the basic physics is perfectly understood. This raises the question: What is the practical use of better understanding pattern formation on short to medium spatial and temporal scales? Perhaps the most important reason for studying pattern formation processes is not in forecasting, but in hindcasting and analysing coastal evolution. Trends in coastal morphology on large spatial and temporal scales generally proceed at a much slower rate than short-term local fluctuations; the former are easily masked by the latter. Interpretation of the short-term local phenomena is necessary for the detection of long-trends from field observations. In addition, many short-term local fluctuations are indirect expressions of the large-scale morphology of the coastal system and its response to trends in forcing. For instance, the presence or absence of beach cusps provides information about wave climate and coastal profile. The observation of spatial and temporal patterns of breaker bars and rip currents provides information relevant for the optimisation of shoreline maintenance, design of more efficient nourishment schemes, etc. Misinterpretation of coastal field observations may seriously impair the effectiveness of coastal management strategies. A skilled coastal manager is capable of 'reading' the beach.

5.3. Wave Theory

In this section we present a more quantitative description of wave propagation in shallow water. The mathematical details are given in Appendix D. This quantitative description is needed to better understand the dynamics of wave-topography interaction discussed in Secs. 5.4 and 5.5.

The orbital motion loses its symmetry

Individual waves entering the shoreface zone from deep water are almost perfectly symmetric; wave-orbital motion in the wave propagation direction mirrors the wave-orbital motion in the opposite direction. A small residual velocity exists only in a very thin layer above the bottom, see Sec. 3.2.1. Otherwise, seabed particles move to and fro under the influence of wave-orbital motion without experiencing a net displacement. However, in the shoreface zone things become very different. The symmetry of the incoming wave is entirely lost, up to a point where the wave finally breaks. This dramatic transformation has great consequences for the net displacement of seabed particles under the influence of wave motion during a wave cycle. In this section we will discuss the most important underlying processes; a more detailed mathematical treatment can be found in Appendix D.

Wave motion as a frictionless process

Before answering the question why wind waves become asymmetric after entering the shoreface, we should first understand why they are symmetric before. Tidal waves, for instance, are already asymmetric outside the shoreface, at depths larger than 20 m. The main reason why wind waves conserve their symmetry at smaller depths than tidal waves is related to the much smaller wave period. Tidal wave asymmetry is mainly due to the influence of frictional energy dissipation; wind wave dynamics is much less influenced by frictional dissipation. Due to the small wave period, the rate at which kinetic and potential energy are exchanged within a wind wave is much greater than the energy dissipated by bottom friction. This contrasts with tidal waves on the continental shelf, where energy dissipation is comparable to the exchange rate between kinetic and potential energy. The period T of wind waves is so short that only a very thin turbulent shear layer can develop at the bottom during a wave cycle. This implies that wave motion can be considered as an essentially frictionless process, which may be described by potential flow.

5.3.1. *Linear Theory*

Basic assumptions

We consider a single-frequency (monochromatic) wave, propagating in the negative x-direction, towards the shore. In the case of potential flow, the horizontal and vertical wave-orbital velocities $u(x, z, t)$, $w(x, z, t)$ in the plane x, z can be related to a flow potential $\Phi(x, z, t)$ by

$$u = \Psi_x, \quad w = \Psi_z. \tag{5.7}$$

The requirement of mass conservation implies that the function Φ has to satisfy the Laplace equation

$$\Phi_{xx} + \Phi_{zz} = 0. \tag{5.8}$$

To solve this equation we need to specify the boundary conditions. At the offshore boundary x_∞ of the shoreface the surface modulation $\eta(x_\infty, t)$ of the incoming wave is assumed to be a sinusoidal function with given frequency $\omega = 2\pi/T$,

$$\eta(x_\infty, t) = a \cos \omega t. \tag{5.9}$$

It is assumed that the wave will break on the shoreface and dissipate all its energy; the corresponding condition at the inshore boundary is the absence of a reflected wave. The boundary condition at the seabed, $z = -h$, states that the velocity component normal to the seabed equals zero. If we assume that the seabed is almost horizontal over distances comparable to the wavelength, this condition is equivalent to zero vertical velocity at the bottom. Finally two

conditions need to be specified at the surface $z = \eta(x, t)$. One condition states
that the velocity normal to the surface equals the surface displacement per unit
time. The second condition specifies that the pressure at the surface equals
the atmospheric pressure. The first surface condition depends on the shape of
the wave surface, which in turn depends on the wave-orbital motion. The two
surface conditions are nonlinear and thus produce a deformation of the initial
sinusoidal wave shape. This deformation depends on the curvature of the wave
surface; the degree of curvature can be characterised by the average magnitude
of the surface slope ak. The wavenumber k is related to the wavelength λ by
$k = 2\pi/\lambda$. If the surface slope is small, $ak \ll 1$, the surface curvature can be
neglected and the initial sinusoidal wave shape remains intact; this is called
linear wave propagation. The surface elevation and the horizontal orbital wave
motion are then given by (D.7)

$$\eta(x, t) = a\cos(kx + \omega t),$$
$$u(x, z, t) = -a\omega\frac{\cosh[k(z + h)]}{\sinh(kh)}\cos(kx + \omega t). \tag{5.10}$$

Deep water wave propagation is independent of depth

It should be noted that the wave-orbital velocity $u(x, z, y)$ decreases from the
surface $z = \eta(x, t)$ to the bottom $z = -h$. This decrease is not caused by bottom
friction, as for the tidal case, but by the requirement of potential flow. According
to the potential flow hypothesis, the vertical gradient of the orbital velocity u is
proportional to the streamline curvature. At the bottom, wave-orbital motion is
horizontal and the streamline curvature zero. The curvature increases towards
the surface and the vertical velocity gradient increases accordingly. It follows
that the orbital velocity u will also increase from the bottom to the surface.

Wind wave propagation in deep water does not depend on depth, in strong
contrast with tidal wave propagation. This follows from the expression of the
the wave propagation velocity c, obtained by solving the linear wave equations
(see D.11)

$$c = \frac{\omega}{k} = \sqrt{\frac{g}{k}\tanh(kh)}. \tag{5.11}$$

In shallow water ($kh \ll 1$) we have, just as for tidal waves, $c \approx \sqrt{gh}$; in deep
water ($kh \gg 1$) c is independent of depth, $c \approx \sqrt{g/k}$. Inspection of (5.10)

shows that the water flux related to wave-orbital motion is also independent of the water depth h.

5.3.2. Wave Deformation

Second order nonlinear theory

The expression (5.11) for the wave propagation speed shows that the wavelength $\lambda = 2\pi c/\omega$ decreases with decreasing depth. Therefore the surface curvature of waves propagating onto the shoreface will increase. At some point, ka becomes on the order of 1, invalidating the assumption of linear wave propagation. The first order correction to the linear approximation can be obtained by expanding the flow potential Φ and the wave equations as a power series in the small parameter $\epsilon \equiv ka$ and retaining the linear terms in ϵ. The solution reads [265]

$$\eta = a\cos(kx+\omega t)+\frac{1}{4}ka^2\left(3\coth^3(kh)-\coth(kh)\right)\cos[2(kx+\omega t)], \quad (5.12)$$

$$u - \frac{1}{2}\frac{ga^2}{ch} - u\omega\frac{\cosh[k(z+h)]}{\sinh(kh)}\cos(kx+\omega t)$$
$$-\frac{3}{4}a^2k\omega\frac{\cosh[2k(z+h)]}{\sinh^4(kh)}\cos[2(kx+\omega t)]. \quad (5.13)$$

The second order nonlinear wave contribution introduces constant terms and terms with double wave period. The cosine terms with single and double period are in phase; this implies that nonlinear terms make wave crests higher and steeper, whereas wave troughs are made more shallow and flat. Wave-orbital velocities in the propagation direction (underneath wave crests) are larger than wave-orbital velocities against the propagation direction (underneath wave troughs).

Wave asymmetry

Wave asymmetry is mainly generated through the nonlinear surface boundary condition

$$w_\perp = \eta_t \quad \text{at } z = \eta(x, z, t), \quad w_\perp \approx w + \eta_x u, \quad (5.14)$$

where $w_\perp$ is the velocity component normal to the wave surface. This condition relates the displacement of the water surface not only to the vertical velocity component, but also to the horizontal component. The horizontal orbital wave motion contributes to vertical displacement of the water surface because of the sloping water surface: Horizontal flow is directed towards the surface shortly before the wave crest arrives, and away from the surface shortly after, see Fig. 5.12. It should be noticed that short-wave asymmetry and tidal wave asymmetry are of a different nature. Nonlinear deformation of the tidal wave leads to a sawtooth shaped wave surface, while nonlinear deformation of wind waves leads to a peaked wave surface. Tidal wave deformation is mainly due to a different propagation velocity of high water relative to low water; this

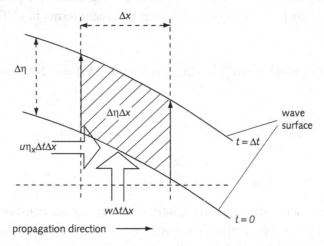

Fig. 5.12. Generation of wave asymmetry. The figure shows part of the wave surface at time $t = 0$, before arrival of the wave crest, and a little later at $t = \Delta t$. The horizontal component u of wave orbital flow has a component perpendicular to the wave surface because of its inclination. Therefore the wave surface is lifted not only by the vertical component of orbital motion (positive velocity w), but also by the horizontal component (positive velocity u). The influence of u on the wave surface motion increases upon propagation into shallow water, because the wave height/wavelength ratio increases and the wave surface curvature increases. The horizontal component u of wave orbital flow is out of phase with water level inclination. It contributes to lifting the wave surface a quarter wave period before arrival of the wave crest and a quarter wave period before arrival of the wave trough; it contributes to lowering the wave surface a quarter wave period after passing of the wave crest and after passing of the wave trough. This results in a sharper rise and fall of the wave crest and in a slower fall and rise of the wave trough. For long waves, such as tidal waves, the wave-peaking process is not significant because of the great wavelength and small water surface slope.

difference relates to the great sensitivity of the tidal propagation speed to water depth. By contrast, short-wave propagation is not strongly dependent on water depth, provided the wavelength λ is smaller than 2π times depth, as shown before. However, in very shallow water, shortly before waves start breaking, the propagation speed of wind waves also becomes strongly depth-dependent. At that stage wind-wave surface takes a sawtooth shape, similar to a tidal wave.

Undertow

Asymmetry of the wave-orbital motion, associated with wave surface asymmetry, causes a net sediment transport. More sediment will be suspended under wave crests than under wave troughs and therefore wave-induced sediment transport does not average to zero over a wave period. This normally leads to shoreward displacement of sediment. However, another second-order nonlinear velocity component influences residual sediment transport in the opposite way. This is the first term in (5.13), which represents the return flow compensating for the excess shoreward mass transport due to a greater water depth during shoreward orbital motion than during seaward orbital motion (Stokes drift). Although derived from linear wave theory, the expression (5.13) gives a

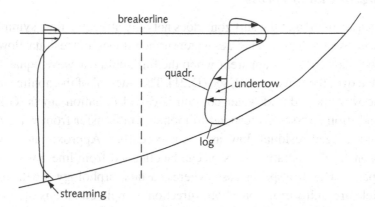

Fig. 5.13. Flow profiles under non-breaking and breaking waves, averaged over the wave period. Near the surface shoreward mass transport is due to the covarying phases of wave-orbital motion and surface elevation. This near-surface shoreward transport is compensated by seaward transport below the wave trough level. In a thin layer at the bottom shoreward drift is due to streaming (see Sec. 3.2.1). In the breaker zone the shoreward mass transport is enhanced, as well as the seaward transport (undertow) lower in the vertical. In the roughness layer close to the bottom the profile is logarithmic; the undertow profile is quadratic [433].

reasonable first order estimate for the depth averaged return current $\bar{U}_0$, even at the stage of wave breaking. Laboratory and field observations indicate [88, 443]

$$\bar{U}_0 = \frac{C_u}{2} \frac{a^2 c}{h^2}. \tag{5.15}$$

The coefficient C_u accounts for the effect of the surface roller in the breaking wave; for irregular waves C_u is found to be of order 1. Below the wave trough the average current is seaward, while shoreward mass transport is concentrated in the upper part of the vertical, between wave trough and wave crest, see Fig. 5.13. In the turbulent wave boundary layer the return flow can be represented by a logarithmic function of depth and above, up to the wave-trough level, it can be represented by a quadratic function of depth. The thickness of the turbulent wave boundary layer in the breaking zone is given by $\delta \approx (\kappa/\omega)\sqrt{\tau_w/\rho}$ [178] with τ_w given by (3.21) [88]. In the wave breaking zone, the maximum return flow velocity occurs close the bottom and attains a value on the order of $2\bar{U}_0$. The return flow under breaking waves is called undertow.

5.3.3. *Wave-Induced Residual Flow*

The longshore uniform case

Wave transformation in the surf zone does not only affect the time symmetry of the incident waves. It also generates water surface slopes and residual flow in the surf zone. This becomes apparent when the full nonlinear wave equations are integrated over the wave period, see (D.21). The integral of the nonlinear terms over the wave period and water column is called radiation stress. Gradients in the radiation stresses correspond to momentum transfer from wave motion to depth-averaged residual flow, as discussed earlier. Approximate analytical expressions of the radiation stresses can be obtained from linear wave theory, see Appendix D.3. In a special case, where coastal morphology and the incident wave field are uniform in longshore direction, simple equations are obtained for the average cross-shore surface slope η_{0x},

$$g\rho D \eta_{0x} + S^{(xx)}{}_x = 0, \tag{5.16}$$

and for the average bottom shear stress $\tau^{(y)}{}_{0b}$ by a longshore current,

$$D\tau^{(y)}{}_{0b} + S^{(xy)}{}_x = 0. \tag{5.17}$$

In these equations $S^{(xx)}$, $S^{(xy)}$ are elements of the radiation stress tensor and D is the total water depth including average wave-induced water-level setup and setdown. For small depth ($kD \ll 1$) the radiation stresses can be approximated by the following first order expressions:

$$S^{(xx)} \approx \left(\frac{1}{2} + \cos^2 \theta \right) E, \quad S^{(yy)} \approx \left(\frac{1}{2} + \sin^2 \theta \right) E,$$

$$S^{(xy)} \approx E \sin \theta \cos \theta, \tag{5.18}$$

where θ is the angle of wave incidence and E is the wave energy given by

$$E = \frac{1}{2} \rho g a^2. \tag{5.19}$$

The energy E in the surf zone is related to the average water depth D, under the assumption of wave saturation (5.6):

$$E = \frac{1}{8} \rho g \gamma_{br}^2 D^2(x). \tag{5.20}$$

Wave set-up

The radiation-stress estimates (5.18–5.19) illustrate some important phenomena which are brought about by wave breaking. The first phenomenon is an increase of water level at the shoreline, called set-up. Set-up is caused by the shoreward decrease of wave amplitude, according to Eq. (D.25). Equation (D.25) can be solved after substitution of (5.18) and (5.20), assuming shore-normal wave incidence $\theta = 0$ and wave saturation. The result is (see D.31)

$$\eta_0 = -\frac{\frac{3}{8} \gamma_{br}^2}{1 + \frac{3}{8} \gamma_{br}^2} h_{br}. \tag{5.21}$$

The wave set-up at the shoreline, η_0, is proportional to the depth h_{br} where waves start breaking. This set-up can attain several decimetres or even one metre under storm conditions. Although the application of linear wave theory in the surf zone may be questioned, the wave set-up determined in this way is in fair agreement with field observations [367]. Due to the presence of wave groups the set-up is not constant, but varies with the period of the wave groups (see Appendix D.2). The high waves in the wave-group centre start breaking at a greater depth than the lower waves at the wave group nodes. The set-up at the wave group centre is larger and the waves propagate further onto the beach. This periodic feature is called surf beat.

Longshore current

The second important phenomenon is the generation of a longshore current in the surf zone. When obliquely incident waves break upon approaching the shore, the radiation stress induces longshore wave-averaged momentum in the surf zone. The strength of the resulting longshore current depends on its frictional momentum dissipation, i.e., on the bottom shear stress $\tau^{(y)}{}_{0b}$. The bottom shear stress can be linearly related to the longshore current velocity V, if it is assumed that the contribution of wave-orbital velocities is much stronger than the contribution of the longshore current. In that case the linear friction coefficient depends on the drag coefficient c_D and on the maximum wave-orbital velocity U, see (D.33). For solving Eq. (D.26) we use again the wave-saturation assumption together with Snell's refraction law $\sin \theta / c =$ constant. The result is (see D.37):

$$V = \frac{5\pi}{32} \frac{\beta \gamma_{br}}{c_D} \sqrt{gh} \sin(2\theta_{br}), \qquad (5.22)$$

where $\tan \beta \approx \beta$ represents the average seabed slope and where θ_{br} indicates the wave-incidence angle at the breaker line. The velocity is proportional to depth and therefore the velocity distribution is triangular in the case of constant slope β: zero velocity outside the surf zone, maximum velocity at the edge of the surf zone, and a gradual velocity decrease to zero at the coast, see Fig. 5.14. In reality, momentum diffusion perpendicular to the shore makes the velocity distribution smoother and less triangular. Observations show that in the case of a barred shoreface the maximum longshore current occurs between the bar and the shoreline [58]. The longshore breaker current is localised in a narrow strip along the coast; it can exceed 1 m/s, with an important lateral velocity shear. This lateral shear may generate instability (meandering) of the longshore current [50, 338, 142].

5.3.4. *Longshore Transport*

The longshore current causes substantial sediment transport along the coast. This is not only due to its strength, but also due to the fact that a lot of sediment is suspended in the surf zone. The longshore transport depends on the wave incidence angle θ_{br} in the same way as the longshore current (5.22),

$$q_{br} = q_0 \sin(2\theta_{br}). \qquad (5.23)$$

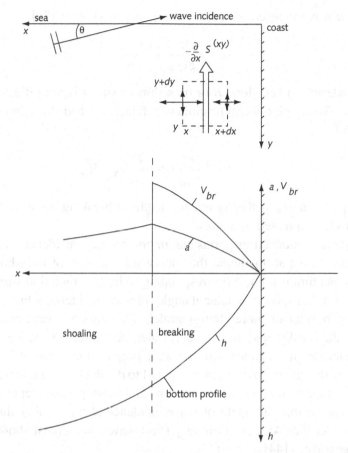

Fig. 5.14. Longshore current V_{br} in the surf zone produced by breaking of obliquely incident waves. The black arrows in the upper panel illustrate the shoreward decrease of wave-orbital velocity and the resulting longshore gradient in the radiation stress (open arrow). The shoreward decrease of wave-orbital velocity is related to the shoreward decrease of wave amplitude, according to the wave-saturation hypothesis, shown in the lower panel. The strength of the longshore current depends locally on the cross-shore gradient of the radiation stress, which is largest at the breaker line and zero at the shoreline. Outside the breaker zone the wave amplitude increases shoreward due to shoaling, but this increase produces no radiation-stress gradient, see Appendix D.4.

In the absence of reliable theoretical descriptions of breaker-induced sediment suspension, estimates of wave-induced longshore transport for practical use are derived from observations. In one of the most commonly used formulas, sediment transport is related to the longshore wave-driven energy flux. This formula for wave-induced longshore transport [m³/s] reads [70]:

$$q_0 = \frac{0.38\, ncE}{g(\rho_{sed} - \rho)}. \tag{5.24}$$

The dependency on bed slope β or on sediment size d is not reflected in this expression. Another empirical formula which includes bed slope and sediment size is [250]:

$$q_0 = 4.10^{-4} \frac{\rho}{\rho_{sed}} \tan \beta \frac{H_{br}^3}{d} \sqrt{g H_{br}}. \tag{5.25}$$

In this expression H_{br} is the significant height of breaking waves and d is the grain diameter expressed in metres.

Longshore sediment transport is maximum at a wave incidence angle of 45° at the breaker line; at this angle the shoreward transport of longshore wave-orbital momentum is highest (corresponding to highest radiation stress $S^{(xy)}$). The longshore transport is smaller at angles of wave incidence which are either more shore-normal or more shore-parallel. The angle of wave incidence is smaller at the breaker line than in deep water, due to refraction. Refraction is related to the depth dependency of the wave propagation speed; it is strongest for waves with a great wavelength compared to depth ($kh \ll 1$), in which case $c \approx \sqrt{gh}$. In the case of a straight coastline and shore-parallel depth contours, it can be shown that the angle of wave incidence at the breaker line, θ_{br}, is always smaller than 45°, even for very short waves and almost shore-parallel deep water waves [144].

A regular order of magnitude of wave-induced longshore transport is $q_{br} \approx 10^6$ m³/year. For certain coasts the angle of wave incidence fluctuates between positive and negative values; in that case the annual net longshore transport is an order of magnitude smaller.

5.4. Shoreline Dynamics

5.4.1. *Large-Scale Shoreline Adaptation*

Shoreline stability

In this section we will investigate how an equilibrium shoreline evolves after perturbation, when a constant uniform wave field is approaching an initially straight coast at a given angle θ_∞ at deep water. We assume that the cross-shore coastal profile is at equilibrium (net cross-shore transport equal to zero), hence longshore transport is the only mechanism for sediment transport in the coastal zone. The unperturbed straight shoreline is in equilibrium; due to the assumed longshore uniformity there is no transport gradient that causes erosion or accretion. Then we perturb the shoreline: We locally shift the shoreline a little forward or backward, keeping the cross-shore profile unchanged (the whole profile is shifted, see Fig. 5.15). The angle of wave incidence relative to the shoreline now varies as a function of the longshore location. This causes a longshore variation of wave-induced longshore sediment transport, producing shoreline adjustment through deposition/erosion. How will the shoreline perturbation evolve; will it remain stable, will it grow or will it decay? We will examine this question in the following.

The shoreline evolution model of Pelnard-Considère

We call h_{cl} the depth at the seaward boundary of the shoreface. Further offshore (at greater depth) the coast is assumed to be inactive: No erosion or sedimentation takes place. The depth h_{cl} is often called 'closure depth'; empirical studies indicate a robust relationship between the closure depth and the mean significant wave height, the wave-height standard deviation, the mean significant wave period and the mean bed-grain diameter [187, 321].

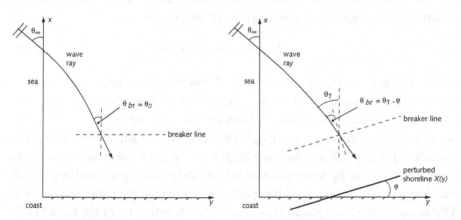

Fig. 5.15. Definition of the angles of wave incidence on a straight and on a perturbed shoreline (rotated by an angle ϕ relative to the unperturbed shoreline).

We use the notation $X(y, t)$ for the shoreline position at location y along the coast at time t; the local advance or retreat of the shoreline, X_t, is given by the following sediment balance [346]:

$$h_{cl} X_t + q_{br y} = 0. \tag{5.26}$$

Indexes t and y denote partial derivatives. The second term, representing the longshore gradient of the wave-induced longshore sediment transport, includes a factor taking into the account the pore-water fraction of the deposited or eroded sediment. Equation (5.26) is often called a 'one-line model'; in more refined 'n-line' models several shoreface strips are distinguished, each with given depth $h_1, h_2, \ldots$ and variable width $X_1, X_2, \ldots$ and mutual sediment exchange by cross-shore fluxes. In the one-line model it is assumed that the cross-shore coastal profile maintains its equilibrium shape while retreating or accreting. This is equivalent to the assumption that cross-shore sediment redistribution takes place at a time scale which is much shorter than the time scale of shoreline adaptation. In that case cross-shore sediment transport can be eliminated from the sediment balance equation by averaging over an appropriate time scale.

The wave-induced longshore transport (5.23) is a function of the angle of wave incidence at the breaker line, θ_{br}, relative to the local orientation of the breaker line, see Fig. 5.15. In the one-line model, the breaker line is assumed to be parallel to the shoreline. The angle θ_{br} depends on the deep-water wave-incidence angle, θ_∞, and on the shoreline orientation $\tan \phi = X_y$, which is a function of the longshore coordinate y. Therefore we write

$$\theta_{br}(y) = \theta_1(y) - \phi(y). \tag{5.27}$$

In the absence of refraction $\theta_1 = \theta_\infty$. However, wave refraction cannot be ignored [145]. The angle θ_1 is not only different from the deep-water incidence angle θ_∞, but also different from the breaker line angle $\theta_{br} = \theta_0$ at the unperturbed straight coast, see Fig. 5.15. Alongshore variation of the coastline orientation ϕ also affects the wave height H_{br} at the breaker line. More wave energy is propagating to portions of the coastline making a small angle to the incident wave front than coastline portions making a high angle, see Fig. 5.16. The wave-induced longshore transport (5.23) therefore takes the form [145]

$$q_{br} = q_1 \sin(2(\theta_1 - \phi)), \tag{5.28}$$

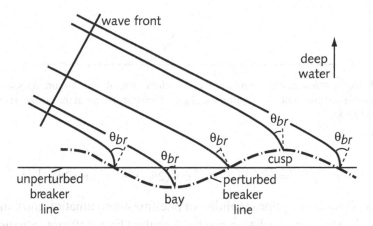

Fig. 5.16. Refraction of wave rays towards a perturbed shoreline in the case of high-angle wave incidence. The strongest convergence of wave rays occurs updrift of the shoreline cusp; at this location wave energy is highest, so at the cusp q_{1y} is negative. The wave incidence angle at the breaker line, θ_{br}, is increasing at the cusp, but this increase is much less than in the absence of refraction. Hence the gradient in the longshore transport at the cusp, $q_{br\,y} = q_{1y}\sin(2\theta_{br}) + 2q_1\theta_{br\,y}\cos(2\theta_{br})$ can be negative and produce accretion.

where not only θ_1, ϕ depend on y but also q_1, because of its dependence on H_{br}.

This equation can be simplified if it is assumed that the wavelength of the perturbation is much larger than the longshore distance of significant wave ray curvature; in this case the refraction of each wave ray only depends on the local shoreline orientation ϕ, i.e., [145]

$$\theta_1 = \theta_1(\phi), \quad q_1 = q_1(\phi). \tag{5.29}$$

This approximation is better satisfied for incident short waves than for swell. We further assume that the coastline perturbation has a much smaller amplitude than wavelength,

$$\phi \approx X_y \ll 1. \tag{5.30}$$

Substitution of the y-derivative of (5.28) in the sediment balance equation (5.26) then yields

$$X_t = D_{CL}X_{yy}, \tag{5.31}$$

Fig. 5.17. Decay of a shoreline irregularity by wave-driven longshore transport. A local shoreline perturbation is diffused along the coast if the angle of wave incidence at the breaker line is much smaller than 45°.

with

$$D_{CL} = -q_{br_\phi}/h_{cl} = \left[2q_1(1 - \theta_{1\phi})\cos(2\theta_{br}) - q_{1\phi}\sin(2\theta_{br})\right]/h_{cl}. \quad (5.32)$$

With the above assumptions (validity of one-line approximation and conditions 5.29, 5.30), shoreline evolution can be described by a diffusion equation. The diffusion coefficient (5.32) incorporates morphodynamic coupling between the incident wave field and shoreline morphology.

Decay of shoreline irregularities

For a wave field approaching an unperturbed coast ($X_y = 0$) at small angles ($|\theta_{br}| \ll 45°$) the refraction terms $\theta_{1\phi}, q_{1\phi}/q_1$ related to shoreline perturbation in (5.32) are smaller than 1. The diffusion coefficient D_{CL} is positive, which implies that a shoreline perturbation will be restored by longshore transport, see Fig. 5.17. A forward or backward shoreline displacement of length l will recover within a period of approximately $\Delta t \approx l^2/D_{CL}$. During periods of heavy wave action (storm events) longshore sediment transport is very intense. Under such conditions it is the dominant transport mechanism in the coastal zone, as assumed in the one-line model. The model predicts that shoreline irregularities present before the storm are straightened out after the storm; this is consistent with field observations [499, 283]. Wave-induced longshore sediment transport under small to moderate wave-incidence angles may thus be considered as an important shoreline stabilising mechanism.

Pocket beach

An illustration of the model is the shoreline morphology of a sandy beach confined between seaward extending piers or cliffs. As longshore transport is interrupted at the piers, morphologic equilibrium requires zero average longshore transport everywhere along the shoreline. This implies that the shoreline will tend to adjust its orientation such that the average wave incidence is always normal to the shoreline. Suppose we have an offshore wave field approaching the coast at a fixed angle θ relative to the piers. The shoreline section which does not lie in the 'shadow' of the piers will then take an orientation perpendicular to the angle of wave incidence at the breaker line (this angle is less oblique than further at sea, due to refraction). The shoreline orientation in the shadow of the piers is determined by wave diffraction around the piers, see Fig. 5.18. Theoretical shoreline curves of pocket-beaches constructed in this way are generally in good agreement with observations [214, 259].

Megascale shoreline cusps

Wave-induced longshore sediment transport has a completely different impact on the shoreline when waves approach the coast at a large angle ($|\theta_\infty| > 45°$). The magnitude of the refraction terms $\theta_{1\phi}, q_{1\phi}/q_1$ related to shoreline perturbation in (5.32) is now of order 1; these terms may therefore change the sign of the diffusion coefficient D_{CL}. This occurs for shoreline perturbations

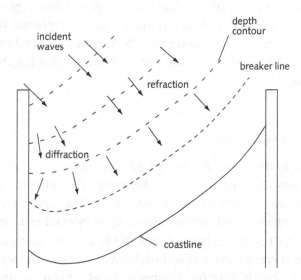

Fig. 5.18. Equilibrium shoreline for constant wave-incidence angle of a 'pocket-beach' enclosed between headlands. The shoreline is perpendicular to the direction of wave incidence, except in the shadow of the headlands.

of great wavelength (on the order of 100 times the surf zone width) [145]. Mathematically this is equivalent to a diffusion process which is reversed in time. With positive diffusion a cuspate shoreline structure will diffuse towards a smooth coastline; with a negative diffusion a smooth coastline will evolve into a cuspate shoreline. The perturbation is amplified because wave rays converge towards the updrift flank of the shoreline cusp (see Fig. 5.16). If the angle of wave incidence is large enough longshore sediment transport will have a maximum upstream of the cusp. Convergence of longshore transport at the cusp produces cusp growth. This behaviour is confirmed by numerical experiments with the one-line model and a wave incidence distribution weighted towards very oblique angles [14]. These numerical experiments show that cuspate structures appear as free instabilities and that the length scale of these structures increases with time. This scale increase is caused by the shadowing effect of large cuspate structures relative to small structures; the large structures mainly determine the shoreline curvature and the resulting angle of wave incidence, thereby limiting the development of smaller structures [14]. The scale of shoreline cusps caused by instability of longshore transport may become very large, exceeding tenths of kilometres. In the case of one prevailing direction of wave incidence the structures migrate in downdrift direction and take the form of downdrift-curved spits. It has been suggested that large-scale cuspate structures occurring along certain coastlines may be explained by this mechanism, for instance, the capes along the US Atlantic coast and the spits along the coastline of the Sea of Azov (see Fig. 5.5), both with a length scale on the order of 100 km.

Conditions for cusp growth

How common is shoreline instability? Longshore currents are produced by radiation stresses resulting from wave breaking. The relevant angle of wave incidence θ therefore refers to the breaker line. Refraction of wave rays starts already further offshore; refraction towards shore-normal wave incidence will be almost completed at the breaker line in the case of long swell waves or gently sloping shoreface. Shoreline instability will therefore occur more easily for reflective coasts than for dissipative coasts. Shoreline instability also requires deep-water wave-incidence angles which are substantially larger than 45°. Dominance of large-angle wave incidence is most probable for elongated water bodies where the fetch is largest in shore-parallel direction. In other cases it is the exception rather than the rule.

5.4.2. Shoreline Patterns

Shoreline straightening

In the previous section we have discussed large-scale long-term shoreline development under the influence of wave-induced longshore transport. Longshore transport generally tends to straighten the shoreline, especially under storm conditions. In the absence of local sediment sources and longshore gradients in the seabed profile, a strong longshore sediment transport cannot be sustained indefinitely. The shoreline will tend to an orientation approximately perpendicular to the average incident wave field — in case of a sufficiently uniform wave field. Observations show that this is indeed the most common shoreline orientation [96].

Temporary patterns

But under milder wave conditions the picture is very different [499, 97, 259, 208, 365]. A great variety of shoreline structures develops, consisting of sequences of cusps, bars and channels. Structures with different spacing, ranging from around ten metres up to a few kilometres, may coexist and different bar orientations relative to the shoreline are possible. A common feature is the presence of so-called 'rip cells', circulations in the surf zone with spatially alternating shoreward and seaward currents. These features strongly depend on wave conditions and they are therefore quite variable and generally short-lived. Their emergence is often a matter of only days, or even less, and they may disappear as quickly. An illustration of nearshore seabed morphology at the Dutch coast is shown in Fig. 5.19. The high longshore variability is characteristic of low-energy summer conditions; in winter most of the longshore variability is removed and the cross-shore pattern of shore-parallel bars (breaker bars) is enhanced [473].

Segregation of shoreward and seaward water motions

Several mechanisms have been identified which may be responsible for the formation of these patterns. Some of these mechanisms will be discussed in

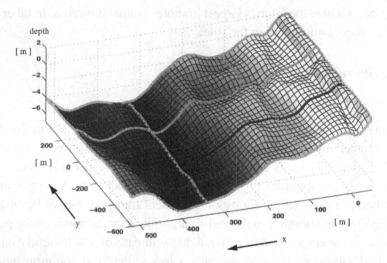

depth
[m]

[m]

[m]

y

x

Fig. 5.19. Nearshore seabed morphology at Egmond (The Netherlands) in summer (17 August 1999), from [473]; notice the different cross-shore and longshore scales. The shore-normal grey line indicates a rip channel. The rip channel ends in a shoal (just landward of the longshore trough), formed by rip-head deposits. Other weaker shore-normal structures are visible as well. In winter the longshore bars are straight with little longshore structure; in summer they are broken up in a sequence of crescentic bars.

below. A common denominator of these mechanisms is their tendency to separate, along the coastline, zones of net shoreward water motion from zones of net seaward water motion. The cross-shore seabed profile of these zones is different. The adjustment of the seabed profile is enhanced by feedback mechanisms: The bed profile is shaped by asymmetry between shoreward-seaward motions such that spatial separation of shoreward-seaward motions is subsequently strengthened.

Shoreline instability

The remarkable regularity of shoreline patterns suggests that resonant wave conditions are responsible for their generation. For a long time, explanations of shoreline patterns have been based on this assumption [48, 259]. It has become increasingly clear, however, that the major reason for the emergence of shoreline patterns is to be found in spontaneous symmetry-breaking, caused by instability of the straight unperturbed shoreline. This means that even an infinitesimal initial disturbance may experience exponential growth through positive feedback with the flow pattern. Three types of shoreline instabilities will be discussed. First we will look at instabilities arising from interaction between the seabed and the incident wave field, then we will consider shoreline instabilities caused by interaction between the seabed and the longshore current and finally we will discuss bed-flow interaction in the swash zone. Our understanding of shoreline pattern generation has been greatly improved during the past decade by the theoretical work of the Catalan morphodynamics school of Falqués and co-workers; many of the ideas exposed in the following pages are based on their work.

Bar generation at normal wave incidence

In our previous discussion of wave set-up (Sec. 5.3.3) we have assumed the absence of residual flow. The radiation stress gradient perpendicular to the shore is exactly balanced by the pressure gradient induced by the average sea surface slope. However, a local disturbance of this balance would immediately lead to a local excess pressure gradient of either the radiation stress or the surface slope. Such an excess pressure gradient drives a flow which affects sediment transport and thereby the bed profile. However, a local change in seabed profile will modify the radiation stress, because the height and orbital velocities of breaking waves depend on depth. The key question to be answered is, whether the modification of the radiation stress opposes or reinforces the original excess pressure gradient. In the first case the process of seabed adjustment will be reversed, in the last case the process will continue.

Feedback for shore-normal bar generation

The feedback process is sketched in Fig. 5.20. It appears that the response of the seabed to a disturbance can yield a positive feedback in the case of normal wave incidence [143]. This can be understood as follows. A low-crested initial shoal in the surf zone situated on an otherwise uniform sloping shoreface will enhance wave breaking and therefore strengthen locally the shoreward radiation stress gradient. The excess radiation stress will drive a current shoreward over the shoal, compensated by a seaward return current at both sides of the shoal. The shoreward current carries sediment from the outer surf zone, where sediment concentrations are high due to intense wave breaking, to the inner surf zone,

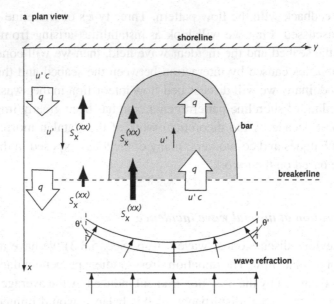

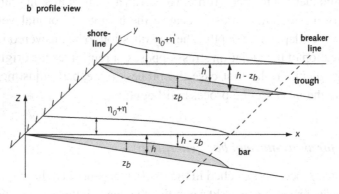

Fig. 5.20. Schematic representation of bar formation at normal wave incidence, plan view in (a) and profile view in (b). Wave breaking enhances the radiation stress $S^{(xx)}$ at the seaward edge of the bar, inducing a shoreward current u_0 over the bar and a shoreward sediment flux q. The suspended load is highest near the breaker line and decreases in landward direction. The resulting convergence of the shoreward sediment flux produces accretion of the bar. The opposite occurs in the trough neighbouring the bar. Wave breaking on the bar is enhanced by refraction of incident waves towards the bar.

where sediment concentrations are lower. The difference between high sediment input from offshore and lower sediment output shoreward will make the shoal grow towards the shore. This growth is enhanced by wave refraction. The incident wave field is refracted to the shoal, according to Snells' law, and therefore enhances the shoreward radiation stress gradient. As a result, the feedback mechanisms generate shore-normal bars extending from the beach throughout the surf zone. Sequences of such shore-normal bars may be observed on natural beaches, in periods of moderate wave intensity [365].

Linear stability analysis

An idealised model illustrating the interaction between seabed profile and incoming wave field has been proposed by Falqués, Caballeria and co-workers [143, 61]. This model is based on the equations describing wave-induced residual flow (D.22–D.24). The initial situation is a plane beach and shoreface, with uniform slope $h_x \approx \beta$. This equilibrium morphology is disturbed by a sequence of bars and troughs with longitudinal spacing $\lambda = 2\pi/k$ and amplitude f, which is of order ϵh, $\epsilon \ll 1$. The bottom perturbation is represented by

$$z_b(x, y, t) = f(x, t) \cos ky. \tag{5.33}$$

The residual flow induced by the perturbation is represented by u', v' and the induced perturbation of the surface inclination by η'_x, η'_y. The total wave-averaged water depth D is modified by $D' = \eta' - z_b$. Momentum is dissipated only through bottom friction; lateral mixing is not considered. The bottom friction term is linearised, assuming that residual velocities are much smaller than wave-orbital velicities. The linear friction coefficient r depends on the amplitude of the wave-orbital velocity, but will be taken constant for simplicity. The perturbed flow is then to first order in ϵ described by the equations

$$g\eta'_x + 2ru'/D + \left([S^{(xx)}_x + S^{(xy)}_y]/\rho D\right)' = 0, \tag{5.34}$$

$$g\eta'_y + rv'/D + \left([S^{(yy)}_y + S^{(xy)}_x]/\rho D\right)' = 0, \tag{5.35}$$

$$(Du')_x + (Dv')_y = 0. \tag{5.36}$$

Residual flow over the shoal

The modification of the radiation stress S' is due to the modification of water depth D' and to modification of the angle of wave incidence. Refraction of the incoming waves causes a deviation θ' from shore-normal incidence. The first order estimates follow from the expressions (5.18), after substitution of wave saturation (5.20),

$$
\left(\frac{S^{(xx)}{}_x}{\rho D}\right)' = \frac{g\gamma^2}{8}\left(2D'_x\left(\frac{1}{2} + \cos^2\theta'\right) + D(\cos^2\theta')_x\right)
$$

$$
\approx \frac{3g\gamma^2}{8}(\eta'_x - z_{bx}),
$$

$$
\left(\frac{S^{(xy)}{}_y}{\rho D}\right)' = \frac{g\gamma^2}{8}\left(2D'_y\sin\theta'\cos\theta' + D(\sin\theta'\cos\theta')_y\right) \approx \frac{g\gamma^2}{8}D\theta'_y,
$$

$$
\left(\frac{S^{(yy)}{}_y}{\rho D}\right)' = \frac{g\gamma^2}{8}\left(2D'_y\left(\frac{1}{2} + \sin^2\theta'\right) + D(\sin^2\theta')_y\right)
$$

$$
\approx \frac{g\gamma^2}{8}(\eta'_y - z_{by}),
$$

$$
\left(\frac{S^{(xy)}{}_x}{\rho D}\right)' = \frac{g\gamma^2}{8}\left(2D_x\sin\theta'\cos\theta' + D(\sin\theta'\cos\theta')_x\right)
$$

$$
\approx \frac{g\gamma^2}{8}(2\beta\theta' + D\theta'_x),
$$

where we have assumed $\theta' \ll 1$. The shoal dips at the seaward flank, i.e., $z_{bx} < 0$. We may assume that the surface perturbation η' is smaller than the bottom perturbation z_b; it then follows that at the shoal $S'^{(xx)}{}_x > 0$. The incident waves are refracted towards the shoal; this implies that at both sides of the shoal $\theta'_y > 0$, or $S'^{(xy)}{}_y > 0$. It then follows from (5.34) that on the shoal $u' < 0$; hence, the flow on the shoal is shoreward.

Growth of the shoal

The implication of shoreward flow over the shoal for accretion or erosion can be derived from the sediment balance equation (5.64),

$$
z_{bt} + \langle\vec{\nabla}.\vec{q}'\rangle = 0, \tag{5.37}
$$

where $\vec{q}'$ is the sediment flux related to the perturbation and where $\langle\ldots\rangle$ stands for averaging over the wave period. We use the formula for suspended-load transport in a wave-dominated environment (3.46) and ignore bottom-slope effects. Then we average over the wave period and ignore contributions from wave asymmetry and undertow, relative to the maximum near-bottom wave-orbital velocity $U_1 \approx ac/h$. This yields, to first order in ϵ

$$\langle \vec{q}' \rangle = \langle \vec{q_s}' \rangle = \alpha(4u', v'), \quad \alpha = \frac{4}{3\pi} \frac{\varepsilon_s \rho c_D}{g w_s \Delta \rho} U_1^3. \tag{5.38}$$

In the case of wave saturation we have $U_1 \approx \frac{1}{2}\gamma_{br}\sqrt{gh}$. If we assume that the longshore component of the perturbed flow, v', is small compared to the cross-shore component, u', then the bottom evolution equation (5.37) can be written

$$z_{bt} + 4\frac{d(\alpha u')}{dx} = z_{bt} + 4\alpha u' \frac{d\ln(\alpha/h)}{dx} = 0. \tag{5.39}$$

For the second equality we have used the continuity equation, which to first order in ϵ reads: $d(hu')/dx = 0$. In the case of wave saturation we have $\alpha/h \propto \sqrt{h}$. The off-shore increase of the wave-stirring factor α is therefore such that the factor $d/dx \ln(\alpha/h)$ is positive. Because at the shoal crest $u' < 0$ the shoal will grow, according to (5.39). This confirms the result anticipated from the previous qualitative discussion.

Shore-normal bars

Gravity-induced down-slope transport introduces an additional positive term in $\langle \vec{\nabla}.\vec{q}' \rangle$, which increases almost quadratically with the longshore wavenumber of the perturbation. This implies that cross-shore bars with a large longshore wavenumber (i.e., small wavelength) are suppressed, in spite of increasing wave refraction. Maximum growth of seabed perturbations occurs for two different types of bars, as shown by numerical experiments of Caballeria *et al.* [61] with the idealised model (they also included the influence of the bar on the location of the breaker line). At low wave intensity, shore-normal bars are found with a spacing on the order of the width of the surf zone or less. These bars extend from the beach throughout the surf zone, see Fig. 5.21. The length of the bars can be attributed to the effect of wave refraction; strong wave refraction causes shoreward flow over the bar (and thus accretion) from the breaker line up to the beach.

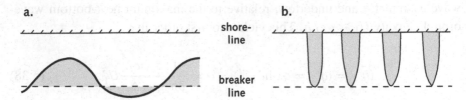

Fig. 5.21. Schematic representation of a crescentic bar-pattern (a) and a pattern of transverse, shore-normal bars (b). The bars are grey, troughs are white. Wave-refraction is more intense in the case of shore-normal bars, because of their smaller longshore wavelength and steeper slope. Therefore shoreward flow over the shore-normal bars is stronger than over the crescentic bars. For this reason the shore-normal bars are better developed close to the shoreline than the crescentic bars.

Crescentic bars and rip channels

At moderate wave intensity shore-normal bars have a greater spacing, on the order of four times the width of the surf zone. These bars do not extend throughout the surf zone, as wave refraction is not strong enough to produce shoreward flow over the bar up to the beach. The bars are limited to the outer part of the surf zone, just landward of the breaker line; their alongshore width is therefore greater than their cross-shore length. The resulting seabed morphology strongly resembles the crescentic bar-pattern which is frequently observed on the shoreface in periods of moderate wave action, succeeding a storm period, see Fig. 5.21. If it is assumed that outside the surf zone the wave-stirring function α ceases to grow as a function of offshore distance, then seaward of the breaker-line an opposite pattern of bars will be found. The troughs between the bars represent initial rip-channels and the shoals seaward of the breaker-line represent rip-flow deposits, also called rip-head bars. Observations show that over time the rip-head bar may become attached to the longshore bar; these sediments may then be transported shoreward and contribute to constriction of the rip channel [54]. Numerical simulations indicate that the morphologic characteristics resulting from initial instability are more or less preserved upon further development of the bar-pattern. This is confirmed by laboratory experiments [186] and numerical model investigations [106] of rip channel development in a coastal system with pre-existing longshore bar. These experiments are consistent with the most common field situation, where rip channels develop under quiet conditions, after periods of strong wave action in which a straight longshore bar has formed. The preferred wavelength of the rip-channel

Fig. 5.22. Rip cells along the coast of North-Holland. Both the outer longshore bar and the inner bar are interrupted by rip channels at regular intervals. Wave breaking at the bars and the resulting water level elevation behind the bars drive the rip current. Waves travelling shoreward through the rip channel of the outer bar break near the shoreline; the resulting water level elevation drives a longshore circulation which forms a bay at the rip location and a shoal behind the outer breaker bar, see Figs. 5.23 and 5.24. Local suppression of wave-induced longshore transport behind the outer bar may also contribute to shoreline accretion behind the bar, in the case of oblique wave incidence.

pattern is on the order of 10 times the distance between the shoreline and the first bar.

Rip cell dynamics

Incoming waves break more strongly at the bar than at the rip channel; the result is a greater set-up of the average water level just behind the bar than in the rip channel. This water level difference drives a residual circulation from behind the bar towards the rip channel, see Fig. 5.24. The momentum advected by this circulation towards the rip channel produces a strong acceleration of the rip current; the rip flow may therefore run far offshore, as a

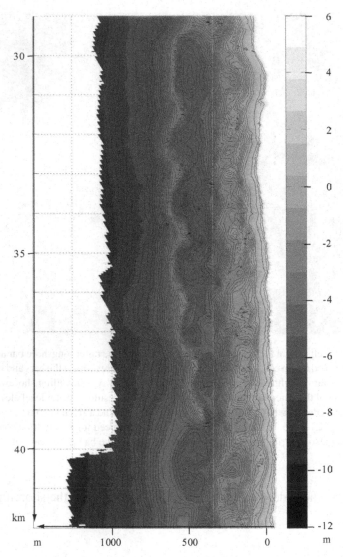

Fig. 5.23. Bathymetric map of the shoreface along the coast of North-Holland (at the right), viewed from south (bottom) to north (top) [421]. It is based on bottom soundings in cross-shore transects with 200 m alongshore spacing. An aerial photograph is shown in Fig. 5.22. The outer bar exhibits a pattern of shoreward pointing cusps opposite to seaward pointing cusps of the inner bar. The cusp spacing is on the order of one to two kilometres and the average distance between the inner and outer bar is on the order of 300 m. Small seabed depressions in the outer bar are visible at both sides of the cusps of the outer bar; these depression likely correspond to rip channels. The alongshore resolution of the bathymetric survey is not sufficient to define with precision the depth and width of the rip channels. According to the model of Falqués *et al.* the cusps correspond to deposits of wave-driven net shoreward flow and the crescents to deposits of net seaward return flow.

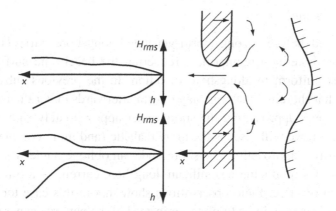

Fig. 5.24. Schematic representation of circulation in a rip cell. Wave breaking at the longshore bars and at the shoreline behind the rip channel increases locally the water level; the resulting water level slopes drive a circulation which feeds the rip current.

narrow meandering jet. However, incoming waves will be refracted to the centre of the rip current. The rip current opposes wave propagation and therefore increases wave steepness; waves will break close to the rip channel [185, 503]. The corresponding radiation stress produces a water level set-up in the channel which counteracts the development of the rip flow. Finally the rip current may take on a slowly pulsating behaviour in response to increased refraction and wave breaking when the current strengthens and decreased refraction and breaking when the current weakens.

Flash rips

We should also mention that there are rip currents which are not related to topography. These currents have a very short lifetime, only minutes, and appear randomly in the surf zone [435, 319]. Several mechanisms involving wave-current interaction have been proposed which might be responsible for the generation of these flash currents. One of these mechanisms relates to the instability of the cross-shore balance between radiation stress and wave set-up, caused by an incipient seaward current which locally attenuates offshore wave heights. This reduces the radiation stress gradient opposing the incipient current, which therefore will 'flash'. The flash is offset by fluctuations in the incident wave field [319]. A similar form of feedback may also contribute to the amplification of topography-related rip currents. This last category of rip currents is more persistent than the flash rips because they are sustained by topography.

Oblique wave incidence

As mentioned before, strong storm-generated longshore currents normally smooth out shoreline-irregularities, rendering the beach and surf zone profiles almost uniform in alongshore direction. In the previous subsection we have seen that shoreline patterns may reform afterwards under the influence of moderate waves approaching the shore under approximately right angles. In this subsection we will discuss the more realistic (and more complex) case of moderate waves approaching the shore under an oblique angle which is sufficiently large for producing a significant longshore current — a situation often occurring in practice. Field observations show that in this case too shoreline patterns may emerge, in addition to, or instead of the patterns generated by the shore-normal component of the radiation stress. These patterns may be produced by feedback of seabed irregularities to the longshore current and to the radiation stress; feedback to wave-refraction also plays a role. The resulting bedforms are bar-shaped, with oblique orientation to the shore, see Fig. 5.25. Such bars are frequently observed in the field; their spacing is typically on the

Fig. 5.25. Oblique, down-drift oriented bars along the beach of Terschelling. The picture was taken at low water.

order of the width of the surf zone. These bars run offshore from the beach, or offshore from the first longshore breaker bar; in the latter case the longshore spacing is greater than in the former [260].

Generation of oblique bars

The generation mechanism of oblique bars that will be discussed here, is similar to the generation mechanism of shoreface-connected ridges, see Sec. 3.4.5. It is triggered by the deflection of the longshore flow when crossing the oblique bar. The shore-normal component of the deflected longshore current is either shore-ward, in the case of an offshore down-drift oriented bar, or seaward, in the case of an offshore up-drift oriented bar. This shore-normal component will trans-port sediment either shoreward or seaward. Convergence of this shore-normal sediment flux at the crest of the bar will produce bar growth and divergence will cause bar decay. In the case of shoreward transport, convergence will occur if the sediment load decreases shoreward. In the case of seaward transport, con-vergence will occur if the sediment load decreases seaward. In theory, both situations are possible, but in practice the former is more likely in the surf zone. If we assume that the sediment load depends on wave stirring, then the decrease of wave height shoreward causes a decrease in sediment load. As a consequence, the seabed will be unstable when affected by a bar-perturbation with down-drift orientation in offshore direction. The principle is shown in Fig. 5.26.

Feedback between wave breaking and refraction

The generation of an offshore down-drift oriented bar also receives positive feedback from the radiation stress, at least at its offshore extremity, close to the breaker line. Increased wave breaking on the bar produces an additional gradient in the shore-normal radiation stress and a corresponding shoreward directed force. This will enhance the shoreward flow over the bar and enhance the shoreward convergence of the sediment flux. Near to the shore, however, wave height and radiation stress are decreased; this decrease counteracts the shore-ward deflection of the longshore current over the bar. Therefore the oblique bar will be best developed in the surf zone, near the breaker line. Wave refraction also yields positive feedback, as it will orient incident waves towards the bar, and therefore enhance wave breaking and radiation stress.

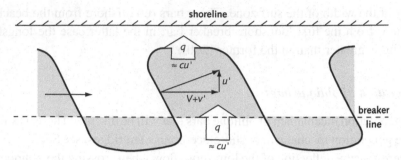

Fig. 5.26. Schematic representation of bar formation by a longshore current (bar is grey: Accretion, trough is white: Erosion). The longshore current is deflected landward over an offshore down-drift oriented bar (the landward velocity component u' is the combined effect of flow continuity, bottom friction torques, vorticity conservation and radiation stress). The bar will accrete if the landward flux of sediment caused by flow deflection is larger offshore than onshore, i.e., if the load of suspended sediment is larger offshore than onshore. This is the case in the surf zone, where a bar may consequently develop. The reverse is true outside the surf zone, where a negative bar (trough) is created. The shoreward flow induced by radiation stress (wave breaking on the bar) is largest near the breaker line; near the shore, the radiation stress on the bar is suppressed and shoreward flow may change to seaward flow. The bar is best developed just landward of the breaker line.

Bar pattern mirrored by the breaker line

The shoreward flow extends over some distance seaward of the surf zone. However, here the sediment load is not decreasing shoreward. The shoreward sediment flux seaward of the breaker line is therefore not converging but diverging. This divergence of the sediment flux produces a trough in the seabed, situated in the seaward extension of the bar. A shoal will form offshore in the extension of the trough neighbouring the bar. The resulting bedform is schematically depicted in Fig. 5.26.

Linear stability analysis

The previous qualitative description is illustrated below by a simple model of the morphodynamic feedback between a seabed-perturbation and the longshore current, proposed by Ribas *et al.* [374]. An originally plane sloping seabed is perturbed by a low, offshore down-drift oriented bar, which deflects the longshore current shoreward. The unperturbed longshore current $V(x)$ is given by (5.22), the shoreward deflected flow is designated u' and the perturbation of the longshore current v'. We assume that these flow velocities are much smaller than the wave-orbital velocity amplitude U_1 (dominance of wave-induced currents over steady currents). For the sediment flux we use the formula (3.46) for wave-dominated suspended-load transport, which in this case can be approximated by (see also 5.51)

$$\langle \vec{q}_s \rangle = \alpha(4u', V + v'), \quad \alpha(x) = \frac{4}{3\pi} \frac{\varepsilon_s \rho c_D}{g w_s \Delta \rho} U_1^3. \tag{5.40}$$

Here we have averaged over the wave period and we have neglected gravity effects. This expression is substituted in the sediment balance equation (5.37). The perturbation of the longshore current v' is eliminated from the expression of the sediment flux gradient by using the continuity equation, which to first order ($|z_b| \ll h$, $|v'|$, $|u'| \ll |V|$) reads

$$h(u'_x + v'_y) + u'h_x - V z_{by} = 0. \tag{5.41}$$

After a few manipulations we find for the sediment balance

$$z_{bt} + \frac{\alpha V}{h} z_{by} = -3\alpha u'_x - \alpha u' \frac{d \ln(\alpha^4/h)}{dx}. \tag{5.42}$$

The first term in the left member represents the time evolution of the bar and the second term stands for longshore migration; the migration velocity is given by $\alpha V / h$. The terms at the right-hand side indicate growth (decay) of the bar, if they are positive (negative) for positive z_b. In the surf zone we have shoreward flow over the bar ($u' < 0$), which is decreasing shoreward ($u'_x < 0$). The wave-orbital velocity decreases in landward direction in the case of saturated wave breaking ($U_1 \approx ac/h \approx \frac{1}{2}\gamma_{br}\sqrt{gh}$). The x-derivative of the factor $\ln(\alpha^4/h)$ is thus positive, as it varies approximately according to h^5. Altogether, the sediment balance equation indicates growth at the bar crest in the surf zone. Just outside the surf zone, in the seaward extension of the bar, the flow is still shoreward, but increasing in landward direction ($u' < 0$, $u'_x > 0$). The wave-stirring factor α now increases in landward direction, thus the gradient of $\ln(\alpha^4/h)$ is negative. This implies that $z_{bt} < 0$: Just outside the surf zone in the extension of the bar a pool will develop. These results corroborate the earlier qualitative description of oblique bar growth.

Field evidence

Observations do not provide hard evidence about the presence and behaviour of transverse bars. With similar longshore currents transverse bars may either be present or absent and the migration of the bars is observed either in downdrift direction or in updrift direction [260]. It therefore seems that in practice longshore currents cannot be solely responsible for the formation of transverse bars.

Tidal influence

The timescale for generation of longshore patterns, such as rip channels and shore-normal or oblique bars, is on the order of one or a few days [260]. The generation process may therefore strongly interfere with tidal motion. Especially at macrotidal beaches the surf zone is swept forth and back in cross-shore direction. Observations show that bar-type longshore patterns develop both at microtidal and macrotidal beaches [65]; no systematic differences have been reported in literature [374]. This supports the idea that tidal motion does not produce a basic modification of the generation mechanism of longshore bar patterns. The instantaneous flow pattern does depend on the tidal phase, however. For instance, observations of rip currents systems show that the rip current is strongest at low tide, when the rip channel depth is smallest, and weakest at high tide [54].

Summary of alongshore patterns in the surf zone

Even if alongshore gradients are absent (or insignificant) in the incident wave field, the alongshore coastal profile may exhibit significant structure. Observed alongshore patterns in the surf zone have been described as crescentic and transverse (normal or oblique) bars corresponding to horizontal circulation cells. In the case of almost shore-normal wave incidence, physical mechanisms responsible for these patterns are related to the instability of seabed morphology through feedback to wave breaking, wave refraction and the resulting wave set-up. In the case of oblique wave incidence, pattern formation is related to the instability of seabed morphology through feedback to shoreward/seaward deflection of the longshore current. The alongshore scale of these patterns is on the order of one hundred meters up to a few kilometres.

These patterns develop under low to moderate incident waves; the patterns are normally wiped out by high-energy waves. We have considered only depth-averaged flow; asymmetry in shoreward/seaward orbital wave motion has been ignored. We may expect that this assumption does not hold for high waves. Wave breaking on shoals in the surf zone produces near bottom undertow; this seaward flow counteracts the shoreward flow produced by the shoreward decrease of radiation stress. This may explain why alongshore patterns do not develop under high-energy conditions, even for shore-normal wave incidence.

In the foregoing we have considered an idealised wave field consisting of monochromatic and unidirectional waves. Realistic wave fields are more complex and generally consist of waves with different periods and directions. The spreading in wave period and wave direction introduces modulations in the wave amplitude in longshore and cross-shore directions (wave groups), with much larger time and space scales than the peak wave period and peak wave length of the incident wave field. There is evidence from field observations and mathematical models that patterns in the incident wave field due to spreading in wave period and wave direction may generate longshore morphologic patterns at the same scale as observed crescentic bars and rip cell patterns [372].

5.4.3. Beach Cusps

Alongshore patterns in the swash zone

Whereas the previous processes are acting mainly in the surf zone, a similar segregation between zones of dominating shoreward flow from zones and dominating seaward flow also occurs on the beach, in the swash zone. This process relates to uprush and downrush flow after final wave breaking and consists of a separation between zones of dominating uprush from zones of dominating downrush [387]. The swash circulation gives rise to the development of cuspate patterns, called beach cusps, with a longshore wavelength of typically ten to several tens of metres, see Fig. 5.27. These patterns exhibit a great regularity in form and spacing and may extend over coastal stretches of several hundreds of metres.

Beach cusp morphology

The morphology of beach cusps is characterised by a sequence of steep, cuspate cross-shore bars (horns), separated by gently sloping embayments. In the developing phase of these structures the swash flow is concentrated at the cusps, while the backwash flow is concentrated in the embayments, see Fig. 5.28. This flow pattern stimulates the development of the beach-cusp pattern. The

Fig. 5.27. Sequence of beach cusps on a shingle beach situated in a small embayment of the exposed, mesotidal coast of Asturias (northern Spain). Top: Wave uprush. Bottom: After downrush.

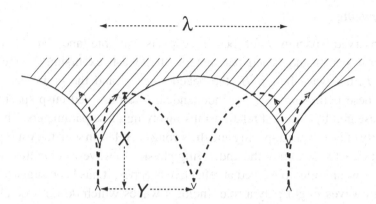

Fig. 5.28. Schematic representation of a beach cusp and the residual flow pattern during the constructional phase. The backwash is diverted to the cusp bays and blocks the following swash towards the bay if the uprush-downrush duration is equal to the wave period. Uprush is then concentrated at the cusp horns.

sediment-laden uprush flow nourishes the horn, especially with the coarsest sediment fraction; at the end of the uprush the swash flow has lost not only its kinetic energy, but also most of its (coarse) sediment. The backwash flow brings sediment in suspension from the embayment, which consequently erodes. The circulation from horn to embayment gets stronger while the cusp structure is building up, generating a positive feedback to beach-cusp development. A beach-cusp pattern may develop in less than one day [303].

Field observations

Beach cusps are most often observed under moderate wave conditions on reflective beaches with a moderate to steep slope and coarse sediment. Beach cusp patterns even occur on beaches covered with boulders. Energetic swash flow and permeable substrate are clearly favourable for beach-cusp development. Beach-cusp circulation is also strengthened when the swash period (period of uprush and downrush) is close to the average period of the incoming waves. In that case the downrush flow counteracts the following swash flow uprush; the uprush is concentrated at the horn and almost entirely absent in the embayments. However, the beach-cusp pattern decays when the incident wave field becomes highly irregular and when the angle of incidence becomes oblique, generating longshore transport. Beach cusps never develop at weakly sloping, dissipative fine-grained beaches.

Cusp spacing

The observed spacing λ of beach cusps is variable and can be related
to several swash and beach parameters: the swash excursion X, the wave
period T, the wave height H, the beach slope β and the grain size d [78,
428]. These parameters are not independent; the observed cusp spacing can
be represented by different relationships involving these parameters. The great
regularity of beach cusp spacing and the strong correlation with the wave period
has first directed search of the underlying physics to wave-forcing theories. As
beach cusps are often observed at reflective beaches, it has been suggested that
reflected waves might play a role. Incident waves which do not lose all their
energy by breaking, but which are partially reflected, may be trapped at the coast
by refraction. Such waves are called edge waves; their occurrence on reflective
beaches has been established experimentally [224]. When such waves propagate along the coast in opposite directions they may form a standing wave.
The standing wave modulates the swash flow of the broken incident wave; the
alongshore wavelength of this modulation is constant and given by [184]

$$\lambda = (g/2m\pi)\beta T^2, \tag{5.43}$$

where $m = 1$ for synchronous edge waves and $m = 2$ for sub-harmonic
edge waves. This relationship yields a fair representation of observed beach
cusp spacing. Nevertheless, edge wave theories have become less credible to
explain beach cusp generation, for several reasons: (1) Both field observations
and numerical models have shown that beach cusps may develop in the absence
of edge waves [205, 485, 303, 80] and (2) there is also no clear experimental
support for the linear dependence of beach cusp spacing on beach slope [428].

Field observations and numerical simulations suggest a linear relationship
between beach cusp spacing λ and swash excursion X [485, 303, 78],

$$\lambda = (1.5 - 1.7)\, X. \tag{5.44}$$

The swash excursion is the distance over which waves run up the beach from
the toe of the cusp. If the effects of friction and percolation are ignored, it can
easily be shown that $X = U^2/(2g\beta)$, where U is the uprush velocity at the toe
of the cusp. If the uprush duration corresponds to half the wave period T, we
have $U = g\beta T/2$ and

$$X = g\beta T^2/8. \tag{5.45}$$

In this case the relationships (5.43) and (5.44) are similar. However, taking into account energy dissipation and percolation, another expression is found for the swash excursion [221],

$$X = 0.4\varphi T\sqrt{gH}, \qquad (5.46)$$

where φ is a coefficient depending mainly on grain size d (expressed in m),

$$\varphi \approx \exp(-0.005\,d^{0.55}).$$

Sunamura [428] found that field data on the spacing of beach cusps are well represented by the relationship

$$\lambda = A\varphi T\sqrt{gH}, \qquad (5.47)$$

where A is a constant with values between 0.7 and 2.

Swash-backwash asymmetry

Numerical investigations [485, 303, 79] have provided evidence that beach cusps may develop as an instability inherent to the uprush-downrush process on a sloping beach. This theory is most widely accepted at present as the primary generation mechanism. The feedback mechanism was described earlier in this section. If uprush and downrush of swash flow takes place at exactly the same location, no structure will develop and sedimentation and erosion will be almost in balance. However, even a very small departure from this symmetric situation to an asymmetric situation where uprush and downrush are slightly shifted relative to each other, will produce a circulation which initiates the genesis of a cusp pattern. The cusp-pattern will further develop by positive feedback.

The feedback model of Dean and Maurmeyer

It was first shown by Dean and Maurmeyer [100] that this feedback process produces a beach-cusp wavelength corresponding to (5.44). We will reproduce the essential features of their model below. Based on observations, they assumed that for optimal morphodynamic feedback the cusp spacing λ corresponds to twice the maximum lateral swash excursion $Y(T)$ produced by deflection of the uprush from the lateral slope of the horn into the embayment during the wave period T. The lateral swash excursion $Y(t)$ can be estimated by comparing the down-slope acceleration experienced by the deflected swash flow on the lateral horn slope to the down-slope acceleration experienced by the uprush flow on the beach slope. We call $z_b(x, y)$ the height of the horn relative to average beach level; the average lateral horn slope is approximated by $z_{by} \approx 4\hat{z}_b/\lambda$, where $\hat{z}_b$ is the average height of the horn relative to the unperturbed beach. The lateral flow deflection thus experiences an acceleration which is $4\hat{z}_b/\lambda\beta$ times the down-slope acceleration experienced by the uprush flow. The lateral excursion $Y(T/2)$ in the same time interval $T/2$ as the uprush flow can therefore be estimated at $Y(T/2) = (4\hat{z}_b X)/(\lambda\beta)$. As the excursion $Y(t)$ increases approximately as t^2 we estimate $\lambda/2 = Y(T) = 4Y(T/2) = (16\hat{z}_b X)/(\lambda\beta)$. The cusp spacing λ is this given by

$$\lambda = \sqrt{32\hat{z}_b X/\beta}. \tag{5.48}$$

Now we assume that the average height of the horn $\hat{z}_b$ can be expressed as a fraction of the average beach elevation over the uprush distance X,

$$\hat{z}_b = 0.5\varepsilon\beta X. \tag{5.49}$$

The parameter ε represents the prominence of the cusp horn relative to the beach profile. Substitution in (5.48) then yields

$$\lambda = 4\sqrt{\varepsilon}X. \tag{5.50}$$

Typical values of the prominence ε are in the range 0.1–0.3 [100, 303]. The cusp spacing predicted by the Dean-Maurmeyer model thus corresponds well with the empirical relationship (5.44). Although the model does not describe the initial emergence of beach cusps, it provides strong evidence for the generation and maintenance of beach cusps through morphodynamic feedback between swash flow and beach morphology.

5.5. Coastal Profile Dynamics

Significance of cross-shore transport

In previous sections we have focused on wave-driven longshore transport as major agent for shoreline evolution. In reality, the evolution of coastal morphology often depends as much, and sometimes even more, on wave-driven cross-shore transport. Although the magnitude of cross-shore transport is several orders of magnitude smaller than the magnitude of longshore transport, the total cross-shore sediment flux may contribute more to the large-scale coastal sediment balance, as it acts along the full length of the coast line. Cross-shore sediment transport not only influences the position of the shoreline, but in particular it determines the average shape of the shoreface profile. In fact, for an alongshore uniform coast the shoreface profile adjusts to cross-shore sediment transport such that in the long term it tends to an equilibrium where the net cross-shore transport is constant along the profile. This constant is zero if we assume that (1) there is no cross-shore transport across the seaward boundary of the shoreface, (2) there is no net sand loss from the upper beach to the dunes and (3) the average sea level is constant.

5.5.1. *Shoreface Profile Models*

Cross-shore transport processes

In this section we will examine the influence of several cross-shore transport processes on the shoreface equilibrium profile. A discussion of the influence of wave breaking is postponed to the next section. We will assume that coastal morphology and incident wave field are uniform in longshore direction. In this case, wave-driven longshore transport will produce no coastal change at all and coastal change is dependent only on cross-shore transport. The most important cross-shore sediment transport processes are [382]:

- Seaward gravity-induced transport. Gravity-induced down-slope sediment transport is incorporated in the semi-empirical expressions (3.46 and 3.45) [19, 20]. In the expression for suspended load, down-slope transport is modelled as a tendency towards auto-suspension; for bedload it is modelled as a tendency towards avalanching. For fine sediment (ratio of settling velocity to orbital velocity less than a few percent), suspension down-slope transport is relatively stronger than bedload down-slope transport [20].
- Seaward transport due to undertow (5.15). Everywhere along the shoreface there is a net seaward flow in the lower part of the water column. This return flow compensates for the shoreward mass transport between the wave-trough and wave-crest levels (Stokes transport); the return flow is strongest in the breaker zone.
- Shoreward transport due to streaming, $u_s \approx -U_1^2/c$ (3.22). Wave propagation provides a net input of momentum to the viscous wave boundary layer, driving a residual forward drift in this very thin layer. Shoreward streaming hardly occurs in the presence of bed ripples.
- Shoreward transport due to wave asymmetry (5.13). Wave propagation is influenced nonlinearly by the curvature of the surface slope; this

effect becomes significant when waves enter shallow water and when the wavelength decreases. Shoreward wave-orbital velocities are therefore higher than seaward wave-orbital velocities. This asymmetry produces a shoreward sediment transport near the bottom, but a seaward transport higher in the vertical [477, 75, 249]. This seaward transport affects in particular the fine sediment fraction, due to suspension lag, see Sec. 3.2.4. However, as the sediment concentration is much higher near the bottom than up in the vertical, the total net sediment transport due to wave asymmetry will normally be shoreward.

Bowen's cross-shore profile model

In the following we will derive the shoreface equilibrium profile from a simple cross-shore transport model, following an approach first outlined by Bowen [49]. This model is based on expressions (3.46) and (3.45) for the wave-averaged cross-shore transport. In these expressions we substitute the different processes contributing to seaward and shoreward transport. The equilibrium requirement of zero net cross-shore transport then yields an equation from which the equilibrium profile $h(x)$ can be solved. We simplify the onshore incident wave field to a single-frequency wave, with amplitude and frequency corresponding to the average wave-field energy (wave amplitude $a = H_{rms}/2$) and to the peak spectral frequency. We assume that the amplitude U_1 of the first-order wave-orbital velocity $u_1 = U_1 \cos(kx + \omega t)$ is much larger than other contributions u', for instance, from streaming or from higher-order wave contributions. In that case the wave-averaged expressions for the sediment transport formulas (3.46) and (3.45) can be approximated by:

$$\langle q_s \rangle = \frac{\varepsilon_s \rho c_D}{g w_s \Delta \rho} \left(4 \langle u' u_1^2 |u_1| \rangle + \frac{\varepsilon_s h_x}{w_s} \langle u_1^4 |u_1| \rangle \right), \qquad (5.51)$$

$$\langle q_b \rangle = \frac{\varepsilon_b \rho c_D}{g \Delta \rho \tan \varphi_r} \left(3 \langle u' u_1^2 \rangle + \frac{h_x}{\tan \varphi_r} \langle u_1^2 |u| \rangle \right). \qquad (5.52)$$

Balance of streaming and gravity

In a first model we assume that sediment is mainly transported as suspended load and that the major onshore and offshore transport mechanisms are shoreward drift by streaming in the bottom shear layer, and down-slope transport by gravity effects. The wave-averaged cross-shore transport then reads

$$\langle q_s \rangle \approx \frac{16}{3\pi} \frac{\varepsilon_s \rho c_D}{g w_s \Delta \rho} U_1^3 \left[u_s + \frac{\varepsilon_s h_x}{5 w_s} U_1^2 \right]. \tag{5.53}$$

We assume that equilibrium corresponds to $\langle q_s \rangle = 0$. Substitution of u_s and U_1 then shows that the bottom slope should fulfil the condition

$$h_x \approx \frac{5 w_s}{\varepsilon_s c} = \frac{5 w_s \omega}{\varepsilon_s g \tan h(kh)} \approx \frac{5 w_s}{\varepsilon_s \sqrt{gh}}. \tag{5.54}$$

The last approximation is valid only for very small depths (typically a few metres or less). The equilibrium bottom slope depends on the fall velocity w_s, i.e., on the sediment grain size d. If we have equilibrium for sediment with median grain size, then we have no equilibrium for the coarse sediment fraction (fall velocity w_s higher than the average). According to (5.53) the coarse sediment fraction will be transported landward ($\langle q_s \rangle < 0$). The fine sediment fraction, by contrast, will be transported seaward. This shows that sediment sorting takes place on the shoreface, with coarser sediment higher in the profile and finer sediment at greater depth. This result is due to the dependence of the suspended load on fall velocity in the transport formula (5.51). The model requires perfect sorting at equilibrium, with a specific grain size at each depth. For uniform sediment, equation (5.54) can be integrated to yield the following off-shore bottom profile

$$h(x) \approx 3.8 \left(\frac{w_s^2}{g \varepsilon_s^2} \right)^{1/3} x^{2/3}. \tag{5.55}$$

We have indicated earlier that many sandy shore-profiles are indeed fairly well described by a 2/3 power of x [99, 101]. The formula (5.55) predicts that the coastal profile will be steeper for coarse sediment (large w_s) than for fine sediment (small w_s). This feature is in agreement with field observations (see Eq. (5.3)). We should expect, however, that the above model is too simple, as, for instance, wave asymmetry cannot be neglected in shallow water. The assumption that suspended load transport mainly depends on the drift velocity in the very thin bottom shear layer may also be questioned. Sediment which is suspended higher in the water column (in particular the fine fraction) is not transported shoreward by streaming, but transported seaward by Stokes return flow (undertow), see Sec. 5.3.2.

Balance of wave asymmetry and gravity

In a more realistic model wave asymmetry is taken into account, in addition to streaming. To this end we include the second order harmonic component of the near-bottom orbital velocity

$$u' = u_s + u_2 \approx -\frac{U_1^2}{c} - U_2 \cos[2(kx + \omega t)], \qquad (5.56)$$

where

$$U_1 = \frac{a\omega}{\sinh(kh)}, \qquad U_2 = \frac{3}{4} \frac{a^2 k\omega}{\sinh^4(kh)}.$$

We disregard suspension lag effects (i.e., net seaward transport higher in the vertical) with the crude assumption that it is incorporated in the down-slope gravity term. Substitution in the suspension transport formula (5.51) yields

$$\langle q_s \rangle \approx \frac{16}{15\pi} \frac{\varepsilon_s \rho c_D}{g w_s \Delta \rho} U_1^3 \left[5u_s - 3U_2 + \frac{\varepsilon_s h_x}{w_s} U_1^2 \right]. \qquad (5.57)$$

The equilibrium profile is now given by

$$h_x \approx \frac{9}{4} \frac{w_s}{\varepsilon_s \sqrt{gh}} \left[\frac{20}{9} + \frac{1}{\sinh^2(kh)} \right]. \qquad (5.58)$$

From (5.58) we can derive that wave asymmetry effects become larger than streaming effects at water depths smaller than $h < 0.01gT^2$, where T is the wave period. For instance, along the Dutch coast under average conditions, this corresponds to $h < 4$ m. If wave asymmetry is dominant over streaming and if $kh \ll 1$, the equilibrium equation can be integrated to yield (see Fig. 5.29)

$$h(x) = 2 \left(\frac{w_s \sqrt{g}}{\varepsilon_s \omega^2} \right)^{0.4} x^{0.4}. \qquad (5.59)$$

The equilibrium profile is again described by a power law of cross-shore distance x. The exponent is smaller than in the previous case, corresponding to a greater bottom steepness near the coast.

Comparison with observed coastal profiles

The model predicts an increase in profile steepness with increasing wave period; this feature is in agreement with field observations [259]. In order to compare

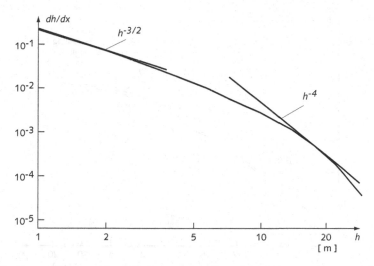

Fig. 5.29. Slope of the coastal profile h_x as a function of depth h, according to (5.58), when neglecting the contribution of streaming. If the depth profile is represented by $h(x) \propto x^n$, then the slope in the double-logarithmic plot $d \ln h_x / d \ln h = 1 - 1/n$. At small depth $d \ln h_x / d \ln h \approx -3/2$, corresponding to $h(x) \propto x^{0.4}$; at large depth $d \ln h_x / d \ln h \approx -4$, corresponding to $h(x) \propto x^{0.2}$.

the model with observed coastal profiles we abandon the assumption $kh \ll 1$. Therefore we plot the bed slope $\beta = h_x$ as a function of depth h, according to Eq. (5.58), in which streaming is neglected; the result is shown in Fig. 5.29. At small depth the bed slope varies according to $h^{-3/2}$, i.e., $h(x) \propto x^{0.4}$. But further offshore, at depths between 10 and 20 metres the bed slope is better represented by h^{-4}, corresponding to $h(x) \propto x^{0.2}$. In Fig. 5.30 the long-term averaged cross-shore profiles are shown for three locations along the Dutch coast. Close to the shore, within the surf zone, the bed slope is approximately constant, $h(x) \propto x^n$, $n \approx 1$. Seaward from the breaker zone the exponent n rapidly decreases to approximately 0.2. This is in fair agreement with (5.58), when streaming is ignored and supports the idea that wave asymmetry is mainly responsible for the sharp increase of the bed slope when moving from the outer shoreface into the breaker zone.

Bedload cannot make equilibrium

As a third example we consider cross-shore sediment transport dominated by bedload, with shoreward transport driven by streaming in the bottom boundary

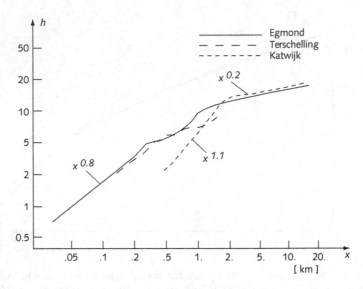

Fig. 5.30. Double logarithmic plot of coastal profiles along the Dutch coast averaged over a period of several decades [490, 473]. The locations can be found in Fig. 5.34. If the profile is written $h(x) \propto x^n$, then the exponent n corresponds to the slope of the double-log profile curve. This slope represents the concavity of the coastal profile (increasing concavity with decreasing n). At large depth, offshore the breaker zone, $n \approx 0.2$; at small depth, in the surf zone, the bottom slope h_x becomes almost constant ($n \approx 1$).

layer. The bedload transport formula may be developed in the same way as the suspension transport formula,

$$\langle q_b \rangle = \frac{3\varepsilon_b \rho c_D}{2g\Delta\rho \tan \varphi_r} U_1^2 \left(u_s + \frac{8}{9\pi} \frac{h_x U_1}{\tan \varphi_r} \right). \qquad (5.60)$$

For the equilibrium profile we then find at a first approximation

$$h_x \approx \frac{9\pi a}{8h} \tan \varphi_r \approx \frac{9\pi}{16} \gamma_{br} \tan \varphi_r, \qquad (5.61)$$

where the last approximation holds for wave saturation. In the surf zone the bottom slope approximates to the avalanching angle φ_r. Such a steep slope is unrealistic. The conclusion is that on the shoreface suspension transport has to dominate over bedload transport. It has also been suggested that suspension transport might dominate during seaward wave-orbital motion, whereas bedload transport dominates for landward wave-orbital motion.

Seaward transport due to long infragravity waves

Interference in the wave spectrum gives rise to wave groups and thereby produces gradients in the radiation stress with the wavelength of the wave-group envelope. These radiation stress gradients induce long infragravity waves [288] (see Appendix D.2 and Fig. D.1). The long waves are bound to the wave group with opposite phase; the maximum of the wave-group envelope coincides with the trough of the long wave, i.e., with seaward orbital velocity of the long wave. Wave stirring of the seabed is strongest at the maximum of the wave envelope; the long wave thus causes a net seaward displacement of sediment. Several authors have reported a significant contribution of long infragravity waves to seaward sediment transport [404, 284, 302].

5.5.2. *Influence of Wave Breaking*

Surf zone processes

In this section we will extend the previous analysis by considering the influence of wave breaking on the cross-shore profile. In the surf zone wave characteristics are strongly modified by wave breaking. These characteristics are poorly described by formulas derived from first order Stokes theory (5.13). The following additional contributions to cross-shore sediment transport have to be considered:

- *acceleration asymmetry*

When waves break, their peakedness and the corresponding velocity asymmetry are reduced. The shape of the waves changes; the front face becomes steeper and the rear face becomes more gentle [137]; this is similar to the saw-tooth distortion of long waves in shallow water. Under the wave front the orbital velocity switches in a very short time interval from seaward to shoreward; asymmetry in maximum seaward/shoreward acceleration is more pronounced than in maximum seaward/shoreward orbital velocity. Observations indicate a strong correlation of fluid accelerations (above a certain critical level) with sediment suspension [189, 235]. This suggests that in the surf zone, shoreward sediment transport induced by wave asymmetry will be stronger than predicted from velocity asymmetry alone; the strongest acceleration asymmetry is observed around the bar crests where waves start to break.

- **wave breaking**

The shoreward near-surface velocity in breaking waves is strongly enhanced; the same applies to the compensating return flow near the bottom (undertow). Plunging breakers produce a strong vortex motion at the plunging point which produces a sand fountain when touching the bottom. During storm periods net shoreward transport by wave asymmetry is offset by seaward sediment transport caused by strong undertow. The strongest seaward transport is observed at the onshore flank of breaker bars.

- **diffusion**

Wave stirring causes intense sediment suspension in the surf zone. Most sediment is suspended around the breakpoint, where wave-orbital velocities are highest. This sediment is transported and diffused away from the break-point by wave-orbital currents. Diffusive fluxes depend on the strength of the diffusion processes (diffusivity) and on the concentration gradient of suspended sediment. Redistribution of sediment in the surf zone is determined by the cross-shore gradient of the diffusive transport; generally most transport is shore-ward [38].

Average surf zone profile

For many sandy beaches the average surf zone profile can be roughly repre-sented by a power law, $h(x) \propto x^{2/3}$, according to observations (5.1). It appears that such a profile satisfies the condition of uniform energy dissipation per unit water volume in the surf zone [99]. The energy dissipation per unit volume within the surf zone for saturated breaking waves is given by (see 5.20)

$$\frac{1}{h}(cE)_x \approx \frac{1}{h}\left(\sqrt{gh}\frac{1}{8}\rho\gamma_{br}^2 gh^2\right)_x = \frac{5}{24}\rho\gamma_{br}^2 g^{3/2}(h^{3/2})_x. \tag{5.62}$$

If we assume that this quantity is independent of x, then the profile $h(x)$ can be obtained by integration; this yields the 2/3 power law. However, there is no apparent physical reason for uniformity of the energy dissipation per unit volume. The energy dissipation observed at laboratory experiments is much higher than predicted by (5.62) around the breaker line and much lower in the surf zone [482].

Longshore bars

The energy-dissipation model yields a smooth surf zone profile. In reality, however, the coastal profile in the surf zone is not smooth; it generally exhibits large undulations, corresponding to bars running approximately parallel to the coastline. These longshore bars migrate in both cross-shore directions; the net perennial migration is typically seaward [490]. Because of this migration the seabed undulations are filtered out when averaging the coastal profile over a period of sufficient length. The required averaging period is on the order of several decades, because bar migration is a slow process. The presence of longshore bars masks the average shape of the surf zone profile. Inspection of the long-term averaged profile reveals that the average surf zone profile has a more complex shape than a simple power-law. Some portions of the profile are not concave, but terraces and convex portions are also present [498, 39].

Field observations

The presence of longshore bars is an almost universal characteristic of the coastal profile at depths up to ten metres or less. The cross-shore spacing of these bars is in the order of one hundred or a few hundred metres and their height is typically a few metres. There is strong evidence that these bars are associated with shoaling and breaking of high energy waves. For that reason they are also called breaker bars. Breaker bars are highly dynamic features; their location and their height can strongly change on time scales of months or even weeks [15]. The bar morphology in the surf zone is particularly sensitive to storms. A seasonal variation of bar morphology is related to alternating periods of high and low wave activity. Breaker bars normally grow and migrate seaward during periods of high wave activity [166]; during periods of low wave activity breaker banks generally decay and migrate shoreward [15]. As mentioned above, the bars often exhibit a long-term trend of seaward migration, see Fig. 5.31 [490]. Breaker bars may extend along the shore over distances of several kilometres or even tens of kilometres. This suggests that they have essentially a two-dimensional character and that their dynamics are primarily related to cross-shore sediment transport. It should be noted, however, that under low-energy conditions breaker bars may break up in structures of smaller longshore extent with a crescentic shape, as discussed earlier.

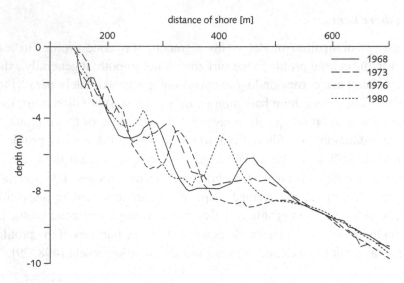

Fig. 5.31. Coastal profiles at Egmond (The Netherlands, tidal range ≈ 2 m) in the period 1968–1980. In the surf zone (depth below approximately 7 m) two or three subtidal bars are present in the coastal profile, all exhibiting net seaward migration. From [473].

Morphodynamic feedback

Observations show that bar formation takes place at cross-shore locations where the vortex motion generated by wave breaking touches the bottom [506]. There is also strong evidence that bar formation produces feedback to wave breaking. It has been observed, for instance, that under breaking waves the undertow is maximum just onshore of the sandbar crest, indicating feedback between bar morphology and undertow [442, 166]. This suggests that bar formation may be considered as a self-organising process in the surf zone. In the following we will adopt this view, and present a model which explains bar formation as the result of nonlinear interaction between wave and seabed dynamics. This model is based on a qualitative description of net wave-induced sediment transport on the shoreface ignoring the details of wave-transformation dynamics on the shoreface.

Inflection of the shoreface profile

Seaward of the breaker zone, the equilibrium shoreface profile follows a smooth curve with shoreward steepening seabed. However, in the zone where waves

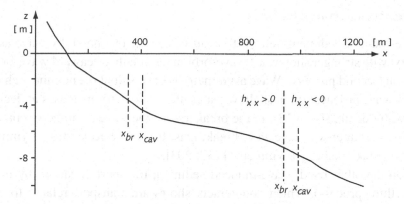

Fig. 5.32. Coastal profile at Egmond averaged over a period of 34 years. The profile changes from concave to convex at two locations. One inflexion point is located at the offshore boundary of the surf zone (depth 7 m) and another some 500 m further onshore at a depth of approximately 4 m. Other long-term averaged profiles along the Dutch coast exhibit a similar behaviour. From [386].

start breaking, the steady shoreward increase of the slope is interrupted. This is true not only for the instantaneous barred profile, but also for the long-term averaged (bar-filtered) profile. At some distance x_{br} from the shore, the bar-filtered profile has an inflexion point ($h_{xx} = 0$) with onshore a locally smaller slope than offshore. Closer to the beach the slope may increase again. This is illustrated in Fig. 5.32 by the long-term averaged coastal profile at Egmond, the Netherlands. This particular shape of the long-term averaged profile can be explained in several ways. In the following we will discuss a hypothesis which also provides a possible explanation of the emergence of longshore bars.

Gravity transport and wave-induced transport

We will assume that cross-shore sediment transport in the surf zone is primarily the result of three competing processes. The first process is gravity-induced down-slope transport, with a magnitude which mainly depends on the steepness of the coastal profile and on wave stirring of the sediment bed. This down-slope transport is counteracted, on average, by wave-induced shoreward transport. If the average wave-induced shoreward transport is strong, a steep average coastal profile is required for equilibrium. If the average wave-induced shoreward transport is weak, equilibrium is reached at a weakly sloping profile. If the average wave-induced transport is directed seaward, the average coastal profile will dip towards the shore.

Wave asymmetry and undertow

Wave-induced sediment transport is caused, on the one hand, by wave asymmetry, with stronger shoreward wave-orbital motion than seaward wave-orbital motion (second process). Wave asymmetry is the result of the nonlinear character of wave propagation in shallow water; the asymmetry increases at decreasing water depth, see (5.13). In the breaker zone the wave asymmetry changes; around the bar crest, where wave breaking is strongest, acceleration asymmetry may overtake velocity asymmetry [137, 203].

On the other hand, wave-induced sediment transport is caused by undertow (third process), which counteracts shoreward transport related to wave asymmetry. Wave breaking produces strong near-bottom return flow and therefore reduces the average wave-induced shoreward transport. This reduction of wave-induced shoreward transport is reflected in the slope of the long-term averaged cross-shore profile. When approaching the breaker zone, the seabed slope no longer increases, but decreases; the coastal profile turns from concave to convex. If the width of the surf zone is sufficiently large (dissipative coasts, with small average seabed slope) waves may reform after breaking. Wave-induced shoreward transport starts increasing again and the coastal profile will recover a concave shape. Close to the beach, in the intratidal zone, an increase of shoreward transport may be caused by swash and aeolian transport; this also produces a landward increase of the seabed slope.

Inflexion point in the average coastal profile

The two assumptions:

- the coastal profile results from a balance between wave-induced and gravity-induced transport,
- the long-term average shoreface depth $h(x)$ increases as a function of offshore distance x with an inflexion point near the breaker line,

have consequences for the stability of the profile. This will be discussed in the following. Therefore the net cross-shore sand transport is written as

$$q = s(U_1)(h_x - f(h)). \qquad (5.63)$$

The first term on the right-hand side, sh_x, represents down-slope transport due to gravity. The second r.h.s. term, sf, represents net wave-induced sediment transport, which is taken positive if shoreward. The factor s represents the effect of wave stirring of the sediment bed and is proportional to the suspended load carried on the average by shoreward/seaward orbital-wave motions. This factor is often parameterised as a n power ($n = 3$–5) of the wave-orbital velocity amplitude U_1; it has a maximum around the breaker line and decreases from there both shoreward and seaward. The transport function f includes different processes that contribute to wave-induced sediment transport; we will assume that f mainly depends on the mean offshore wave amplitude a and on the local depth h. One component of f represents transport driven by wave asymmetry; it is proportional to the relative difference in average suspended load between shoreward and seaward wave-orbital motion. Another component represents the effect of wave breaking; it is proportional to the undertow velocity. The inflexion point at the breaker line $x = x_{br}$ corresponds to $h_{xx} = 0$; at this point the shoreward wave-induced transport function f is at its maximum, $f_h \equiv df/dh = 0$. The seaward decrease of f is due to seaward decreasing wave asymmetry; the shoreward decrease is due to shoreward increase of undertow, which is stronger than the shoreward increase of wave asymmetry.

Wave-transformation process

It is not accurate to represent wave-induced transport as just a function of local depth, because wave transformation processes (dissipation, distortion, shoaling, breaking) during shoreward propagation play a role [210, 441, 366]; for different coastal profiles the function f is not necessarily the same. However, the dependence of wave-induced transport on the offshore depth profile is not relevant for the following discussion. We will consider only small local depth changes relative to the equilibrium profile and we will assume that their influence on f does not depend significantly on wave transformation processes further offshore. A more general model without this restriction has been formulated by Plant, Ribas and co-authors [352, 373].

Response to a local depth change

According to the previous assumption, seaward of the inflexion point a local decrease of depth produces an increase in wave-induced transport, because the increase of wave asymmetry will be more important than the increase of undertow. Inversely, landward of the inflexion point a local decrease of depth produces a decrease of wave-induced transport. Seaward of the inflexion point x_{br} we find the location $x = x_{cav}$, where the increase of wave-induced transport with decreasing depth, f_h, is at its maximum. If the coastal profile is in equilibrium $x = x_{cav}$ corresponds to the location where the concavity of the coastal profile is greatest (minimum of h_{xx}/h_x). At this location a small local depth perturbation produces an increase in wave-induced transport which is the same at the onshore and the offshore flank of the perturbation. Seaward of $x = x_{cav}$ a local depth perturbation produces an increase in wave-induced transport which is greater at the onshore than at the offshore flank, while landward of $x = x_{cav}$ the inverse will occur. We will see that this is relevant to seabed stability and bar-formation. The locations x_{br} and x_{cav} not only depend on depth, but also on wave conditions far offshore. Under conditions with high waves x_{br} and x_{cav} are located further offshore than under conditions with low waves.

Perturbation in the zone of shoreward increasing concavity

We consider the long-term averaged coastal profile as representative of a morphologic equilibrium situation without breaker bars. We now investigate the stability of this equilibrium by perturbing the profile with a low longshore bar. First we consider the case of a bar situated at $x > x_{cav}$, i.e., at the concave offshore portion of the shoreface, seaward of the location of maximum concavity. The water depth is locally decreased at the bar and in this portion of the coastal profile a decrease in depth produces an increase in f: Shoreward wave-induced transport exceeds gravity-induced down-slope transport at the bar crest. The increase of shoreward wave-induced transport is greater at small depth, i.e., at the inner side of the bar (onshore of the crest), than at large depth, i.e. at the outer (offshore) side of the bar. We also assume that wave stirring decreases seaward of the surf zone. The result is a divergence of sediment transport at the bar. The bar will decay and we may conclude that in the zone of shoreward increasing concavity the coastal profile is stable against perturbation. The bar-migration depends on two competing factors. Wave stirring will cause seaward transport as a result of gravity (wave stirring and slope decay stronger at the onshore bar flank than at the offshore flank) and wave-induced transport causes shoreward transport. If the latter effect dominates the bar will move shoreward while decaying, see Fig. 5.33.

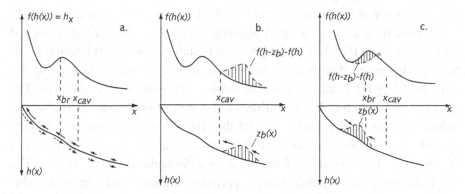

Fig. 5.33. Generation of breaker bars. The figures (a)–(c) explain the influence of a small bed perturbation on wave-driven cross-shore transport, neglecting variations in wave stirring for simplicity. Figure (a): The unperturbed long-term average coastal profile, corresponding to an equilibrium between wave-driven shoreward transport (black arrows) and gravity-induced seaward transport (dotted arrows). The concavity of the profile first increases shoreward due to increasing wave asymmetry. Then the concavity decreases and even changes to convexity as the shoreward wave-induced transport decreases due to increasing undertow. Figure (b): The profile is perturbed at its concave part, in the zone where the concavity increases shoreward (outside the surf zone). The perturbation enhances wave asymmetry and this increase is stronger at the onshore flank of the perturbation than at the offshore flank (dashed arrows). The result is landward divergence of sediment transport and net shoreward migration; the perturbation will decay. Figure (c): The profile is perturbed in the zone where the convexity increases shoreward. In this case wave-driven shoreward sand transport decreases due to wave breaking and undertow. This decrease is enhanced by the bed perturbation (dashed arrows). Shoreward transport is more decreased at the onshore flank than at the offshore flank of the perturbation. Sediment transport converges at the perturbation; the perturbation grows and migrates seaward.

Perturbation in the zone of shoreward decreasing concavity

Next we consider the case of a small bar situated in the zone landward of maximum concavity ($x < x_{cav}$); here the concavity of the coastal profile decreases in landward direction and concavity may even turn in convexity. Again water depth is locally decreased at the bar, but now the increase of wave-induced shoreward transport is smaller at the landward flank of the bar than at the seaward flank. The reason is that the influence of undertow on the variation of the wave-induced transport with decreasing depth is greater than the influence of wave asymmetry. If situated just landward of x_{cav}, the bar produces an increase in shoreward wave-induced transport, which is larger at the seaward flank of the bar than at the landward flank. If situated further landward, in the convex part of the profile, the bar produces a decrease in shoreward wave-induced transport, which is larger at the landward flank of the bar than at the seaward flank. In both situations sediment transport converges at the bar. The bar will grow and we may conclude that the coastal profile is unstable against perturbation in the zone of shoreward decreasing concavity or shoreward increasing convexity. This instability is a mechanism for the initiation of longshore breaker bars. The essential assumption is that the influence of undertow on the variation of wave-induced shoreward transport is greater than the influence of wave asymmetry. Growth of the breaker bar will shift the breaker line and x_{br} in the direction of the bar crest. The migration of the bar is again influenced by two factors. The bar flank closest to the breaker line experiences highest wave stirring and this bar slope will decay faster; the bar moves moves away from the breaker line. Wave-induced transport causes landward migration in the zone of decreasing concavity and seaward transport in the zone of increasing convexity.

Tidal influence

The location x_{cav}, which separates zones of seabed stability and seabed instability, depends on tidal level and on offshore wave conditions. With rising tide x_{cav} will shift shoreward and at falling tide it will shift seaward; for strong tides and gently sloping coastal profiles this shift will be very important. The time scales for bar migration, bar growth or bar decay are substantially larger than the tidal period. The bar generation mechanism may therefore remain qualitatively similar when averaging over a large number of tidal periods. In a quantitative sense, tides have a significant influence. The conditions for bar growth and bar migration in a particular profile segment are variable during the tidal period; if at a particular tidal phase the rate of growth and the migration rate are strong and positive, then they may be weak or even negative at another tidal phase. Therefore it may be anticipated that growth rate and migration rate of longshore bars are weaker in a macrotidal situation than in a microtidal situation, if all other conditions are the same. This is consistent with bar behaviour observed for different tidal regimes [276, 263].

Seaward transport under storm conditions

Wave-induced transport is also modulated by alternating wind conditions. During periods of high waves the zone of strongest wave-induced landward transport ($f_h \approx 0$) will shift seaward because waves are breaking earlier and undertow develops in deeper water. The generally observed occurrence of coastal erosion under storm conditions suggests that wave-induced landward transport is reduced by intense wave breaking. This reduced (or even reversed) wave-induced transport cannot neutralise gravity-induced seaward transport, because the equilibrium seabed slope is too steep. The net cross-shore sand transport is thus directed offshore and the coastline will retreat.

Shoreward transport during calm weather

The general tendency of coastal accretion observed during periods of low wave activity can be explained by the landward shift of the zone of strongest wave-induced landward transport. In this zone the equilibrium profile is often more convex than concave and gravity-induced seaward transport is relatively weak. The net cross-shore sand transport is thus directed onshore.

Sediment stirring by breaking waves

The wave-stirring factor in the model (5.63) depends on the amplitude U_1 of the wave-orbital velocity. However, this assumption ignores the influence of breaking waves on sediment resuspension. The increased seabed stirring by waves breaking on a bar will normally lead to a higher suspended concentration at the onshore flank than at the offshore flank. In this case sediment transport diverges at the bar crest if the bar increases shoreward wave-induced sediment transport; convergence of sediment transport at the bar crest occurs in the opposite case. Increased sediment suspension by breaking waves therefore stimulates decay of landward migrating bars and growth of seaward migrating bars.

Mechanisms of wave-induced sand transport

Seabed instability and growth of longshore bars result in the model from feedback of the initial bar formation to the balance between cross-shore gradients in wave-induced shoreward transport (related to wave asymmetry) and wave-induced offshore transport (related to undertow). The exact nature of the wave asymmetry and undertow feedback has not been specified; it is derived from the observed shape of the equilibrium profile (variation in concavity and convexity). Models which relate shoreward wave-induced transport only to asymmetry in shoreward/seaward maximum wave-orbital velocity underpredict shoreward sand bar motion [382, 442, 166]. Predictions are improved when acceleration-asymmetry effects are taken into consideration [137, 203]; these effects are implicitly included in the wave-induced transport function f. There are also other mechanisms which may enhance shoreward transport in the surf zone. Diffusion processes, for instance, produce shoreward sediment transport because suspended sediment concentrations generally decrease from the breaker line shoreward. This transport mechanism is disregarded in most existing models. For coastal profiles of sufficient steepness this mechanism can stimulate bar development [39].

Neglect of longshore transport

A comprehensive theory for cross-shore sediment transport in the surf zone has not yet been established. The previous models for explaining the long-term average cross-shore profile are very crude; different models can produce similar cross-shore equilibrium profiles. We have ignored the longshore sand transport component; gradients of longshore transport may strongly affect coastal erosion or coastal accretion, for instance, transport gradients related to rip channels or transverse bars. These transport patterns are not persistent, however, and there-fore less relevant for the long term average sand balance. Longshore currents also provide a mechanism for the generation of highly oblique bars in the surf zone [374]; these bars make only a small angle to the shoreline (see 5.4.2) and therefore look similar to breaker bars. However, the longshore current is the result and not the cause of wave breaking offshore and so are the bars generated by the longshore current. Hence, this mechanism cannot explain bar generation at the breaker line.

Analytical approach

The results of the previous qualitative description can also be derived by using a more analytical approach, following suggestions of Plant and Ribas *et al.* [352, 373]. We start from the sediment balance equation

$$z_{bt} + q_x = 0, \tag{5.64}$$

where $z_b(x, t)$ is a small time-dependent perturbation of the equilibrium seabed profile $h_{eq}(x)$. The shape of the perturbation is assumed to resemble a longshore bar with height of order $\epsilon \ll 1$. The wave-averaged sediment transport is given by (5.63), with $h = h_{eq} - z_b$ and $q_{eq} = 0$. Substitution in (5.64) and Taylor development to first order in ϵ yields

$$z_{bt} + (sf_h - s_h f)z_{bx} = sz_{bxx} - (sf_h)_h h_x z_b, \qquad (5.65)$$

where $h = h_{eq}$ and $s_h = s_{U_1} U_{1h}$. For evaluating U_{1h} the wave amplitude may be considered as approximately constant seaward of the breaker line and landward one may use the wave-saturation assumption. The second term on the left-hand side represents migration of the perturbation; the migration speed

$$(sf_h - s_h f) = \frac{1}{sf}\left(\frac{f_h}{f} - \frac{s_h}{s}\right) \qquad (5.66)$$

is the result of a balance between, on the one hand, relative increase or decrease of wave-induced transport and on the other hand, upslope-down-slope asymmetry of gravity-induced bar decay due to relative increase or decrease of wave stirring. If the variation of the wave-stirring factor s along the coastal profile is less important than the variation of the wave-induced transport, then the migration direction depends on the response of wave-induced transport to the presence of the bar. Landward of x_{br} (in the surf zone) this response produces a decrease of wave-induced transport, due to increased undertow relative to wave asymmetry, which results in seaward migration of the bar. Seaward of x_{br} the wave-induced transport increases, due to increased wave asymmetry at the bar; hence the bar will migrate landward.

The first term on the right-hand side of (5.65) represents decay of the perturbation under the influence of gravity effects. The last term stands for either decay $(sf_h)_h > 0$ or growth $(sf_h)_h < 0$ of the perturbation resulting from feedback to wave-induced transport. We have decay (negative feedback) if the increase of shoreward wave-induced transport at the onshore bar flank is smaller than at the offshore flank or if the decrease of shoreward wave-induced transport at the onshore bar flank is greater than at the offshore flank. This will occur seaward of the location x_{cav} where sf_h is minimum, which corresponds approximately to the location x_{cav} where concavity of the equilibrium seabed slope is greatest ($sf_h = sh_{xx}/h_x$ = minimum). Landward of x_{cav} the perturbation will grow (positive feedback, $(sf_h)_h < 0$).

Multiple bars

At many dissipative sandy coasts there is not just a single breaker bar, but a sequence of two or three shore-parallel bars. The cross-shore spacing of these bars is typically on the order of 50 to 300 metres. There is no unique explanation for the formation of such a bar sequence. In some cases the bar sequence seems to be related to the pattern of long standing waves, resulting from partial reflection at the shoreline of long infragravity waves [93, 197]. These waves may result, as indicated earlier, from radiation-stress gradients within wave groups of the incident wave field. This generation mechanism is questionable, however, because the wave spectrum is often too broadband for creating nodes and antinodes at fixed locations.

Wave reformation

Another explanation is based on multiple wave breaking. According to this theory the outer bar is associated with the initial breaking of incoming waves. Waves are reformed in the trough after the bar and will break a second time, closer to the shore. The validity of this hypothesis is probably restricted to weakly sloping, dissipative coasts. An example is the Dutch North Sea coast; here the coastal profile exhibits two inflexion points, a first inflexion point at the offshore boundary of the surf zone and a second some 500 metres further onshore (see Fig. 5.32). We may expect that the presence of a well-developed outer bar will influence the development of inner bars. The reason is that, in the through shoreward of the outer bar, wave height is insufficient for breaking. Waves reform and increase their asymmetry, while undertow remains relatively small. Sediment transport then increases in shoreward direction and landward of the outer bar the coastal profile assumes a concave shape. In this zone the growth of an inner bar will be inhibited. An inner bar may only develop closer to the shore, at smaller depths, where waves start breaking again.

Outer bar control

Observations at the Dutch North Sea coast show that in a multiple bar system, the development of an inner bar starts only when the outer bar has migrated far enough offshore and then decays [490]. According to the present theory, the outer bar starts as an instability of the shoreface profile under the influence of wave-driven sediment transport. Growth of the outer bar inhibits the development of a bar over some distance onshore. Seaward migration of the outer bar creates space for development of inner bars and for their seaward migration. The outer bar thus constrains the development of a multiple bar system. At coastal zones with a steep profile (reflective coast) the width of the surf zone much smaller than at dissipative coasts. The development of a bar near the breaker line then inhibits the growth of other bars in the remaining onshore portion of the surf zone. This may explain the absence of multiple bar systems in steep coastal profiles.

Bruun rule

Models based on the concept of an equilibrium profile implicitly satisfy the Bruun rule [57]. The Bruun rule states that the nearshore coastal zone always tends to an equilibrium profile, also when sea level is rising. As the seabed slope is normally steeper nearshore than offshore, an initial equilibrium profile will be out of equilibrium at a sudden rise of the sea level; the concave parts of the initial equilibrium profile will be too steep. In the case of a longshore uniform coast the profile will adapt by cross-shore redistribution of sand. Profile adaptation will occur over the depth range where wave-driven cross-shore transport is active. The active zone extends from the shoreline to the closure depth h_{cl}, i.e., the depth at which wave-induced cross-shore sediment transport is insignificant. Typical values of the closure depth range between 10 and 20 m [187]. The sand for profile adaptation is withdrawn from the shore, because the excess steepness of the profile favours seaward down-slope transport over shoreward transport. This produces a shoreline retreat Δx given by

$$\Delta x \approx \Delta \eta / \beta_{cl}, \tag{5.67}$$

where β_{cl} is the average slope of the coastal profile between the upper beach and the closure depth h_{cl} and $\Delta \eta$ is the long-term sea-level rise. Direct verification of the Bruun rule is difficult due to short-term coastline fluctuations and due to the presence of longshore gradients. Average observed erosion trends broadly agree with the Bruun rule in coastal zones where longshore gradients in sediment transport are of minor importance [59]. For instance, the net annual erosion along the whole European sedimentary coastline (length 45 000 km) is estimated at 15 km^2 [140]. Taking $\beta_{cl} \approx 0.01$ as a rough estimate of the average profile slope and considering an average annual sea-level rise of $\Delta \eta \approx 0.002$ m the Bruun rule yields a net coastal erosion of 9 km^2. However, local deviations of observed coastal erosion from the Bruun rule are much larger.

Table 5.1. Characteristics of the Holland coast [489, 421]; all the figures represent long-term mean values. Longshore sediment transport is the annual mean of
northward (positive) and southward (negative) wave-induced transports; at the central portion of the Holland coast northward and southward contributions are almost
equal.

Waves	height	1–1.5 m	period	5–6 s
Tide	amplitude	0.5–1 m	excursion	6–10 km
Longshore drift	velocity	3–6 cm/s	sand transport	$0–2 \times 10^5 \, m^3/yr$
Shoreface	grain size	200–350 μm	slope	0.01–0.002
Surf zone	width	400–800 m	longshore bars	1–4 (number)

Inverse application of the Bruun rule

The Bruun rule should, in principle, remain valid in the case of a falling sea
level is instead of a rising sea level. Return to an equilibrium profile then would
require at most shoreface locations an increase of the seabed slope instead of a
decrease. Because under a sudden fall of the sea level the seabed slope is locally
too flat, wave-induced shoreward transport will on the average dominate over
down-slope gravity-induced sediment transport. This net shoreward transport
will produce a seaward advance of the shoreline. The same response may be
expected when the shoreface seabed is suddenly raised, without change of sea
level. Such a raise of the shoreface seabed can be achieved by shoreface sand
nourishment. The Bruun rule then implies that the shoreface will respond by
restoring the equilibrium profile through an advance of the shoreline. In the
following section we will examine the coastal response to shoreface nourishment and try to answer the question: Is it possible indeed to combat shoreline
retreat by shoreface nourishment?

5.5.3. Response to Perturbation: Shoreface Nourishment

The Holland coast

The Holland coast is a strip of 120 km length between the Rotterdam Waterway
at Hook of Holland and Texel Inlet at Den Helder, see Fig. 5.34. It borders the
Southern Bight of the North Sea; a sand beach and dunes are present almost all
along. At Scheveningen and IJmuiden the coastline is interrupted by harbour
moles; at IJmuiden the moles extend a few kilometres into the sea. At many
places the Holland coast has retreated during the last centuries at an average

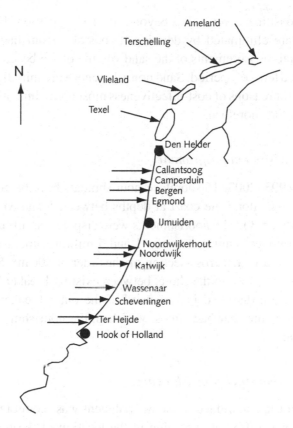

Fig. 5.34. Locations along the Dutch coast where shoreface nourishments have taken place in the period 1993–2003.

rate up to about 2 m/yr (see Fig. 4.46); retreat is largest at the northern part near Texel Inlet. The convex central part of the Holland coast is almost stable, or even slightly accreting. Some characteristics of the Holland coast are indicated in Fig. 5.1. The coastline has been fixed at a few locations by dikes or seawalls; a 5 km coastal stretch in the north, which is protected by a dike for almost 200 years, is now protruding almost 100 metres seaward relative to the adjacent unprotected coastline.

Shoreline maintenance policy

In 1990 a coastal policy was adopted to maintain the average position of the Dutch coastline at the position of that time. Sand is dredged far offshore (seaward from the 20 m depth contour) and brought to the coast at locations where the average coastline is retreating beyond the 1990 position. Fluctuations in the coastline are eliminated by defining its position from linear regression, based on annual measurements of the sand volume of the beach and the upper shoreface over a ten-year period. Sand nourishments were initially deposited on the beach, but for reasons of cost-effectiveness most nourishments are presently effectuated on the shoreface.

Shoreface nourishment programme

In the period 1993–2003, 15 shoreface nourishments have been carried out at different locations along the coast at depths between 5 and 8 m below mean sea level, see Fig. 5.34. The nourishments were disposed as artificial longshore berms, with volumes ranging between 1 and 5 million cube, lengths ranging from 1 to 6 km and with a cross-section on the order of $500 \, m^2$. Some nourishments were placed close to the shore, between existing breaker bars, but most nourishments were disposed just seaward of the outer breaker bar. In some cases, where the outer bar had almost vanished, the nourishment was located at this outer bar.

Monitoring of shoreface nourishments

The response of the shoreface to the nourishment was monitored in all cases. The response was different according to the location of the nourishment (in particular the cross-shore position), but certain features were systematically observed [421, 465, 181]: the cross-shore profile of the artificial berm always takes a shape similar to that of the existing breaker bars; very soon after disposal of the berm a trough is formed at its landward flank. The sand volume landward of the initial nourishment location is always increased during the first year after the nourishment. In some cases an average landward displacement of the artificial berm is observed, but in all cases there is a significant average shoreward displacement of the coastal profile landward of the artificial berm. However, the observed increase of the sand volume landward of the berm is larger (and in some cases even much larger) than the increase resulting from this landward shift. Sand losses are observed in coastal profiles outside the nourishment zone, in particular at the updrift side relative to longshore transport. After a few years the artificial berm is displaced in the dominant longshore transport direction; it is also diffused in cross-shore and longshore direction and the coastal profile finally assumes a shape similar to the situation before nourishment. Although the tidal amplitude is not the same at different locations along the coast, there is no evidence for substantial differences in shoreface response related to the tidal amplitude.

Observed bar response

The response to shoreface nourishment depends in particular on the position of the artificial berm relative to the existing longshore bars.

An artificial berm placed in between existing bars is redistributed in less than a year over the existing bars. The existing bar system is not only amplified but also modified; the bar just landward of the nourishment experiences a substantial shoreward migration.

An artificial berm placed at the location of a fading outer bar (or just seaward of it) develops into a new outer bar, which remains in place for at least several years without substantial decay or migration. The seaward migration of the first landward bar in front of the artificial bar is stopped and turned into shoreward migration, but just outside the nourishment location the seaward migration of this bar continues (even slightly enhanced). After some time the extremities of the artificial bar may become connected to the seaward migrating parts of the first landward bar. The artificial bar is then integrated into the natural bar pattern. The first stages of this bar response are illustrated in Fig. 5.35.

Integration in existing bar pattern

The shoreface responds to the introduction of an artificial longshore bar by its integration in the existing bar system; this response is most pronounced when the artificial bar is placed in between existing bars. When the artificial bar replaces a fading outer bar it remains a distinct feature of the shoreface profile during a much longer time; apparently there is no strong negative feedback to a location which fits into the natural bar pattern. The stability or slight landward displacement of the artificial outer bar is atypical, however, compared to the general seaward migration trend of natural bars. Different factors may play a role, which are discussed qualitatively below.

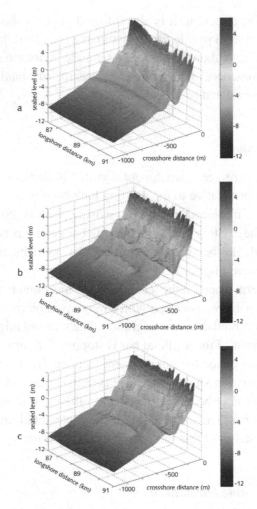

Fig. 5.35. The shoreface at Katwijk (see Fig. 5.34) interpolated from bottom soundings along cross-shore transects with 200 m spacing; the seabed level is indicated in metres relative to mean sea level with different grey tones, from [421]. The shoreface was nourished during Spring 1999 with an artificial bar disposed against the offshore flank of the outer bar. (a) Shoreface before nourishment in August 1998. Three longshore bars are visible: The outer bar is fading, the middle bar is well developed, and a new bar is developing at the shore. (b) Shoreface one year after nourishment. Soon after disposal the artificial bar migrated landward and became asymmetric. The onshore flank was steepened; a trough formed at the onshore bar front and around the bar extremities. The middle bar was lowered and the landward trough was partially filled. A cuspate shoreline protrusion formed in front of the nourishment, with a small shoreline retreat at both sides. (c) Shoreface two and a half years after nourishment. The artificial bar was still present without great change. The middle bar had migrated seaward, except in front of the nourishment. The longshore uniformity of the trough landward of the middle bar was restored. The cuspate shoreline advance was straightened and was developing into a new longshore bar.

Interpretation

According to the simple cross-shore transport model (5.63), the bar migration speed is given by (5.66), where $s(h)$ is the wave-stirring factor and $f(h)$ the average wave-induced transport function (positive if directed landward). The wave-stirring gradient, s_h, is typically positive in the surf zone (highest wave stirring at the breaker line), causing landward migration if net wave-induced transport is shoreward. The gradient in the wave-induced transport function, f_h, is positive if the shoreward increase of seaward transport due to undertow is stronger than the increase of shoreward transport due to wave asymmetry. However, an offshore artificial berm reduces wave breaking over some distance landward. One may then expect that shoreward of the berm, increase of wave asymmetry will dominate over increase in undertow, i.e., $f_h < 0$. The artificial offshore berm thus contributes to shoreward migration of the first bar at its landward side and to shoreward increase of shoreward transport. This effect, together with strong undertow at the onshore flank of the artificial berm due to wave breaking, provides an explanation for the formation of a trough just landward of the nourishment.

Effect of finite berm length

The finite length of the artificial offshore berm also accounts for the different behaviour in comparison to natural longshore bars. Two processes in particular seem to play a significant role: Gradients in the radiation stress $S^{(xx)}$ and gradients in the radiation stress $S^{(xy)}$, which both have been discussed in Sec. 5.4.

Wave breaking at the berm produces a shore-normal gradient in the radiation stress $S^{(xx)}$, which drives a residual circulation with shoreward velocities over the berm and a seaward return current around the berm extremities. This circulation moves the berm shoreward, in particular the central portion of the bar. The berm shape becomes slightly convex, with a trough at the extremities, see Fig. 5.35(b). This response is similar to the mechanism responsible for the development of rip cells, see Sec. 5.4.2. This mechanism is predicted to be most effective if the longshore berm length is about 5 times the distance to the shore.

In the case of oblique wave incidence, wave breaking at the artificial offshore berm produces close to the shore, in front of the berm, a local minimum in the shoreward gradient of the radiation stress $S^{(xy)}$. This corresponds to a minimum of the longshore current and the longshore sediment transport between the bar and the shoreline, see also Sec. 5.4.1. Hence, local shoreline accretion does not only result from shoreward transport fed by the nourishment, but also from convergence of longshore transport. This may explain the shoreline cusp observed in front of the nourishment, see Fig. 5.35(b). It also explains the loss of sand observed in coastal profiles updrift of the nourishment zone relative to the average longshore transport direction.

5.5.4. *Conclusion*

From the observed shoreface response to perturbation by an artificial shore-parallel berm, we may conclude:

- The shoreface response is primarily determined by wave breaking at the artificial berm. Wave breaking at the artificial berm affects wave-induced cross-shore sediment transport and affects sediment transport produced by gradients in the radiation stresses;
- The shoreface response is qualitatively consistent with the behaviour predicted by idealised models;
- Shoreface nourishment is an effective method to combat shoreline retreat at a dissipative coast. Wave energy is used for moving sediment towards to the shore in front of the nourishment. If sand is easily available offshore, shoreface nourishment is a cost-effective alternative for beach nourishment.

Chapter 6

Epilogue

6.1. Progress in Understanding

Spectacular progress has been achieved in the understanding of coastal morphodynamics during the past 50 years. Much of this progress is due to better knowledge of the dynamic balances responsible for the characteristic morphologic patterns observed in the coastal environment. For many situations the underlying physical processes have been identified and physical-mathematical descriptions have been developed. A major breakthrough is the understanding that coastal morphology and coastline change are not just a passive response to the action of waves and currents. This is due to the nonlinear mutual dependency between water motion and coastal morphology; this nonlinear dependency introduces new spatial and temporal variability and so creates additional complexity in coastal behaviour. Describing coastal dynamics in terms of nonlinear feedback processes has been an essential step to better understand coastal morphology and coastline change.

The sea is not just an eroding force

The sea is often perceived as primary responsible of coastal erosion. However, if this is true, then how to explain the presence of a sandy beach even in cases where the offshore seabed is deprived of sand? And how to explain the tendency of a beach to be restored, at least partially, after a severe storm with heavy coastal erosion? How to explain the landward displacement of coastal barriers when the sea-level rises? How to explain that non-floating materials (shells, for instance) can wash ashore? How to explain the development of coastal dunes, the development of coastal sandbanks and the development of tidal flats?

We presently know and understand that the interaction between water motion and seabed morphology generates transport components favouring sediment accumulation in the coastal zone, in particular for the coarse fraction. In many cases the sea acts more as an accretive force than as an eroding force.

Water motion and morphology are dynamically interdependent

However, onshore sediment transport and sand accumulation in the coastal zone is not a uniform or random process. It produces specific morphologic structures, like sandbanks, dunes, ripples, bars, spits and cusps, often according to a rhythmic pattern. Onshore sediment transport does not explain in itself the development of these patterns. Pattern formation is inherent to the nonlinear interaction of seabed morphology, water motion and sediment transport. The existence of positive feedbacks between perturbations of the morphology and water motion plays an essential role. These feedbacks can generate instability of the existing morphology and disrupt its symmetry characteristics. A linear relationship between morphology and water motion would not produce any other patterns than those present in the external conditions, while in reality the morphologic patterns exhibit wavelengths and fluctuations at spatial and temporal frequencies which are completely absent from the external conditions.

Symmetry breaking

Temporal symmetry breaking is illustrated by wave propagation in shallow water. Wave asymmetry (landward orbital wave motion stronger than seaward orbital motion) contributes to landward sediment transport and therefore produces a shoreward decrease of water depth. This decrease of water depth in turn enhances wave asymmetry. The self-strengthening process of mutual feedback between wave propagation and seabed topography continues until the steepness of the coastal profile is sufficient for compensation of wave-induced transport by gravity-induced down-slope transport. Shoreward sediment transport through wave asymmetry is also counteracted by wave breaking close to the shore; undertow currents generated by wave breaking may even reverse the direction of cross-shore sediment transport and produce a barred coastal profile which locally dips towards the shore. The physical processes involved in these different dynamic balances are not yet fully understood.

An example of symmetry breaking in the spatial domain is the circulation pattern produced by emerging seabed structures, which disturb the initial spatial flow symmetry. This circulation pattern may induce growth of these seabed structures, so providing a positive feedback to symmetry-breaking flow circulation. The development of rip cells in the surf zone and the development of beach cusps are examples of self-strengthening spatial symmetry breaking.

Initial development, linear instability

Idealised models have been very successful for studying the initial response to perturbation of coastal morphology, using a technique called linear stability analysis. This approach has provided basic insight in the processes which initiate the development of morphologic patterns. This insight has served to distinguish between coastal behaviour due to internal dynamics on the one hand and coastal adaptation to changes in the external conditions on the other hand. This is essential for the interpretation of morphological data and the detection of trends. We can also better understand why there are certain similarities or certain differences between coastal environments; this understanding is essential to make predictions based on analogies. Interpretation of field observations by using idealised models has been one of the major fields of progress during the past decades.

6.2. Working with Nature

Understanding the capacity of the sea to shape the coastal zone through interaction with morphology has practical implications for coastal management. It enables the development of strategies for working with nature, which in the long run are more efficient than interventions that oppose natural processes. Coastal engineering applications have not been discussed in great detail; a few examples are mentioned below:

- Shoreline maintenance. An eroding coast can be stabilised by sand nourishment of the upper shoreface. This strategy uses the capacity of waves to move sand in shoreward direction and is in many cases a less expensive alternative of beach nourishment.
- Maintenance of navigation channels. Dredged material is not removed from the estuary but disposed in a way to stimulate the capacity of tidal currents to maintain the navigation channel at greater depth. One strategy is to stimulate ebb dominance. This is achieved by disposing dredged material against

the channel banks, such that at HW the channel width is decreased and the intertidal area increased. This geometry promotes tidal asymmetry with enhanced currents during ebb relative to flood. The maintenance costs are reduced because tidal currents help carrying sediment to the sea; this may more than compensate the expense of extra dredging due to bank erosion.

- Concentration of flow through a single channel branch. Initial deepening or filling of one branch of a two-channel system may induce an ongoing accretion process of the shallow channel branch leading to its final closure, without additional intervention. In general a significant initial intervention is required for changing the current morphologic evolution of the channel system.
- Land reclamation. The sediment retention capacity of tidal basins can be enhanced by stimulating vegetation growth. Marsh vegetation traps fine sediment and prevents erosion until the tidal flat level is raised to the highest astronomical tide level. The resulting decrease of high-water basin width affects tidal asymmetry in such a way that flood currents are enhanced with respect to ebb currents, thereby stimulating further sediment import.

6.3. Future Challenges

Coastlines are ever changing

Coastal morphology never reaches a stationary state; this is one of the major difficulties when analysing coastal behaviour. Cliff coasts of soft rock never become stable when the sea-level rises; fine sediments released by cliff erosion are washed away by currents and deposited in sheltered inland basins and in deep offshore water where the seabed is not significantly stirred by waves and currents. Coasts built of mobile sediment can erode or accrete, but normally they will not become stationary. There are several reasons:

- The external forces strongly fluctuate at time scales of days, weeks and seasons. These fluctuations are reflected in the coastal morphology, especially at small to intermediate spatial scales.
- Fluctuations of coastal morphology also occur under average stationary forcing. Such fluctuations are related to the nonlinear interaction between different coastal subsystems, for instance the tidal flood and ebb deltas; this interaction generates cyclic or chaotic oscillations between different

morphologic configurations. A typical example is the cyclic or quasi-cyclic dynamics of the channel-shoal configuration in tidal inlets.

- Coastline change results from the presence of sediment sinks or sediment sources. Sediment supply by high river loads may produce ongoing accretion, while aeolian transport may produce coastline retreat by moving sediment inland.
- A long-term trend in coastal evolution is due to sea-level rise and to subsidence or uplift of the earth's crust. The morphologic adaptation time scale largely depends on sediment supply and sediment transport capacity in the coastal system.
- The coastline adapts to human interventions that produce structural changes in the pattern and the strength of currents, waves and tides acting on the coast.

The continuous change of coastal morphology implies that it is almost impossible to interpret coastal observations without historical knowledge. The larger the considered spatial scale, the longer is the relevant history. The survey costs with conventional observation techniques are often prohibitive. However, new remote sensing techniques, for example video camera and marine radar [207, 30], hold the potential for a breakthrough in morphologic field surveys covering large temporal and spatial ranges. This technological innovation is crucial to advance understanding of feedback processes in coastal systems.

Fully developed morphology

Although successful for explaining the initial development of morphologic patterns, linear stability analysis with idealised analytical models is inadequate for studying the behaviour of morphologic patterns in later stages of development. With increasing complexity, analytic methods loose their transparency and feasibility. Laboratory and field observations together with numerical models then offer the best opportunities to progress. However, it is much less easy to distil generic understanding from this approach than from analytic models. Numerical models are well suited for testing hypotheses, for analysing the sensitivity to certain parameters and for interpreting field observations. A major difficulty is related to uncertainty about the representation of the basic physical processes; this uncertainty limits the capability to produce reliable predictions for situations where the numerical model has not been validated with observations. The technique of complex principle component analysis, using empirical orthogonal functions [209] has been applied with success to analyse dynamic properties of highly complex morphologic situations [280, 489, 385]. Further development of these techniques is an important research line to advance understanding of the dynamic balances that govern the behaviour of fully grown morphologic patterns.

Outer deltas

Outer deltas or ebb tidal deltas are a typical example of interacting morphologic structures; these structures result from current-topography interaction, tide-topography interaction and wave-topography interaction. There are many qualitative descriptions of the behaviour of outer deltas based on field observations [334, 335, 151, 402]. From these descriptions some general characteristic features appear, but the basic nature of the feedbacks operating in these systems and the competition between different processes leading to the observed quasicyclic behaviour are not yet well understood. Some attempts have been made to formulate idealised models [170], but they heavily rely on empirical relationships. Outer deltas have received only minor attention in this book. The complexity of these systems is such that even numerical models are not yet capable to incorporate all the major physical processes. Outer deltas link the dynamics of the coastal zone with the dynamics of inland tidal basins. Understanding the dynamics of outer deltas is therefore essential to understand large-scale long-term behaviour of barrier coasts with tidal inlets.

Long-term coastal evolution, large-scale feedback

Long-term large-scale evolution refers to the formation and fate of entire coastal systems, such as coastal barrier systems, river tidal deltas and shelf seas. This evolution is very slow and can hardly be observed in real time, except in situations of very high sediment supply, for example in the Yellow River delta. One of the major difficulties is that long-term evolution is steered by small residual processes. There is in general great uncertainty about the nature and strength of the interactions and feedbacks responsible for long-term residual evolution. Geological reconstructions yield important evidence, but part of the morphologic evolution remains hidden due to erosion processes. Geological reconstructions therefore have to be completed by modelling. Some attempts have been made for the Dutch coastal zone [460, 77]. A way to progress is the development of morphodynamic models at aggregate scales, which describe the large-scale dynamics of coastal subsystems and the interaction between these subsystems. This approach requires a thorough understanding of aggregate-scale dynamics, which is still in a developing stage [87].

Biotic influences

Sedimentation and erosion processes are often strongly influenced by biotic activity. The general tendency of biotic activity is to stabilise the seabed and to stimulate sedimentation, but the inverse effect may also occur. There is evidence that in some coastal environments biotic influences have played a significant role in the long-term fate of coastal systems. For example, geologic recon-structions in The Netherlands indicate that vegetation in tidal fresh and brack-ish water zones has been a major cause of gradually progressing sedimentary infill, finally leading to the complete closure of former tidal basins [33]. This would probably not happen in the present tidal basins, where dike construction has stimulated the formation of deep tidal channels which prevent the seaward progression of fresh water marshes. Nevertheless, this example illustrates the potentially important influence of interaction between biotic and physical pro-cesses in the coastal zone. As yet this interaction is still poorly understood and there are no reliable predictive models. The lack of understanding also hampers the development and implementation of eco-engineering techniques as a tool for the sustainable management of coastal systems.

Field observations under high-energy conditions

Reality is far more complex than idealised models. Although the capability to simulate coastal processes with numerical models has tremendously increased over the past decades, the uncertainty in predictions has not decreased accord-ingly. Better observational evidence is crucial for reducing these uncertainties.

Laboratory experiments and field observations show a very strong increase of sediment suspension with increasing bottom shear stress. Shear stress and turbulence generated by currents may lift sediment particles high in the water column, whereas wave-induced shear stress mainly produces an increase of concentration near the bottom [472]. The shear stress exerted on the bottom by currents and waves can become so high that the top layer of the seabed starts to behave as a fluid sediment layer which is entrained by the near bottom current (sheet flow). Estimating the magnitude of sediment transport under high energy conditions therefore requires detailed knowledge of the structure of flow and sediment profiles near the bottom. There is ample evidence that major morphologic change takes place under high energy conditions: spring tides, high river floods and storms. However, field measurements of flow velocities and sediment concentrations under such conditions are scarce and even almost absent. Better techniques to measure in situ under these conditions would greatly contribute to validate sediment transport models and to achieve more reliable predictions of long-term coastal behaviour.

Sediment management

Human interventions are a major cause of coastline change. Sometimes the coastline is modified directly by engineering works, but more often coastline changes are the indirect consequence of interventions affecting sediment supply by rivers or interventions affecting the average flow pattern or local wave climate. Coastal development has often reduced the space for natural dynamics of inland seas, lagoons, tidal basins and deltas; this has consequences not only for the availability of sediment but also for the capacity of natural processes to import sediment from the sea. Other causes of reduced availability and mobility of sediment are related to retention of fluvial sediment and sea sand mining. At the long term this may threaten the capacity of the coastal system to adapt to sea-level rise. Therefore it is important that strategies are developed to manage sand resources in the coastal zone, especially for coastal zones with limited stocks of offshore sand.

The need to join research efforts

The costs of programmes for investigating these basic knowledge gaps are considerable. The strong fragmentation of the coastal research community is an obstacle for the development of research programmes at the required scale. Joining research capacities among coastal research institutes for designing and carrying out such programmes is a major challenge for the global coastal research community in the next decades.

Appendix A

Basic Equations of Fluid Motion

A.1. General Nature of the Basic Equations

Newton's law

The equations describing water motion in seas and rivers are derived from Newton's law. Although at first sight these equations look quite straightforward, their solution for real situations is very complex. The reason lies in the nonlinear nature of these equations; according to Newton's law the velocity of a fluid parcel depends on its trajectory. Nonlinearity also arises from the free surface, where the fluid motion has to satisfy conditions which depend on the location of this boundary. The nonlinearity of the equations of motion, together with viscous interaction at the molecular scale, means that water motions at different spatial and temporal scales are linked to each other. The topography of river and sea basins is structured over an almost infinite range of scales, from the scale of the individual sediment grain to the scale of the entire basin; the hydrodynamic response to this topography interrelates all these scales. The equations can therefore only be solved after simplification. Such simplifications imply assumptions additional to Newton's law, which are valid only under certain conditions. Assumptions conflicting with Newton's law are avoided (as far as possible) by validation from observations.

Notations

Different notations will be used, to keep the appearance of the equations as simple as possible. This refers in particular to the use of spatial coordinates and the notation of partial differentiation. The spatial coordinates and velocities will be noted

$x_1 = x$, $u_1 = u$ longitudinal coordinate and velocity,

$x_2 = y$, $u_2 = v$ transverse (lateral) coordinate and velocity,

$x_3 = z$, $u_3 = w$ vertical coordinate and velocity.

We choose a right-handed coordinate system with an upward vertical axis. Partial differentials of any function f will be noted

$$f_{,1} \equiv f_x \equiv \partial f / \partial x, \quad f_{,2} \equiv f_y \equiv \partial f / \partial y, \quad f_{,3} \equiv f_z \equiv \partial f / \partial z.$$

We will use the convention that if an index i, j, etc. appears twice in a product term then summation over the repeated index is implied, for instance $u_{i,i} \equiv \vec{\nabla}.\vec{u}$. A simple index $(i, j, \ldots)$ indicates the vector component following the x_i axis.

Small-scale limit

The equations describing water motion and transport of substances are balance equations based on conservation properties of the balance variable f; this variable may represent, for instance, mass, momentum, vorticity, energy, suspended matter or sediment, etc. In practice it is not possible to describe the spatial structure for the full range of scales down to the molecular scale; the variable f is therefore defined as an average over a given spatial and temporal domain (the model scale) which should be taken smaller than the scales at which we want to resolve its variation. The difference between f and its real counterpart in nature is designated f', with $\langle f' \rangle = 0$, where the brackets stand for averaging over the temporal model scale. The model scale cannot be chosen arbitrarily, but depends on the scales of the physical processes which are not represented explicitly in the model. If these processes have a periodic character the model scale should be a multiple of the spatial or temporal periodicity; if these processes are aperiodic the model scale should be larger than the spatial or temporal correlation scales. The model scale is the lower limit at which the dynamics of the system is explicitly resolved. In practice there is also an upper limit; the validity of the balance equations does not extend to the global scale but is restricted to given boundaries at which the behaviour of the system need to be specified by boundary conditions.

Balance equations

In 3D space the balance equation has the following form

$$f_t + \Phi_{i,i} = -\Psi_{i,i} + P. \tag{A.1}$$

This equation describes the change of a variable f as a result of input, output, production and destruction. Implicit within this equation is the scale at which f

is described and at which the different balance terms are modelled. The various terms have the following meaning:

- First term on the left-hand side: Change of f in a unit volume per unit of time, averaged over temporal and spatial scales of small-scale fluctuations.
- Second term on the left-hand side: Gradient of advection of f, i.e., difference between input and output of f for a unit cell per unit of time. Often $\Phi_i = u_i f$, where u_i is also defined as an average over the model scale. (But if, for instance, f is the energy of a propagating wave, then u_i has to be replaced by the wave-group velocity.)
- First term on the right-hand side: Gradient of transport by fluctuations of the velocity field related to processes on a smaller scale than the model scale. If $\Phi_i = u_i f$ then Ψ_i is given by $\Psi_i = \overline{\langle u_i' f' \rangle}$. If the small-scale processes fluctuate on time scales which are much smaller than the model scale and if these fluctuations are uncorrelated at the model time scale (random walk), then Ψ_i can be described as a diffusion process, $\Psi_1 = -K_1 f_{,1}$ etc. The diffusion coefficients K_i cannot be obtained from (A.1) and should be specified by an additional 'closure' relationship.
- Second term on the right-hand side: Local production or destruction of f.

As the balance equation (A.1) applies within a limited domain, a solution can only be obtained if boundary conditions are specified. These are, on the one hand, conditions specifying f in the entire domain at an initial time t_0, and on the other hand, conditions which either specify f at the boundary or specify the flux of f through the boundary at any time.

Nonlinearity

The balance equation is often a nonlinear equation. The transport term $\Phi_i = u_i f$ is nonlinear if u_i and f are mutually dependent variables. This is the case, for instance, if f represents the momentum ρu_i, the salinity S or the seabed topography Z_b. Other nonlinearities may arise from the diffusion term and the production-destruction terms.

Reduction of dimensions

Balance equations may also be formulated in a 2D space (by averaging over depth or width) or in a 1D space (averaging over a cross-section A). In the latter case the balance equation takes the form

$$(A\overline{\overline{f}})_t + (A\overline{\overline{u}} \cdot \overline{\overline{f}})_x = A\overline{\overline{P}} - \Psi_x. \tag{A.2}$$

In this equation Ψ is the transport due to processes in the 'hidden' (transverse) dimensions. This is called dispersive transport, $\Psi = \overline{\overline{A(u - \overline{\overline{u}}) \cdot (f - \overline{\overline{f}})}}$. Under certain conditions this dispersive transport can be approximated by a gradient-type transport, with dispersion coefficient D,

$$\Psi \approx -AD\overline{\overline{f}}_x, \tag{A.3}$$

see Sec. 4.7.1.

A.2. Water Motion in Three Dimensions

Momentum balance

Application of Newton's law to fluids yields the momentum balance equation

$$(\rho u_i)_{,t} + (\rho u_j u_i)_{,j} = -p_{,i} + \tau^v{}_{ij,j} + \rho F_i. \tag{A.4}$$

In the right-hand side of this equation the first term represents the pressure gradient, the second term the viscous stresses between fluid parcels and the third term external forces acting on the water mass, other than forces transmitted through shear stress. This last term corresponds, for instance, to the attraction force of the earth, $-g\rho$, and gravitational forces exerted by celestial bodies. These latter forces are only significant in very deep oceanic basins and will be left out of consideration from now. We are not interested in resolving water motion at the spatial and temporal scale of turbulence. Therefore the equations are averaged over temporal scales characteristic for turbulence; this yields

$$(\rho u_i)_{,t} + (\rho u_j u_i)_{,j} = -p_{,i} + \tau_{ij,j} - g\rho\delta_{i3}. \tag{A.5}$$

The stress terms τ_{ij} now incorporate stresses produced by viscosity and stresses produced by turbulence (Reynolds stresses); the latter are much larger than the former (except in a very thin layer at the bottom), so we may write

$$\tau_{ij} = -\langle\overline{\rho u'_i u'_j}\rangle. \tag{A.6}$$

Turbulent fluctuations in the density ρ have been disregarded in this equation.

Boussinesque approximation

Temporal and spatial scales of turbulence are not particularly small; in shallow coastal waters they may span the whole water column and attain periods exceeding 10 minutes. Turbulent stresses are often related to local velocity gradients, although such an assumption is a rough simplification. Following the Boussinesque approximation a turbulent viscosity N is defined such that

$$\tau_{ij} = \rho N (u_{j,i} + u_{i,j}).$$ (A.7)

The mass balance equation reads

$$\rho_{,t} + (\rho u_j)_{,j} = 0.$$ (A.8)

Combining this with the previous equations we obtain the Boussinesque equations

$$u_{i,t} + u_j u_{i,j} + p_{,i}/\rho + g\delta_{i3} = (N(u_{j,i} + u_{i,j}))_{,j},$$
$$u_{j,j} = 0.$$ (A.9)

In Eqs. (A.7) and (A.9) variations in density have been ignored; they are generally much smaller than variations in velocity.

A.3. Horizontal Flow Equations

Hydrostatic approximation

If the horizontal scale of water motion is much larger than the vertical scale we may ignore horizontal momentum diffusion as it is much smaller than vertical momentum diffusion and we may ignore vertical accelerations in the vertical momentum balance (hydrostatic pressure assumption). The vertical momentum balance can then be integrated and we find for the hydrostatic pressure

$$p = p_s + \int_z^{\eta} g\rho \, dz,$$ (A.10)

where η is the surface elevation relative to a horizontal reference level and p_s is the atmospheric pressure. The horizontal momentum balance equations read

$$u_{i,t} + u_j u_{i,j} + p_{,i}/\rho = (N u_{i,3})_{,3}.$$ (A.11)

These equations can be integrated to yield balance equations for the depth averaged velocity momentum. In the derivation we ignore terms $(u_i - \overline{u_i})^2$ as they are much smaller than $\overline{u_i^2}$. The result is

$$(D\overline{u}_i)_{,t} + \overline{u}_j(D\overline{u}_i)_{,j} + p_{,i}/\rho + \tau_{ib} - \tau_{is} = 0, \qquad (A.12)$$

$$\eta_{,t} + (D\overline{u}_j)_{,j} = 0. \qquad (A.13)$$

In these equations $D = h + \eta$ is the total water depth, τ_{ib} is the x_i-component of the bottom shear stress and τ_{is} the x_i-component of the shear stress at the water surface.

A.4. Earth's Rotation

Coriolis acceleration

Due to the rotation of the earth a centrifugal force, perpendicular to the earth's rotation axis, acts on ocean and sea water. This centrifugal force is only partly compensated by gravity; it has a component acting along the surface of the earth towards the equator. If water is at rest this tangential component of the centrifugal force is balanced by a sea surface slope. When seawater is brought in motion, the balance is broken and water motion experiences Coriolis acceleration. This acceleration is zero at the equator, maximally right-turning at the North Pole and maximally left-turning at the South Pole. The Coriolis acceleration is the additional force to which a water particle is subjected, as soon as a flow velocity (u, v) relative to the surface of the earth is superposed on the East-West directed rotation velocity of the earth surface. This rotation surface velocity is given by $U_A = \Omega r_A \cos \psi$, where Ω is the radial earth rotation frequency, r_A is the earth radius and ψ is the inclination relative to the equator. In the presence of earth's rotation, a flow u parallel to the equator from West to East will experience an acceleration toward the equator as a result of increased centrifugal force

$$\frac{dv}{dt} = \sin \psi \frac{(U_A + u)^2 - U_A^2}{r_A \cos \psi} \approx fu. \qquad (A.14)$$

The approximation is very accurate because $|u| \ll |U_A|$. From (A.14) we find the expression $f = 2\Omega \sin \psi$ for the Coriolis acceleration. A current directed away from the equator ($v = r_A \partial \psi / \partial t$) experiences an acceleration in

E-W direction; this current implies, on the one hand, a decrease of the surface rotation velocity U_A and, on the other hand, a gradient of momentum transport associated with U_A

$$\frac{du}{dt} = \frac{\partial U_A}{\partial t} + v \frac{\partial U_A}{\partial y} \approx -fv. \qquad\qquad (A.15)$$

Earth's rotation has a significant influence on stationary currents or on slowly varying currents, such as tidal motion. For currents varying on time scales much shorter than the Coriolis time scale $1/f$, such as wind waves or swell, the effect of earth's rotation can be ignored.

Momentum balance for large spatial and temporal scales

Often the influence of earth's rotation can also be ignored in certain tidal rivers, estuaries or other narrow coastal systems. The Coriolis acceleration induced by a decrease or increase of the current velocity is counteracted by a cross-flow inclination of the water surface. The surface inclination induced by earth rotation $\eta_y = -fu/g$ is often small and dominated by the effects of river or channel meandering. However, in wide basins Coriolis acceleration has to be taken into account.

Earth's rotation needs to be included in the momentum balance equations (A.12). These equations can be rewritten in a simpler form by using the mass balance equation to eliminate the depth from the acceleration terms. After substitution of (A.14, A.15) and dropping the overbars of the depth-averaged velocities the equations (Navier-Stokes equations) finally read

$$u_t + uu_x + vu_y - fv + g\eta_x + \frac{\tau_b^{(x)} - \tau_s^{(x)}}{\rho D} = 0, \qquad\qquad (A.16)$$

$$v_t + uv_x + vv_y + fu + g\eta_y + \frac{\tau_b^{(y)} - \tau_s^{(y)}}{\rho D} = 0. \qquad\qquad (A.17)$$

In these equations another notation of the partial derivatives has been used and $\tau_b^{(x)}$, $\tau_b^{(y)}$ are the x, y-components of the bottom shear stress and $\tau_s^{(x)}$, $\tau_s^{(y)}$ the x, y-components of the surface shear stress.

A.5. Vorticity Balance

For studying residual circulation or other spatial structures of the flow velocity field, it is often more practical to use the vorticity balance instead of the momentum balance. The vorticity balance describes angular momentum conservation in a fluid and can be derived from the momentum balance. The horizontal depth-integrated vorticity and potential vorticity are defined as

$$\zeta = v_x - u_y, \quad \zeta^{pot} = (\zeta + f)/H, \tag{A.18}$$

where u, v are depth-integrated velocities. For potential flow ($u_i = \Phi_{,i}$) the vorticity ζ is zero, for circular rotational currents the vorticity equals twice the angular velocity. The potential vorticity ζ^{pot} is the sum of flow vorticity and planetary vorticity (Coriolis acceleration) divided by depth. The vorticity balance can be obtained from the depth-averaged momentum and mass balance equations using the curl operator. In this way the pressure gradient $p_{,i}$ is eliminated and the result reads

$$\zeta_t^{pot} + u\zeta_x^{pot} + v\zeta_y^{pot} + \frac{1}{H}\left(\left(\frac{\tau_b^{(y)}}{\rho H}\right)_x - \left(\frac{\tau_b^{(x)}}{\rho H}\right)_y\right) = 0. \tag{A.19}$$

In the absence of friction the left-hand side of equation (A.19) can be written as a time differential d/dt along the flow trajectory,

$$d\zeta^{pot}/dt = 0. \tag{A.20}$$

Equation (A.20) states that, in absence of bottom friction, potential vorticity is conserved along a flow trajectory.

Appendix B

Tidal Propagation in One Dimension

B.1. 1D Tidal Equations

Channelised tidal flow

Flood and ebb flow in tidal inlet systems, such as tidal basins, estuaries and tidal rivers, is generally concentrated in a main channel. Tidal propagation also follows this main channel. This implies that tidal propagation in inlet systems can be described approximately by a set of one-dimensional equations where the longitudinal coordinate x follows the channel axis. The balance equations for mass and momentum describe how a water level change, imposed by external conditions at the inlet (astronomic tide, wind setup or setdown), propagates landward and how a reflected wave propagates seaward. The balance equations are established for each infinitely short segment of the inlet system. The momentum balance equation is the equivalent of Newton's Law and describes flow acceleration or deceleration due to a pressure gradient produced by a water level difference over the segment and due to frictional momentum transfer to the bottom. The mass balance equation states that a difference between inflow and outflow produces an equivalent change of water volume in a segment through raising or lowering the water level.

One-dimensional flow schematisation

For deriving these equations assumptions have to be made concerning the flow distribution over the basin cross-section $A(x, t)$. The basic assumption is that for each cross-section a distinction can be made between two parts: (1) A momentum-conveying part (channel section) of the cross-section, where almost all along-channel flow is concentrated, and (2) a storage part, corresponding to banks and tidal flats, where water hardly flows in the along-channel

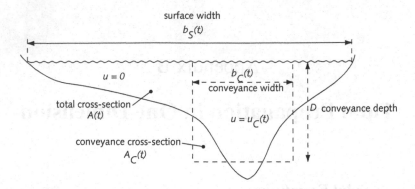

Fig. B.1. Schematisation of the cross-section and definition of notations.

direction, see Figs. B.1 and B.2. This is not always an accurate representation of the actual flow pattern, but it is generally sufficient for describing the major characteristics of tidal propagation [115]. The conveyance section has a cross-section $A_C(x, t)$, which is the product of a representative conveyance width $b_C(x, t)$ and a representative conveyance depth $D(x, t)$ at each time t. From the one-dimensional equations alone it cannot be derived which part of the cross-section should be considered as conveying and which part as storage area. For practical computations estimates are made based on measuring campaigns or 2D and 3D numerical flow models. As a rule of thumb we take as conveyance section that part of the cross-section which is deeper than 1 to 2 metres [115].

1D flow variables

We consider a cyclic tide with period $T = 2\pi/\omega$. For describing water level and flow velocity we introduce the following quantities:

$Z_s(x, t)$ is the instantaneous water level relative to a horizontal reference level,
$Z_b(x)$ is the seabed level relative to the same reference level,
$\eta(x, t)$ is the tidal component of the instantaneous water level Z_s; the tidal average of this quantity is by definition equal to zero,
$h(x)$ is the tidal average of the instantaneous water depth D,
I is the tidally averaged water surface slope,

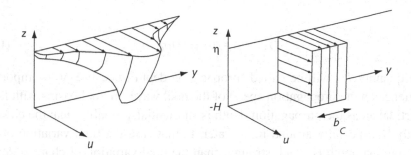

Fig. B.2. 1D-schematisation (right) of the flow velocity distribution (left).

$b_S(x, t)$ is the total cross-sectional width at the water surface,
$Q(x, t)$ is the instantaneous discharge,
$u_C(x, t)$ is the average flow velocity in the conveyance section A_C.

These quantities are related to each other as follows

$$D = Z_s - Z_b = h + \eta, \quad h = \langle D \rangle = \langle Z_s \rangle - Z_b, \quad I = \langle Z_{sx} \rangle,$$
$$D_t = Z_{st} = \eta_t, \quad A_t = b_S \eta_t,$$
$$Q = A\bar{\bar{u}} = A_C u_C = D b_C u_C. \tag{B.1}$$

Angle brackets $\langle .. \rangle$ designate tidal averages and the double line over the flow velocity u stands for averaging over the total cross-section.

1D mass balance equation

The flow is characterised by two variables, η and u_C. They can be derived from the two equations for mass balance and momentum balance. The mass balance (continuity) equation reads

$$(A_C u_C)_x + b_S \eta_t = 0. \tag{B.2}$$

The first term describes the volume change due to inflow and outflow, the second term describes the volume change due to water level change. The mass balance equation may also be written as

$$D_S u_{Cx} + \left(\frac{b_{Cx}}{b_C} + \frac{D_x}{D} \right) D_S u_C + \eta_t = 0, \tag{B.3}$$

with

$$D_S = A_C/b_S = Db_C/b_S. \tag{B.4}$$

The quantity D_S will be called 'propagation depth', because of its important influence on the propagation speed of the tidal wave. In tidal basins with large intertidal areas the propagation depth is substantially smaller than the channel depth D, especially around high water. In that case the tidal variation of the propagation depth is much stronger than the tidal variation of channel depth; tidal variation of the propagation depth is responsible for the high nonlinearity of the mass balance equation and for distortion of the tidal wave when propagating through the basin.

1D momentum balance equation

The momentum balance equation reads

$$(A_C u_C)_t + (A_C u_C^2)_x + Q_t^{trans} + g A_C Z_{sx} + \frac{b_C}{\rho} \tau_b = 0. \tag{B.5}$$

The first term describes the momentum change, the fourth term is the surface slope pressure gradient and the last term represents momentum transfer to the bottom. The second and third terms describe the inflow-outflow balance of along-flow and cross-flow momentum respectively,

$$(A_C u_C^2)_x + Q_t^{trans} = (A\overline{\overline{u^2}})_x.$$

Momentum exchange with storage areas

The term Q_t^{trans} describes the momentum carried to and from the storage zone

$$Q_t^{trans} = (A_t - A_{Ct})u^{ex}, \tag{B.6}$$

where ρu^{ex} is the momentum carried from the channel to the storage zone when $\eta_t > 0$ and the momentum carried from the storage zone to the channel when $\eta_t < 0$. Especially in the second case this momentum is smaller than the average momentum ρu_C carried by the flow in the channel.

The momentum balance can be written in a more convenient way by developing the product terms and using the mass balance equation; this yields

$$u_{Ct} + u_C u_{Cx} + g Z_{sx} + c_D \frac{|u_C| u_C}{D}$$

$$= \frac{A_t u_C}{A_C} \left(1 - \frac{A_{Ct}}{A_t} \right) \left(1 - \frac{u^{ex}}{u_C} \right). \tag{B.7}$$

We will ignore the term at the right-hand side of this equation; this is a reasonable approximation only if the tidal amplitude is very small compared to the depth or if the momentum exchanged between the channel and the storage zone is not very different from the momentum conveyed in the channel. This last assumption is more justified for flood than for ebb; it implies that during ebb the bottom friction term does not fully represent the total momentum loss. In tidal basins with a great width-to-depth ratio (> 100), transverse mixing is a slow process relative to the tidal timescale. In these basins momentum loss on the tidal flats affects primarily the flow along the channel bank and to a lesser degree the flow at the centre of the channel.

In the following we will consider the momentum balance equation

$$u_t + u u_x + g(I + \eta_x) + c_D \frac{|u| u}{D} = 0 \tag{B.8}$$

instead of (B.7). This means that the cross-sectional geometry only plays a role in the mass balance equation and not in the momentum balance equation. From now on we leave out the index C from the flow velocity.

B.2. Scale Analysis

Scaled variables

Scale analysis helps simplifying the tidal equations, based on estimates of the relative orders of magnitude of the various terms. The relative order of magnitude is expressed in dimensionless parameters. Scale analysis is also the starting point for investigating the influence of small nonlinear terms on tidal propagation, by developing the solution in a power series of the corresponding small parameter. The following scales are introduced

$$u = [u]u^*, \quad \eta = [\eta]\eta^* = \epsilon[h]\eta^*, \quad t = [t]t^*, \quad x = [x]x^*,$$

$$h = [h]h^*, \quad D = [h]D^*, \quad D_S = [h_S]D_S^*, \quad I = [I]I^*. \tag{B.9}$$

All variables labelled with $*$ are of absolute magnitude order 1 (locally or instantaneously the value may be much smaller than 1, but never much larger). The square brackets [...] express the order of magnitude of an unscaled variable; these values are constant. The inverse tidal frequency is the dominant time scale in the tidal equations, i.e. $[t] = \omega^{-1} = T/2\pi$, if we assume that linear terms are larger than quadratic or higher order terms. However, several length scales can be defined: $[x_u]$ for the length scale on which the tidal velocity varies, $[x_\eta]$ for the length scale on which the tidal amplitude varies, $[x_b]$ for the length scale on which the channel width varies and $[x_h]$ for the length scale on which the channel depth varies.

Scaled flow equation

The mass and momentum balance equations in their scaled form read

$$F\frac{[h_S]}{\epsilon[h]}D_S^*u_{x_u^*}^* + F\frac{[h_S][x_u]}{[h][x_\eta]}\frac{D_S^*}{D^*}u^*\eta_{x_\eta^*}^*$$

$$+F\frac{[h_S]}{\epsilon[h]}\left(\frac{[x_u]}{[x_b]}\frac{b_{Cx_b^*}^*}{b_C^*} + \frac{[x_u]}{[x_h]}\frac{D_{x_h^*}^*}{D^*}\right)D_S^*u^* + \eta_{t^*}^* = 0, \qquad \text{(B.10)}$$

$$FGu_{t^*}^* + F^2Gu^*u_{x_u^*}^* + \frac{[x_\eta][I]}{\epsilon[h]}I^* + \eta_{x_\eta^*}^* + F^2G\frac{c_D[x_u]}{[h]}\frac{|u^*|u^*}{D^*} = 0, \quad \text{(B.11)}$$

where

$$F = \frac{[t][u]}{[x_u]}, \qquad G = \frac{[x_u][x_\eta]}{\epsilon g[h][t]^2}$$

are dimensionless parameters. In a tidal river with almost uniform geometry we have $[x]/[x_h]$, $[x]/[x_b] \ll 1$, where $[x_u] \approx [x_\eta] \approx [x]$ is a measure of the distance over which the tide propagates in the tidal inlet system. For tidal amplitudes on the order of $\epsilon = O[10^{-1}]$ one may to a first approximation neglect the second term on the left-hand side of both equations. From the continuity equation it then follows that $F \approx \epsilon[h]/[h_S]$. The ratio of the frictional and inertial terms R/S is then equal to $R/S = \epsilon c_D[x]/[h_S]$. If we assume that $R/S \ll 1$, then the momentum balance equation requires $FG \approx 1$.

Together with the mass balance equation this gives $[x] \approx [t]\sqrt{g[h_S]}$ and $R/S = \epsilon c_D[t]\sqrt{g/[h_S]}$.

Relative magnitude of the friction term

In tidal basins the relative tidal amplitude ϵ is typically larger than 0.1, the friction coefficient c_D larger than 0.002 and the propagation depth $[h_S]$ smaller than 10 m. The ratio R/S is then larger than 1 and typically on the order of 5. Hence the friction term cannot be neglected in the momentum balance equation and should be of the same order of magnitude as the surface-slope pressure gradient, i.e. $F^2 G c_D[x_u]/[h] \approx \epsilon$. In that case we find for the ratio R/S between frictional and inertial terms $R/S = (g\epsilon^2 c_D^2[t]^2/[h_S])^{1/3}$. For values of c_D and $[h_S]$ in the same range as before we still find $R/S > 1$; the ratio is typically between 2 and 3. This implies that the dynamics of tidal propagation in tidal inlet systems is normally ruled by bottom friction; inertial terms in the momentum balance play a secondary role. The third term in the momentum balance corresponds to the average surface slope and is by definition constant throughout the tide; it can only balance with the tide-averaged value of the nonlinear terms in the momentum balance, in particular with tide-averaged friction. If tidal flow is much stronger than river flow, instantaneous friction is much stronger than tide-averaged friction and the third term only has a minor contribution to tidal propagation.

Uniform geometry

In the case of uniform geometry an approximate description of tidal propagation is given by

$$D_S u_x + \eta_t = 0,$$

$$u_t + g\eta_x + c_D\frac{|u|\,u}{D} = 0. \tag{B.12}$$

In this description terms of order ϵ have been ignored, except if they are multiplied by $[h]/[h_S]$ or by R/S. The most important terms in the tidal equations are underlined. Nonlinearity of tidal propagation is primarily caused by the tidal variation of D_S in the first term of the mass balance and by the tidal variation of the depth in the frictional term in the momentum balance. In frictionally dominated basins with large intertidal zones these contributions to nonlinear propagation are more important than the other neglected nonlinear terms in (B.10, B.11).

Strongly converging basins

In most tidal inlet systems the geometry may not be considered constant over the whole length of tidal intrusion; in many basins of the type river tidal inlets the length scale on which geometry varies is of the same order or even smaller than the length scale of tidal intrusion. Tidal basins with a substantially smaller geometric than tidal length scale, $[x_h], [x_b] \ll [x_u], [x_\eta]$, are said to be 'strongly converging'; often it is the channel width which strongly decreases in landward direction, more than the channel depth. In such basins the first two terms in the mass balance equation (B.10) may be neglected to first order with respect to the last three ones. This implies that the factor F is given by $F = \epsilon[h][x_b]/[h_S][x_u]$. As $F \ll 1$ the second term in the momentum balance equation can be neglected relative to the first one. By substitution we find that the ratio R/S of frictional and inertial terms in the momentum balance equals $R/S = \epsilon c_D[x_b]/h_S$. This ratio is often larger than 1, in particular for shallow tidal basins of at least 10 km length with large tidal flats. In this case the dominant terms (underlined) in the tidal equations are

$$D_S u_x + \left(\frac{D_x}{D} + \frac{b_{Cx}}{b_C} \right) \underline{D_S u} + \underline{\eta_t} = 0, \quad u_t + g\underline{\eta_x} + c_D \frac{|u|u}{D} = 0. \quad \text{(B.13)}$$

In these equations the smaller nonlinear terms have been ignored.

B.3. Linear Tides in a Uniform Channel

Small amplitude and negligible friction

The tidal equations can be linearised in the case of very small amplitude-to-depth ratio $a/h = \epsilon$, negligible friction ($c_D = 0$) and longitudinally uniform cross-section ($[x_b] = [x_h] = \infty$). We start from the scaled equations (B.10, B.11), where, in this case, $F = \epsilon h/h_S$ and $FG = 1$. This last equality implies $[x]/[t] = \sqrt{gh_S}$. We further assume $[x_u] = [x_\eta] = [x]$ and we ignore the mean surface slope I, because in frictionless flow there is no balance for a residual down-slope acceleration. The scaled tidal equations then read

$$D_S^* u^*_{x*} + \epsilon \frac{D_S^*}{D^*} u^* \eta^*_{x*} + \eta^*_{t*} = 0,$$
$$u^*_{t*} + F u^* u^*_{x*} + \eta^*_{x*} = 0. \tag{B.14}$$

We have $D^* = 1 + \eta^*$ and the propagation depth D_S^* is approximated by a linear function of η^*

$$D_S^* = 1 + \epsilon p \eta^*, \quad p = \frac{1}{\epsilon} \frac{D_S^+ - D_S^-}{D_S^+ + D_S^-}, \tag{B.15}$$

where D_S^+ is the propagation depth at high water and D_S^- the propagation depth at low water. Substitution in (B.14) yields

$$(1 + \epsilon p \eta^*) u^*_{x*} + \epsilon \frac{1 + \epsilon p \eta^*}{1 + \epsilon \eta^*} u^* \eta^*_{x*} + \eta^*_{t*} = 0,$$
$$u^*_{t*} + \frac{\epsilon h}{h_S} u^* u^*_{x*} + \eta^*_{x*} = 0. \tag{B.16}$$

We now develop the scaled variables as a power series of the small dimensionless parameter $\epsilon = a/h$ as follows,

$$\eta^* = \eta^{(0)} + \epsilon \eta^{(1)} + \epsilon^2 \eta^{(2)} + \cdots,$$
$$u^* = u^{(0)} + \epsilon u^{(1)} + \epsilon^2 u^{(2)} + \cdots. \tag{B.17}$$

If we substitute these expansions in the tidal equations (B.16) and collect terms of the same order in ϵ we obtain a series of linear equations from which the functions $\eta^{(0)}, u^{(0)}, \eta^{(1)}, u^{(1)}, \ldots$ can successively be solved.

First order tidal wave equations

The linear tidal equations correspond to the balance of first-order terms in ϵ

$$u^{(0)}_{x*} + \eta^{(0)}_{t*} = 0,$$
$$u^{(0)}_{t*} + \eta^{(0)}_{x*} = 0. \tag{B.18}$$

In dimensional terms this is equivalent to

$$h_S u_x + \eta_t = 0,$$

$$u_t + g\eta_x = 0, \tag{B.19}$$

where we have dropped the superscript $^{(0)}$ and used the notation $h_S = \langle D_S \rangle$ for the tide-averaged propagation depth. By cross-differentiation we can eliminate u from these equation; this yields the wave equation

$$\eta_{tt} = c^2 \eta_{xx}, \quad c = \sqrt{gh_S}, \tag{B.20}$$

where c is the wave propagation speed. The boundary condition at the inlet is a sinusoidal tidal oscillation of the water level with period $T = 2\pi/\omega$,

$$\eta(x = 0, t) = a\cos(\omega t). \tag{B.21}$$

Progressive tidal wave

If the tidal inlet system is a channel of infinite length, a tidal wave will propagate into the channel with constant amplitude a and wavenumber $k = 2\pi/L$, where L is the tidal wavelength,

$$\eta(x, t) = a\cos k(x - ct). \tag{B.22}$$

The wavenumber k is related to the wave propagation speed c by $kc = \omega$. The tidal flow velocity can be obtained by substitution in (B.28). We find

$$u(x, t) = \frac{ac}{h}\cos k(x - ct), \tag{B.23}$$

where $k = \omega/c$ is the wave number.

Stokes drift

In a frictionless propagating tidal wave the surface wave and the flow velocity are in phase. Largest flood flow occurs at high water and largest ebb flow at low water. The average water level during flood flow is therefore higher than during ebb flow, see Fig. B.3. This implies that more water is transported in flood direction than in ebb direction. This residual discharge $\langle b_C(h + \eta)u \rangle$

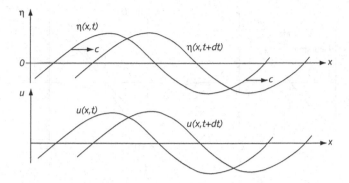

Fig. B.3. Representation of a frictionless propagating wave.

corresponds to Stokes drift. It is a term of order ϵ in the mass balance and will appear explicitly in the second-order solution of the tidal equations.

Standing tidal wave

Tidal inlet systems have a finite length; as the channel cross-section is uniform, the tidal wave will be reflected at the channel end $x = l$. The tide is then a superposition of an incoming and a reflected wave; together these waves form a standing wave (see Figs. B.4 and B.5)

$$\eta(x, t) = \frac{1}{2} a_l [\cos(k(x - l) - \omega t) + \cos(k(x - l) + \omega t)]$$

$$= a \frac{\cos(k(x - l))}{\cos kl} \cos \omega t,$$

$$u(x, t) = \frac{1}{2} \frac{ac}{h_S \cos kl} [\cos(k(x - l) - \omega t) - \cos(k(x - l) + \omega t)]$$

$$= u_0 \frac{\sin(k(x - l))}{\sin kl} \sin \omega t, \tag{B.24}$$

where

$$h_S = \frac{A_C}{b_S}, \quad a_l = \frac{a}{\cos kl}, \quad u_0 = u(0, 0) = -\frac{ac \tan kl}{h_S}.$$

The velocity at the landward boundary $x = l$ is zero. Reflection of the tidal wave produces an increase of the tidal amplitude when approaching the landward

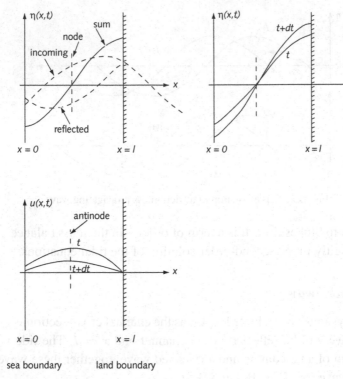

Fig. B.4. Representation of a standing wave as the sum of an incoming wave at $x = 0$ and a reflected wave at $x = l$ in the absence of friction.

boundary. The tidal variation of flow velocity is 90 degrees in advance of the tidal water level variation; flood flow corresponds to rising water and ebb flow to falling water. The mean water level during flood flow is the same as during ebb flow.

Resonance

Resonance will occur if the basin length is equal to a quarter of the tidal wave length,

$$l = L/4 = \pi/2k. \tag{B.25}$$

As the factor $\cos(kl)$ then becomes infinitely small, the tide becomes infinitely large. It is clear that bottom friction cannot be neglected in this case.

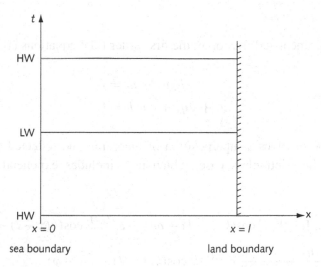

Fig. B.5. Location of HW and LW as a function of time for a standing wave in a tidal basin of length l in the absence of friction.

Linear friction

A more realistic description of tidal propagation requires inclusion of friction in the momentum balance. We first consider the linear case and replace the quadratic expression for frictional momentum dissipation by a linear expression

$$\tau_b = \rho r u, \tag{B.26}$$

where r is the linearised friction coefficient. This linear expression represents a reasonable approximation of the friction experienced by the main tidal component if tidal flow is much stronger than river flow. In the presence of significant river flow the quadratic expression generates additional tidal components which are absent in the linear expression. In the absence of river flow the coefficient r can be related to the friction coefficient c_D by requiring that the quadratic and linear expressions yield an identical energy dissipation, on average, over the tidal period

$$r \approx c_D \frac{\langle |u|^3 \rangle}{\langle |u|^2 \rangle} = \frac{8}{3\pi} c_D u_{max}, \tag{B.27}$$

where u_{max} is the tidal velocity amplitude.

Damped tidal wave

Inclusion of linearised friction in the first-order tidal equations (B.28) yields

$$hsu_x + \eta_t = 0,$$
$$u_t + g\eta_x + ru/h = 0. \tag{B.28}$$

The solution involves a superposition of incoming and reflected tidal waves, similar to the frictionless case, which now includes exponential damping functions

$$\eta = \frac{1}{2}a_l[e^{-\mu(x-l)}\cos(k(x-l) - \omega t) + e^{\mu(x-l)}\cos(k(x-l) + \omega t)],$$

$$u = \frac{1}{2}\frac{a_l}{hs}\frac{\omega}{\sqrt{k^2 + \mu^2}}[e^{-\mu(x-l)}\cos(k(x-l) - \omega t - \varphi)$$
$$- e^{\mu(x-l)}\cos(k(x-l) + \omega t + \varphi)]. \tag{B.29}$$

Here we have used the following notations

$$a_l = \frac{a}{\sqrt{\cos^2 kl \cosh^2 \mu l + \sin^2 kl \sinh^2 \mu l}}, \tag{B.30}$$

$$\binom{k}{\mu} = \frac{\omega}{\sqrt{2ghs}}\sqrt{\pm 1 + \sqrt{1 + \beta^{-2}}}, \quad \cos\varphi = \frac{k}{\sqrt{k^2 + \mu^2}}, \quad \beta = \frac{h\omega}{r}. \tag{B.31}$$

The coefficient β is a measure of the strength of the inertial acceleration relative to frictional damping in the momentum balance equation. For a resonant tidal basin length, $kl = \pi/2$, the tidal amplitude a_l and flow velocity u remain finite.

Radiation of tidal energy

Apart from bottom friction, outward radiation of tidal energy at the inlet also causes energy loss from the basin to the open sea [317]. This energy radiation modifies the ocean tide near the basin entrance; modelling tidal propagation in a basin close to resonance requires inclusion of tidal motion in the offshore vicinity of the inlet [230]. Far from resonance this effect may be ignored.

B.4. Tidal River with Strong Friction

Tidal diffusion

In many natural tidal channels friction is dominant over inertial acceleration in the momentum balance equation. The linear friction coefficient r is typically larger than 0.002, the average water depth h smaller than 10 m and $r/h \gg \omega$. In that case we have

$$k \approx \mu \approx \sqrt{\frac{r\omega}{2ghh_S}}, \quad \varphi = \pi/4. \qquad (B.32)$$

The tidal velocity $u(x, t)$ has a phase advance on the surface wave $\eta(x, t)$ of about $1/8$ tidal period.

This result can also be obtained by a method proposed by LeBlond [271]. We start directly from the tidal equations in which inertia is neglected relative to friction

$$D_S u_x + \eta_t = 0,$$

$$g\eta_x + c_D|u|u/D = 0. \qquad (B.33)$$

The surface elevation η can be eliminated by cross-differentiation, if D_S is assumed independent of x. We then find an equation in the single variable u,

$$u_t = \Xi u_{xx}, \quad \Xi = \frac{gDD_S}{2c_D|u|}. \qquad (B.34)$$

In the case of dominating friction, tidal propagation is described by a diffusion equation, with diffusion coefficient Ξ. The tide apparently does not propagate as a wave, but it rather diffuses into the river. Periods around slack water are, however, not well described by the tidal diffusion equation. During these periods friction is not dominant and the tide propagates approximately as a wave. If we assume that the wave-dominant periods are short then tidal wave propagation is reasonably represented by the diffusion model. The tidal diffusion coefficient varies strongly during the tidal period, in particular because of its dependence on the flow velocity. The tidal diffusion equation is thus nonlinear; the nonlinearity will causes tidal wave distortion. The tidal wave loses its sinusoidal character when diffusing upstream the river. This implies that the symmetry between ebb and flood will be broken. If we ignore the t- and x-dependence of Ξ, the diffusion equation (B.34) becomes a linear equation with constant coefficients. The solution has the form

$$u = -u_R + u_T, \quad u_T \propto e^{i\kappa x - i\omega t}. \tag{B.35}$$

Here $-u_R$ is the tide-averaged flow velocity, $u_R = -\langle Q/A_C \rangle$, u_T is the tidal flow component and κ is a complex wave number. The complex notation implies that in the final solution only the real part is retained. If friction is sufficiently strong we may neglect the tidal wave reflected at the landward end of the basin; in that case we only retain the solution for which the real part of κ is positive. Substitution in (B.33) with the boundary condition $\eta(x = 0, t) = a \cos(\omega t)$ yields the complete solution

$$\eta = ae^{-kx} \cos\left(k\left(x - c\left(t - \frac{T}{8}\right)\right)\right),$$

$$u = -u_R + u_T, \quad u_T = \frac{1}{\sqrt{2}} \frac{a}{h} ce^{-kx} \cos(k(x - ct)). \tag{B.36}$$

In contrast to frictionless tidal propagation, the tidal water level variation, η, has a phase lag of $T/8$ relative to tidal flow u_T. The solution (B.36) is identical to (B.29) for strong friction, $\beta \ll 1$.

Approximate linearity for tidal phases around maximum tidal flow

The solution (B.36) can be considered a crude approximation of the tidal wave only for those tidal phases at which the diffusion coefficient Ξ does not strongly vary as a function of t and x. The first condition is approximately satisfied during a short period just preceding maximum flood and during a short period after maximum ebb. During the first period (designated by a superscript $^+$) D and $|u|$ are both increasing and in the second period (designated by a superscript $^-$) D and $|u|$ are both decreasing, see Fig. B.8. For these two tidal phases the complex wave number κ is given by $\kappa^\pm = (1 + i)k^\pm$. The real part k and the propagation velocity c follow from substitution in (B.33),

$$k^\pm = \sqrt{\frac{r^\pm \omega}{g D_S^\pm D^\pm}}, \quad c^\pm = \frac{\omega}{k^\pm} = \sqrt{\frac{g\omega D_S^\mp D^\pm}{r^\pm}}, \quad r^\pm = c_D|-u_R + u_T^\mp|. \tag{B.37}$$

In tidal rivers flood flow is counteracted by river flow, see Fig. B.6. If u_R and $|u_T|$ have similar magnitude the flow velocity is very small in the flood period of

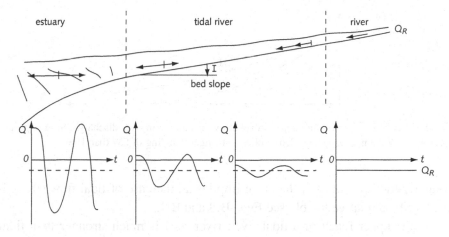

Fig. B.6. Schematic representation of a tidal river and discharge curves along the river.

constant Ξ. In this case friction is dominated by inertia; the tide will behave as a frictionless propagating wave and the Eqs. (B.33) do not adequately describe tidal propagation.

The second condition, Ξ independent of x, cannot be satisfied because it is incompatible with damping of the tidal wave. The implications will be discussed in the next section.

Distortion of the tidal wave

In large portions of a tidal river friction is dominant over inertia, not only during ebb flow, but also when the tidal flood component is at its maximum, see Fig. B.6. However, the maximum flow velocity at ebb, $|u^-| = |-u_R + u_T^-|$, is substantially larger than the maximum flow velocity at flood $|u^+| = |-u_R + u_T^+|$. In that case $k^+ \ll k^-$ and damping of the tidal wave is therefore much stronger for tidal phases around maximum ebb than for tidal phases around maximum flood, see Fig. B.7. The propagation velocity $c^+ = \omega/k^+$ around maximum flood flow is larger than the propagation velocity $c^- = \omega/k^-$ around maximum ebb flow. This is an important cause of tidal wave distortion in tidal rivers; the period of tidal rise is shortened relative to the period of tidal fall. The phase lag between flow velocity and water elevation implies that flood flow coincides to a great extent with tidal rise and ebb flow coincides to a great extent with tidal fall. A shorter period of tidal rise and a

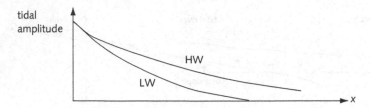

Fig. B.7. Tide propagation into a uniform shallow channel with river discharge. Slower propagation of LW compared to HW also implies a stronger damping of LW than HW.

longer period of tidal fall therefore implies an increase of tidal flow strength $|u_T|$ at flood relative to ebb, see Figs. B.8 and B.9.

In the upper reach of a tidal river, river flow is much stronger than tidal flow. In that case the friction coefficient r is almost constant and the periods of constant Ξ correspond approximately to the tidal phases of high and low water.

B.5. Nonlinear Tides in a Uniform Channel

Frictionless tidal propagation

The linear tides which are solution of the first-order tidal equations represent a very rough approximation of tidal propagation. In reality the tidal wave is distorted in shallow water due to the nonlinearity of several terms in the mass and momentum balance. This nonlinearity is the cause of tidal asymmetry; flood and ebb flow are not mirror images of each other, but are different in strength an duration. Tidal asymmetry has important consequences for sediment transport in tidal inlet systems. In this section we will investigate the influence of nonlinear terms in the tidal equations on distortion of the tidal wave without the assumptions of the previous section. As before, we will first consider the frictionless case and a uniform channel of infinite length. We start form the scaled tidal equations (B.16), substitute the expansion (B.17) and select terms of order ϵ

$$u_{x^*}^{(1)} + \eta_{t^*}^{(1)} = -p u_{x^*}^{(0)} \eta^{(0)} - u^{(0)} \eta_{x^*}^{(0)},$$

$$u_{t^*}^{(1)} + \eta_{x^*}^{(1)} = -\frac{h}{h_S} u^{(0)} u_{x^*}^{(0)}. \tag{B.38}$$

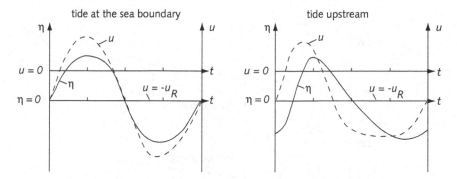

Fig. B.8. Tide propagation into a uniform shallow channel with river discharge. Schematic representation of tidal wave distortion due to stronger frictional damping during ebb than during flood. Tidal rise is faster than tidal fall and the tidal velocity component has a greater amplitude at flood than at ebb.

In dimensional form this is equivalent to

$$h_S u_x^{(1)} + \eta_t^{(1)} = -\frac{h_S}{h}\left(p u_x^{(0)} \eta^{(0)} + u^{(0)} \eta_x^{(0)}\right),$$

$$u_t^{(1)} + g\eta_x^{(1)} = -u^{(0)} u_x^{(0)}, \tag{B.39}$$

with boundary condition $\eta^{(1)}(x = 0, t) = 0$ and no reflected wave, according to (B.21). For the right-hand side of these equations we substitute the first-order solution (B.22, B.23). The homogeneous part of the equations is solved first and then a particular solution of the complete equations is added; coefficients follow from the boundary conditions. For the particular case $h_S = h$, $p = 1$ we find the well-known result [265] (sum of first and second-order solution)

$$\eta(x, t) = a \cos\theta + \frac{3}{4}\frac{a^2}{h}kx \sin 2\theta, \tag{B.40}$$

$$u(x, t) = \frac{a}{h}c\left[\cos\theta - \frac{a}{h}\left(\frac{1}{2} + \frac{1}{8}\cos 2\theta - \frac{3}{4}kx \sin 2\theta\right)\right], \tag{B.41}$$

where $\theta \equiv k(x - ct)$. The condition $\epsilon \ll 1$ is equivalent to $x \ll h/ak$. These formulas clearly display tidal asymmetry; the tidal rise speeds up and tidal fall slows down as the tide propagates. The same applies to the flow velocity: When the tide turns from ebb to flood the low-velocity period is shorter than for the turning of the tide from flood to ebb. Tidal asymmetry is related to differences in wave propagation speed during the tidal period, as will be shown below. From the function (B.40) an expression can be derived for the propagation speed at

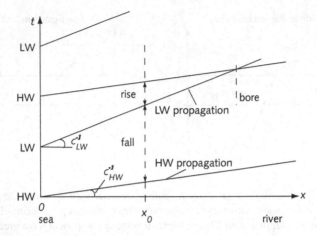

Fig. B.9. Schematic representation of tidal wave propagation for HW and LW tidal phases. LW-propagation is slower than HW-propagation due to stronger frictional damping at LW than at HW.

high water (HW, designated by superscript $+$) and low water (LW, designated by superscript $-$). Therefore we determine the time at which the water level is at its maximum or minimum for every location x. The corresponding times $t^{\pm}(x)$ are solutions of the equation

$$\eta_t = 0 = a\omega \sin \theta - \frac{3a^2}{2h}\omega kx \cos 2\theta \approx a\omega \sin \theta \mp \frac{3a^2}{2h}\omega kx. \qquad \text{(B.42)}$$

The last approximation follows from the condition $akx/h \ll 1$ and therefore $|\theta^+| \ll 1$ for HW and $|\pi + \theta^-| \ll 1$ for LW. Substitution of $\theta^{\pm} = k(x - ct^{\pm})$ yields

$$t^+ \approx \frac{x}{c}\left(1 - \frac{3a}{2h}\right), \quad t^- \approx \frac{T}{2} + \frac{x}{c}\left(1 + \frac{3a}{2h}\right).$$

This is equivalent to (assuming $a/h \ll 1$)

$$c^{\pm} \approx c\left(1 \pm \frac{3a}{2h}\right). \qquad \text{(B.43)}$$

The propagation speed at HW is higher than the propagation speed at LW. This explains why the period of rising tide becomes shorter and the period of falling tide longer as the tide propagates through the channel, see Fig. B.9.

Wave-propagation method

The same result can be found in a simpler way, by the 'wave-propagation method'. For short periods around high water (HW) and low water (LW) the tidal equations for a progressive, undamped tidal wave are approximately linear; around HW, water level and flood flow are at their maximum, so we have $\eta_t \approx 0$, $u_t \approx 0$. The same holds for a short period around LW. As the tidal amplitude is independent of x, the x-derivatives of water level and flood flow also vanish at HW and LW. Therefore product terms involving depth and flow velocity may be linearised by taking the values corresponding to HW (maximum flood) and LW (maximum ebb) for depth and flow velocity.

For $h_S = h$, $p = 1$ the tidal equations read (B.3, B.8)

$$Du_x + u\eta_x + \eta_t = 0,$$
$$u_t + uu_x + g\eta_x = 0. \tag{B.44}$$

During short intervals around HW and LW, u and D are almost constant and equal to

$$D^{\pm} \approx h \pm a, \quad u^{\pm} \approx \pm a\sqrt{g/h}.$$

As before, the superscripts $^+$ and $^-$ indicate HW and LW. After substitution in the nonlinear terms of the tidal equations we seek a solution around HW and LW of the following form

$$\eta, u \propto e^{ik^{\pm}(x - c^{\pm}t)}. \tag{B.45}$$

From this we get an expression for the tidal wave propagation velocity at HW and LW respectively:

$$c^{\pm} \approx \left(1 \pm \frac{3a}{2h}\right)\sqrt{gh},$$

which is identical to the result found with the perturbation method (B.43). As the tidal equations have been solved only in short intervals around HW and LW the wave propagation method only provides information on the amplitude and the phase of the tidal wave around HW and LW. A first-order approximation of the complete tidal curve can be obtained by interpolation between HW and LW.

The influence of friction on tidal propagation

For a more realistic description of tidal propagation, frictional effects cannot be ignored. We restrict the present discussion to a uniform channel of infinite length and we will use a linearised representation of the friction term (B.26). We have already noted that the tidal wave is damped by bottom friction; the tidal amplitude decreases as a function of x. We will assume, however, that the propagation depth D_S is independent of x; this implies, according to (B.15), that p is also independent of x. The tidal equations (B.3, B.8) then read

$$h_S u_x + h_S p\, u_x \eta/h + h_S\, u\eta_x/h + \eta_t = 0,$$
$$u_t + u u_x + g\eta_x + ru/h - ru\eta/h^2 = 0. \tag{B.46}$$

We solve these equations to first order in $\epsilon = a/h$, by the same perturbation expansion method as previously, see equations (B.17) and (B.39). This is most easily done by eliminating the first-order velocity contribution $u^{(1)}$ from the equations (B.46) by differentiating respectively with respect to t and x. This yields, to first order in ϵ,

$$\eta_{tt}^{(1)} - gh_S \eta_{xx}^{(1)} + \frac{r}{h}\eta_t^{(1)} = -\frac{rh_S}{h^2}\big[(p+1)\, u_x^{(0)}\eta^{(0)} + 2u^{(0)}\eta_x^{(0)}\big]$$
$$- \frac{h_S}{h}\big[p\, u_x^{(0)}\eta^{(0)} + u^{(0)}\eta_x^{(0)}\big]_t + h_S(u^{(0)}u_x^{(0)})_x. \tag{B.47}$$

At the inlet we impose a sinusoidal tide, $\eta(x = 0, t) = a \cos \omega t$, no reflected wave and no river discharge, $\langle A_C u\rangle = 0$. The calculation is lengthy but straightforward; the result is

$$\eta = \frac{a^2}{h}\frac{2kh_S - \beta h\mu}{4\mu h_S\sqrt{1+\beta^2}}(1 - e^{-2\mu x}) + ae^{-\mu x}\cos(kx - \omega t)$$
$$+ \frac{a^2}{4h}(p+3)\big(e^{-2\mu' x}\cos(2k'x - 2\omega t) - e^{-2\mu x}\cos(2kx - 2\omega t)\big)$$
$$+ \frac{a^2\beta}{2h}(p+1+h/h_S)\big(e^{-2\mu' x}\sin(2k'x - 2\omega t)$$
$$- e^{-2\mu x}\sin(2kx - 2\omega t)\big),$$

$$u = \frac{a\omega}{h_S} \left[-\frac{a\cos\varphi}{2hK}e^{-2\mu x} + \frac{e^{-\mu x}}{K}\cos(kx - \omega t - \varphi) \right.$$
$$-\frac{a}{2hK}(p+2)e^{-2\mu x}\cos(2kx - 2\omega t - \varphi)$$
$$+\frac{a}{4hK'}(p+3)e^{-2\mu' x}\cos(2k'x - 2\omega t - \varphi')$$
$$-\frac{a\beta}{2hK}(p+1+h/h_S)e^{-2\mu x}\sin(2kx - 2\omega t - \varphi)$$
$$\left. +\frac{a\beta}{2hK'}(p+1+h/h_S)e^{-2\mu' x}\sin(2k'x - 2\omega t - \varphi') \right]. \quad \text{(B.48)}$$

The parameters are given by

$$p = \frac{h}{a}\frac{D_S^+ - D_S^-}{D_S^+ + D_S^-}, \quad \beta = \frac{\omega h}{r},$$

$$\tan\varphi = \frac{\mu}{k}, \quad \tan\varphi' = \frac{\mu'}{k'}, \quad K^2 = k^2 + \mu^2, \quad K'^2 = k'^2 + \mu'^2,$$

$$\binom{k}{\mu} = \frac{\omega}{\sqrt{2gh_S}}\sqrt{\pm 1 + \sqrt{1 + \beta^{-2}}},$$

$$\binom{k'}{\mu'} = \frac{\omega}{\sqrt{2gh_S}}\sqrt{\pm 1 + \sqrt{1 + \frac{1}{4}\beta^{-2}}}.$$

A similar solution has been presented by Van de Kreeke and Iannuzzi [458], who have shown good agreement between the analytical solution and numerical simulations.

Propagation speed of HW and LW

The expressions (B.48) are rather complex and do not provide straightforward insight into the behaviour of the tidal wave. Therefore we will examine the HW and LW propagation speed (c^+), resp. (c^-), which can be obtained from (B.48) by solving the equation $\eta_t = 0$. The solution of this equation yields the HW and LW trajectories $t^\pm(x)$. The calculation is simplified by restricting it to an x-domain much smaller than the tidal wavelength and the tidal damping length, $(kx)^2$, $(\mu x)^2 \ll 1$. We look for solutions $t^\pm(x)$ satisfying for

HW: $(kx - \omega t^+)^2 \ll 1$ and for LW: $(\pi + kx - \omega t^-)^2 \ll 1$. The result is

$$c^\pm = \left(\frac{dt^\pm}{dx} \right)^{-1} = \frac{\omega}{k} \left[1 \pm \frac{a}{hk} \left((p+3)(k-k') \right. \right.$$
$$\left. \left. + 2\beta \left(p + 1 + \frac{h}{h_S} \right) (\mu - \mu') \right) \right]. \tag{B.49}$$

Propagation speed for weak friction

For small friction $(r/h\omega < 1)$ we have $k \approx k'$ and $\beta(\mu - \mu') \approx k/4$. We find for the HW and LW propagation speed

$$c^\pm \approx \sqrt{gh_S} \left[1 \pm \left(\frac{a}{h} + \frac{a}{h_S} + \frac{D_S^+ - D_S^-}{D_S^+ + D_S^-} \right) \right]. \tag{B.50}$$

This expression is similar to (B.43). The last term within brackets () will usually be positive and in that case the HW-propagation speed is larger than the LW-propagation speed. For tidal basins with large tidal flats $(b_S^+ \gg b_S^-)$ the propagation depth $D_S = Db_C/b_S$ is larger at LW than at HW and the term within brackets may become negative; the HW-propagation speed will then be lower than the LW-propagation speed. The physical explanation is that the HW propagation is slowed down by filling of the tidal flat areas.

Propagation speed for strong friction

For strongly dominating friction $(r/h\omega \gg 1)$ we have $k \approx \mu \approx \sqrt{2}k' \approx \sqrt{2}\mu' \approx \sqrt{r\omega/2ghh_S}$. Substitution in (B.49) yields

$$c^\pm = \sqrt{\frac{2g\omega hh_S}{r}} \left[1 \pm \left(1 - \frac{1}{\sqrt{2}} \right) \left(\frac{D_S^+ - D_S^-}{D_S^+ + D_S^-} + 3\frac{a}{h} \right) \right]. \tag{B.51}$$

For strong friction the propagation speed is influenced by tidal flat filling in the same way as for weak friction. This is not surprising, because tidal flat filling is a kinematic effect which to first order is independent of friction. However, water flow is friction dependent and the average tidal propagation velocity therefore also depends on friction.

Comparison with the wave-propagation method

A similar result can be obtained by the wave propagation method. We start from the tidal equations (B.3, B.8) for a uniform channel and linear friction

$$D_S u_x + D_S/D \eta_x u + \eta_t = 0, \tag{B.52}$$

$$u_t + u u_x + g \eta_x + ru/D = 0. \tag{B.53}$$

We only consider short time intervals around HW and LW where D_S, D are replaced by their values at HW and LW, respectively D_S^+, D^+ and D_S^-, D^-. We consider these values as constants and then solve the linear equations to order a/h. For weak friction the result is identical to (B.50). For strong friction the result is

$$c^{\pm} = \sqrt{\frac{2g\omega h h_S}{r}} \left[1 \pm \frac{1}{2} \left(\frac{D_S^+ - D_S^-}{D_S^+ + D_S^-} + \frac{a}{h} \right) \right]. \tag{B.54}$$

The influence of tidal flats on the propagation speed is expressed in almost the same way as in (B.51), but otherwise the expressions (B.51) and (B.54) are almost a factor 2.5 different. The reason is that $D_S^{\pm}$, $D^{\pm}$ are not really constants at HW and LW, but that they are functions of x, due to frictional damping. This shows that the wave propagation method is only valid in situations where the tidal amplitude is not strongly varying as a function of x.

Tidal asymmetry for strong friction

The influence of tides on morphology is much greater in the case of strong friction than in the case of weak friction. Therefore we will analyse (B.48) more in detail for the strong-friction limit $r/h\omega \gg 1$. The solution (η_2, u_2) for the second harmonic tidal component M4 then takes the simpler form

$$\eta_2 = \frac{p'\hat{\eta}^2}{h} kx(\cos 2\theta - \sin 2\theta), \tag{B.55}$$

$$u_2 = \frac{\hat{U}\hat{\eta}}{4h} \left[(-p' + 2) \cos \left(2\theta + \frac{\pi}{4} \right) + 2p'kx \cos \left(2\theta + \frac{\pi}{2} \right) \right], \tag{B.56}$$

where we have assumed $(kx)^2 \ll 1$ and where we have used the notations

$$\hat{\eta} = ae^{-\mu x}, \quad \hat{U} = \frac{a\omega}{h_S K} e^{-\mu x}, \quad \theta = \omega t - kx, \quad p' = (2 - \sqrt{2})(p + 3).$$

The expression (B.55) shows that for $p' > 0$ the harmonic tidal component M4 causes tidal wave peaking by the term $\cos 2\theta$ (more pronounced HW and less pronounced LW) and faster tidal rise than tidal fall by the term $\sin 2\theta$. The inverse occurs for $p' < 0$, i.e., $p < -3$, corresponding to a basin with very large intertidal areas. Calling $b_S^{\pm}$ the total basin at resp. HW and LW and assuming little tidal variation of the channel width b_C, then

$$p \approx 1 - \frac{h}{a} \frac{b_S^+ - b_S^-}{b_S^+ + b_S^-}. \tag{B.57}$$

The condition $p < -3$ corresponds to an intertidal width $b_S^+ - b_S^- > 4a(b_S^+ + b_S^-)/h$, a situation hardly occurring in practice. The expression (B.56) shows that for $p' \gg 1$ the harmonic tidal component M4 causes at the mouth ($x = 0$) a higher ebb velocity than flood velocity, but further landward ($kx > 0.35$) a higher flood velocity than ebb velocity.

Local excitation and damping of higher harmonics

Equation (B.47) explains how the first harmonic overtide M4 is generated and how it propagates. The r.h.s. of this equation represents the generation terms. For weak friction ($r/h\omega \ll 1$) the last two generation terms are much larger than the first two; the M4-tide is then mainly due to the positive correlation between water depth and tidal velocity in the mass-balance and also due to gradients of momentum advection in the momentum balance. For strong friction ($r/h\omega \gg 1$) these terms are subordinate to the first two terms in the r.h.s.; in that case the M4-tide is mainly due to ebb-flood asymmetry in the friction term of the momentum balance. This asymmetry is again related to the positive correlation between the tidal fluctuations of water depth and flow velocity. More M4 is generated in the situation of strong friction than in the situation of weak friction.

The l.h.s. of Eq. (B.47) represents the propagation of the M4-tide. For weak friction only the first two terms are important; the M4-tide then propagates as a wave. This wave is reflected at the basin boundaries and may become resonant if the length scale of the basin is close to 1/8 times the M2 tidal wavelength.

In the case of strong friction the first term at the l.h.s. can be neglected relative to the other two. The M4 tidal equation then reads

$$\eta_t^{(1)} - \Xi\eta_{xx}^{(1)} = -\frac{h_S}{h}\left[(p+1)\,u_x^{(0)}\eta^{(0)} + 2u^{(0)}\eta_x^{(0)}\right]. \tag{B.58}$$

The l.h.s. is a diffusion equation, with diffusion coefficient $\Xi = ghh_S/r$. The M4-tide may still be reflected at the basin boundaries, but it will not resonate because the wave is diffused and damped while propagating. The equations (B.47) and (B.58) show that the M4-tide is locally generated through the depth-dependency of the friction term and that it is dissipated through diffusion when propagating away from the location of generation. The diffusion coefficient Ξ is inversely proportional to the friction coefficient r; however, the length scale $[x]$ is inversely proportional to the wavenumber k, which is proportional to the square root of r in the case of strong friction. The diffusion term $\Xi\eta_{xx}^{(1)}$ therefore will not decrease with increasing friction. The harmonic overtides are sometimes dealt with as freely propagating waves. In the case of strong friction this is a misleading interpretation, and strong friction is precisely the relevant situation for sediment transport and morphology.

B.6. Tidal Wave in a Uniform Basin of Finite Length

Partially standing wave

In a uniform tidal basin of finite length the tidal wave is reflected at the landward boundary. The resulting tide has the character of a damped, partially standing wave. If frictional damping is very strong the tidal amplitude decreases in landward direction; in the case of weak friction the amplitude increases due to reflection. In general the tidal amplitude will not strongly vary along the basin. In that case the tidal equations may be solved using the wave-propagation method. We start from the tidal equations (B.12) in which only the most important linear and nonlinear terms are retained and where the friction is linearised

$$D_S u_x + \eta_t = 0, \tag{B.59}$$

$$u_t + g\eta_x + ru/D = 0. \tag{B.60}$$

Again we focus on time intervals around HW and LW where the depth D and the propagation depth D_S can be replaced by constants, $D^\pm$, $D_S^\pm$. The solution of the linearised equations is given by

$$\eta = \frac{1}{2}a_l[e^{-\mu(x-l)}\cos(k(x-l)-\omega t)$$
$$+ e^{\mu(x-l)}\cos(k(x-l)+\omega t)], \tag{B.61}$$
$$u = \frac{1}{2}\frac{a_l}{D_S}\frac{\omega}{\sqrt{k^2+\mu^2}}[e^{-\mu(x-l)}\cos(k(x-l)-\omega t - \varphi)$$
$$- e^{\mu(x-l)}\cos(k(x-l)+\omega t + \varphi)], \tag{B.62}$$

where a_l is the tidal amplitude at the landward boundary. These expressions are approximately valid only during the HW and LW tidal phases; for each tidal phase we have a different wave number, propagation velocity and phase, $k^\pm$, $c^\pm$, $\varphi^\pm$ respectively, see (B.31). In these expressions the time origin is chosen such that $t = 0$ coincides with HW at the landward basin boundary $x = l$. Since the tidal phases HW and LW are close to slack water ($u \approx 0$) the neglected nonlinear terms in the mass and momentum balance equations, $D_S u \eta_x/D$ and uu_x, are of minor influence on the HW and LW propagation.

Propagation speed of HW and LW

The propagation velocities of HW and LW throughout the basin can be determined from the expression for $\eta(x, t)$ along the HW and LW trajectories. The difference in propagation speed between HW and LW causes a difference between periods of rising and falling tide. The difference between the ebb and flood flow durations can be determined from the expression of $u(x, t)$. This asymmetry between ebb and flood yields an expression for the difference between maximum flow velocities for ebb and flood, which can be related to a difference of sediment transport during the ebb and flood periods. The derivation is given below.

The times of HW and LW at any location can be determined from the times of HW and LW at the basin end $x = l$, by solving $\eta_t = 0$. From the equation for η it follows that a solution should satisfy

$$e^{-\mu(x-l)}\sin(k(x-l)-\omega t) = e^{\mu(x-l)}\sin(k(x-l)+\omega t),$$

or

$$\tan(\omega t) = -\tan(k(x-l))\tanh(\mu(x-l)).$$

We consider basins which are short compared to the tidal wave length, i.e., $(\mu l)^2 \ll 1$, $(kl)^2 \ll 1$. In this case the result reads

$$\omega t \approx -k\mu(x-l)^2, \tag{B.63}$$

implying

$$t_{HW}(x) - t_{HW}(0) \approx \frac{k^+ \mu^+}{\omega}[l^2 - (x-l)^2],$$

$$t_{LW}(x) - t_{LW}(0) \approx \frac{k^- \mu^-}{\omega}[l^2 - (x-l)^2]. \tag{B.64}$$

The HW and LW trajectories $t_{HW}(x)$, $t_{LW}(x)$ are shown in Fig. B.10 for fictitious values of $k^\pm$, $\mu^\pm$.

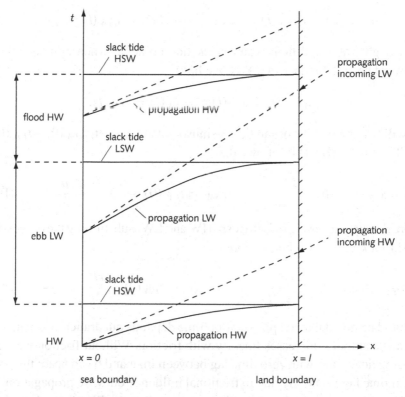

Fig. B.10. Location of HW, LW, and slack waters HSW and LSW as a function of time for a partially reflected damped wave in a tidal basin of length l.

Time of high water slack tide and low water slack tide

To determine the periods of ebb and flood, we have to determine the times HSW (high slack water) and LSW (low slack water); for this we solve the equation $u(x, t) = 0$,

$$e^{-\mu(x-l)} \cos(k(x - l) - \omega t - \varphi) = e^{\mu(x-l)} \cos(k(x - l) + \omega t + \varphi),$$

or

$$\tan(\omega t + \varphi) = \tanh(\mu(x - l)) / \tan(k(x - l)).$$

For a tidal basin much shorter than the tidal wavelength we find $\tan(\omega t + \varphi) \approx \mu/k$. The time t is approximately independent of x, implying that slack water occurs approximately at the same time everywhere in the basin (see Fig. B.10),

$$t_{HSW}(x) \approx t_{HSW}(l) = t_{HSW}, \quad t_{LSW}(x) \approx t_{LSW}(l) = t_{LSW}.$$

Since high water and high water slack tide coincide when approaching the landward basin boundary $(x \to l)$, we have

$$t_{HSW}(l) = t_{HW}(l), \quad t_{LSW}(l) = t_{LW}(l).$$

We will assume that the ocean tide is sinusoidal at the inlet, $t_{HW}(0) - t_{LW}(0) = \frac{1}{2}T$. Then from (B.64) it follows that

$$t_{HSW} \approx t_{HW}(0) + \frac{k^+ \mu^+}{\omega} l^2, \quad t_{LSW} \approx t_{LW}(0) + \frac{1}{2}T + \frac{k^- \mu^-}{\omega} l^2, \quad \text{(B.65)}$$

where the superscripts $\pm$ refer to resp. HW and LW. Substituting the expressions (B.31) for k and μ, we finally obtain

$$t_{HSW} \approx t_{HW}(0) + \frac{rl^2}{2g} \frac{1}{D^+ D_S^+}, \quad t_{LSW} \approx t_{LW}(0) + \frac{rl^2}{2g} \frac{1}{D^- D_S^-}. \quad \text{(B.66)}$$

It should be noted that the propagation time depends quadratically on the basin length l and that it increases linearly with friction. Without friction we have a standing tidal wave with zero time lag between inlet and basin head; the propagation time lag is entirely due to frictional influence on wave propagation. The quadratic dependence of propagation time on distance illustrates that friction induced wave propagation is basically a diffusion process.

Ebb-flood asymmetry

We will now derive the maximum flow velocity difference between ebb and flood from the difference between the periods of ebb and flood. This difference will then be related to the sediment transport difference between ebb and flood. We will assume that river discharge is negligibly small compared to tidal fluxes. We then have

$$P = \int_{floodperiod} A_C u \, dt = - \int_{ebbperiod} A_C u \, dt. \qquad (B.67)$$

in which P is the tidal prism (water volume entering during flood by tidal motion). If we assume that the flood and ebb velocity curves can be represented by a positive and a negative part of a sinus function, then it follows from the above the flow velocity can be approximated by the following function (see Figs. B.10 and B.11)

$$t_{LSW} - T < t < t_{HSW} : \quad u = u_{flood} \sin \left(\frac{\pi(t - t_{LSW} + T)}{t_{HSW} - t_{LSW} + T} \right),$$

$$t_{HSW} < t < t_{LSW} : \quad u = u_{ebb} \sin \left(\frac{\pi(t - t_{LSW})}{t_{LSW} - t_{HSW}} \right). \qquad (B.68)$$

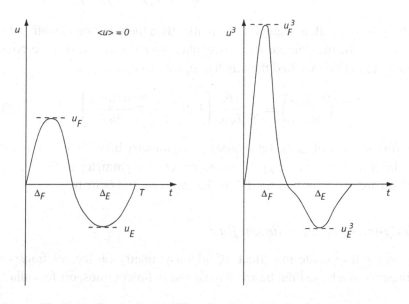

Fig. B.11. Sketch of net sediment flux due to ebb-flood asymmetry.

We will assume that the conveyance section $A_C(t)$ varies linearly with the tidal level $\eta(t)$

$$A_C \approx A_{0C} + A_{1C}\eta/a. \tag{B.69}$$

If the width of the conveyance section is more or less constant during the tidal period ($b_C \approx b_C^+ \approx b_C^-$), then $A_{0C} \approx hb_C$, $A_{1C} \approx ab_C$. Substitution of (B.68) and (B.69) in (B.67) yield estimates for u_{flood} and u_{ebb}

$$\begin{pmatrix} u_{flood} \\ u_{ebb} \end{pmatrix}$$
$$= P \left[\frac{A_{0C}}{\pi}(T \pm 2\Delta) \pm \frac{A_{1C}}{4\pi}\frac{((T-\Delta)^2 - \Delta^2)T}{\Delta(T-\Delta)} \sin\frac{\pi\Sigma}{T} \sin\frac{\pi\Delta}{T} \right]^{-1},$$
$$\tag{B.70}$$

where $\Delta = -\Delta_{EF}/2$ is half the difference between flood and ebb duration and $\Sigma = 2\Delta_S$ is twice the average phase difference between velocity and water level variation at the inlet,

$$\begin{pmatrix} \Sigma \\ \Delta \end{pmatrix} = \frac{l^2}{\omega}(k^+\mu^+ \pm k^-\mu^-) \approx \frac{rl^2}{2g}\left(\frac{1}{D^+D_S^+} \pm \frac{1}{D^-D_S^-}\right). \tag{B.71}$$

For tidal basins with a length much smaller than the tidal wavelength we have $\Delta, \Sigma \ll T$. If, in addition, the conveyance width is assumed to be constant throughout the tide we find the much simpler expressions

$$\begin{pmatrix} u_{flood} \\ u_{ebb} \end{pmatrix} = \frac{\pi}{T}\frac{P}{b_C h}\left[1 \pm \frac{\Delta_{EF}}{T} \mp \frac{\pi^2 a\Delta_S}{2hT}\right]. \tag{B.72}$$

This formula describes the two causes of asymmetry between maximum flood and ebb flow: The term Δ_{EF}/T is the effect of asymmetry in flood and ebb duration and the term $\pi^2 a\Delta_S/hT$ is the effect of Stokes drift.

Tidally-induced net sediment flux

Now we will evaluate the effect of tidal asymmetry on the net transport of sediment through the tidal basin. We use the sediment transport formula

$$q \propto b_C|u|^{n-1}u. \tag{B.73}$$

Upon substitution of (B.68) we find at the inlet $x = 0$

$$\langle q \rangle \propto \left(1 - \frac{\Delta_{EF}}{T}\right) u^n_{flood} - (1 + \frac{\Delta_{EF}}{T}) u^n_{ebb}. \tag{B.74}$$

We substitute (B.72), still assuming $\Delta_{EF}, \Delta_S \ll T$, and find to first order in $\Delta_{EF}/T, \Delta_S/T$,

$$\langle q \rangle \propto \frac{\Delta_{EF}}{2T} - \frac{n}{n-1} \frac{\pi^2 a \Delta_S}{4hT}. \tag{B.75}$$

We now substitute (B.71) and use the approximation $|D^+ D_S^+ - D^- D_S^-| \ll D^+ D_S^+ + D^- D_S^-$; this yields

$$\langle q \rangle \propto \left(1 - \frac{n\pi^2 a}{8(n-1)h}\right) D^+ D_S^+ - \left(1 + \frac{n\pi^2 a}{8(n-1)h}\right) D^- D_S^-. \tag{B.76}$$

Using the approximation $D^{\pm} = h \pm a$, this can also be written

$$\langle q \rangle \propto \left(1 - \frac{n\pi^2 a}{8(n-1)h}\right) \left(1 + \frac{a}{h}\right)^2 \frac{b_C^+}{b_S^+}$$
$$- \left(1 + \frac{n\pi^2 a}{8(n-1)h}\right) \left(1 - \frac{a}{h}\right)^2 \frac{b_C^-}{b_S^-}. \tag{B.77}$$

The Eqs. (B.76, B.77) relate the net sediment import or export in a tidal basin to its geometric characteristics. The proportionality coefficient is a variable, and depends on basin geometry; however, this dependence does not influence the net sand transport direction. In the case of sufficient sediment supply tidal basins are close to equilibrium, i.e., $\langle q \rangle$ is small or zero. A perturbation of an equilibrium geometry which affects $D^+ D_S^+$ or $D^- D_S^-$ will produce a net sediment transport with a sign that does not depend on the variation of the proportionality coefficient.

Influence of Stokes drift

The factors between brackets, $(1 \pm (n\pi^2 a)/(8(n-1)h)$, express the influence of Stokes drift on the net tidal transport. For tidal basins with a small amplitude-to-depth ratio ($a/h \le 0.1$) these factors are close to 1. If the amplitude-to-depth ratio is larger, the Stokes effect favours seaward transport of sediment over landward transport of sediment.

B.7. Strongly Converging Tidal Basins

Mass and momentum balance equations

In strongly converging tidal inlet systems the geometrical length scale is much smaller than the tidal length scale $k^{-1} = L/2\pi$; this is the opposite of previously discussed uniform basins where the geometrical length scale is much larger. The equations describing tidal wave propagation in this case are, according to the scaling analysis (B.13),

$$\left(\frac{D_x}{D} + \frac{b_{Cx}}{b_C}\right) D_S u + \eta_t = 0,$$

$$g\eta_x + r\frac{u}{D} = 0. \qquad (\text{B.78})$$

In these equations the friction term has been linearised and it has been assumed that this term is much larger than the inertial term u_t. We call the topographic convergence length L_b; over a distance L_b the cross-sectional area decreases by a factor e. We assume that the basin length is several times larger than L_b; in that case reflection of the tidal wave at the landward boundary will be hardly significant. The tidal equations (B.78) show a remarkable difference with the tidal wave propagating into a uniform basin without friction. In the latter case the tidal velocity and surface elevation are in phase. In a strongly converging friction-dominated basin the tidal velocity and the surface elevation are 90° out of phase, as for a frictionless standing wave.

Exponentially converging basins

The tidal equations (B.78) have a simple solution for the special situation of an exponentially converging basin with uniform depth. In this case the tidal equations read

$$-\frac{D_S}{L_b}u + \eta_t = 0,$$

$$g\eta_x + r\frac{u}{D} = 0. \qquad (\text{B.79})$$

If the depth and the propagation depth are assumed constant, $D = h$, $D_S = h_S$ we find

$$\eta(x, t) = a \cos k(x - ct), \quad u(x, t) = -\frac{aL_b\omega}{D_S} \sin k(x - ct), \quad \text{(B.80)}$$

where the wavenumber k and the propagation velocity c are given by

$$k = \frac{rL_b\omega}{ghh_S}, \quad c = \frac{ghh_S}{rL_b}. \quad \text{(B.81)}$$

In spite of strong frictional momentum dissipation the tidal wave is not damped when propagating landward. Frictional damping is exactly balanced by convergence of landward propagating tidal energy; tidal systems with this characteristic are called 'synchronous'. Because of the tidal variation of depth D and propagation depth D_S the equations (B.79) are nonlinear.

Propagation speed of HW and LW

To investigate the influence of this nonlinearity on the tidal propagation velocity we can use the wave propagation method; this is permitted because of the constant tidal amplitude. Equations (B.79) are approximately linear around HW and LW, where D and D_S are replaced by their HW and LW values. It immediately follows that the wave propagation speed at HW (c^+) resp. LW (c^-) is given by

$$c^\pm = \frac{gD^\pm D_S^\pm}{rL_b}. \quad \text{(B.82)}$$

The times of high and low water increase linearly with x,

$$t_{HW}(x) = t_{HW}(0) + \frac{rL_bx}{gD^+D_S^+}, \quad t_{LW}(x) = t_{LW}(0) + \frac{rL_bx}{gD^-D_S^-}. \quad \text{(B.83)}$$

Because tidal velocity and surface elevation are 90° out of phase, the times of HW and LW coincide with the times of slack water HSW and LSW, respectively. Therefore we can compare (B.83) with (B.66), for the case of a uniform basin. The two expressions are very similar. We may conclude that tidal asymmetry depends on basin topography (depth, tidal flat area) in the same way for uniform and for strongly convergent basins.

Tidally induced sediment flux

An important difference between the uniform and converging basins is the absence of Stokes drift in the latter case. For strongly converging friction dominated basins tidally driven sediment transport is entirely determined by tidal asymmetry due to differences in HW and LW propagation speed. Instead of (B.76) we now have

$$\langle q \rangle \propto D^+ D_S^+ - D^- D_S^-. \tag{B.84}$$

Tidal bore

If the intertidal area is relatively small compared to the total basin surface, $b_S^+ - b_S^- \ll b_S^+$, we have $D_S^+ D^+ > D_S^- D^-$. In that case the high water wave propagates faster than the low water wave; the tidal wave thus assumes an asymmetric form for which the period of tidal rise is shorter than the period of tidal fall. The tidal component of the flow velocity at flood will be larger than that at ebb. This tidal asymmetry may sometimes lead to the development of a tidal bore, a breaking tidal wave in which HW has overtaken LW. In a uniform tidal river this may happen in theory, but in practice the tide is damped before the tidal bore has developed. In a funnel-shaped tidal river the tidal amplitude increases rather than decreases; tidal bores are observed only in funnel-shaped rivers. Figures B.12 and B.13 show a tidal bore at spring tide in the Amazon.

Moderately converging basins

The tidal equations (B.79) are a very crude simplification of tidal dynamics. Neglecting the term $D_S u_x$ in the mass balance assumes that the convergence length L_b is much smaller than the inverse wavenumber $1/k = L/2\pi$ and neglecting the inertial term u_t in the momentum balance assumes that the friction coefficient r/h is much larger than the tidal frequency ω. The first condition is better satisfied for basins of large depth, while the second condition is better satisfied for basins of small depth. Examples of estuaries belonging to the category of strongly dissipative and strongly convergent tidal basins are the Conwy, Hoogly, Khor, Ord, Tamar and Thames estuaries [161, 268].

For tidal inlets which are moderately converging and moderately friction dominated the terms $D_S u_x$ and u_t in the mass balance equation and the

Fig. B.12. Picture of a tidal bore in the Amazon River near the mouth of the Araguairi River. Note the disturbances propagating behind the front. From [296], reproduced with permission of Scientific American.

momentum balance equation cannot be ignored,

$$-\frac{D_S}{L_b}u + D_S u_x + \eta_t = 0,$$

$$u_t + g\eta_x + r\frac{u}{D} = 0. \tag{B.85}$$

If the depth D and propagation depth D_S are taken constant ($D = h$ and $D_S = h_S$) this set of equations is linear; it can be solved by eliminating η through cross-differentiation and by substituting $u \propto \exp i(\kappa x - \omega t)$. The wavenumber κ is a complex number, $\kappa = k + i\mu$. If we assume $|\mu| \ll k$, then the wavenumber k can still be approximated by (B.81). The damping (or amplification) coefficient μ is given by

$$\mu \approx L_b(k^2 - k_0^2), \quad k_0 = \omega/\sqrt{gh_S}. \tag{B.86}$$

The first term in the brackets is the contribution of $D_S u_x$ and produces damping of the tidal wave; the second term in the brackets is the contribution of u_t and produces amplification. If the tidal basin is strongly converging (L_b is small) and friction is not too strong the expression in the brackets is negative and the tidal wave is amplified when propagating. Such a landward increase of tidal

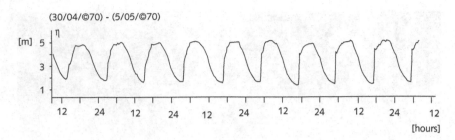

Fig. B.13. Sequence of tidal bores in the Amazon recorded at Ponto do Guara, south of the mouth of the Araguairi River, in the period 30–4–1970 to 5–5–1970. Redrawn after [173].

amplitude is indeed observed in many converging estuaries [161]. Tidal curves in the Gulf of St. Lawrence are shown in Fig. B.14 as an example.

Linearisation of the friction term

In all the analytical models the bottom shear stress has been linearised. However, the linearised friction coefficient is in reality time-dependent and it equals zero at slack water. This may cast some doubts on the validity of the propagation method, where tidal asymmetry is related to different propagation speeds of the tidal wave in the periods around HW and LW. In fact, high and low slack water (HSW and LSW) almost coincide with HW and LW in frictionally dominated, strongly convergent basins and in short, weakly converging basins. Tidal propagation therefore is almost frictionless at HW and LW. The friction-dependent propagation velocity may be considered as a reasonable approximation only some time before and after slack water, but not during slack water. Numerical simulations show that the different propagation velocity in the short periods around the slack water tidal phases will not strongly modify the overall asymmetry of the tidal curve, but it does affect the shape of the tidal curve around slack water. This point has been investigated by Lanzoni and Seminara [268], who have shown that the tidal equations (B.79) with nonlinear friction produce physically unrealistic peaking of the tidal wave around HW and LW. Removal of this anomaly requires introduction of other ignored terms. The term $D_S u_x$ in the mass balance equation, in particular, contributes to smoothing the tidal curve around slack water. This indicates that models with linearised friction should be interpreted with great care. The simple linear models or quasi-linear

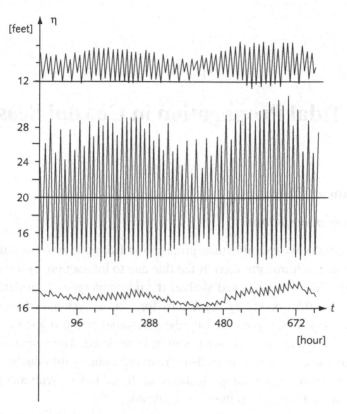

Fig. B.14. Strong tidal amplification in the funnel-shaped Gulf of St. Lawrence and decay in the upstream tidal St. Lawrence River. Water level recordings during the month October 1976; the vertical scale is in feet and the horizontal scale in hours. Upper curve: Port aux Basques, at the inlet of the Gulf; middle curve: Quebec, at the head of the Gulf; lower curve: Trois Rivieres, in the tidal river. Redrawn after [271].

models with idealised basin topography do not aim to reproduce tidal curves in real field situations with any precision; the models are intended only to reproduce the gross characteristics of the tidal wave, and, in particular, the degree of tidal asymmetry.

Appendix C

Tidal Propagation in Coastal Seas

C.1. Linear Tide

Simplifying assumptions

In this appendix we examine tidal propagation from ocean to coastal zone and in particular the transformation of the tide due to interaction with shelf topography. On wide semi-enclosed shelves the tide undergoes considerable modification. The tidal boundary conditions for coastal systems situated on such shelves are not well represented by the sinusoidal ocean tides; the distortion of the tidal wave in the shelf sea has to be considered. The discussion of tidal propagation in coastal seas is made more transparent by introducing a number of simplifications. These simplifications are listed below, with a comment on the conditions under which they are applicable.

- Neglect of tide-generating forces by celestial bodies. This is justified because tide-generating forces are extremely small compared to the tidal momentum dissipation in shallow coastal seas.
- Neglect of vertical fluid accelerations. This is justified because the horizontal scale of motion exceeds the vertical scale by several orders of magnitude.
- Neglect of the interaction of tidal flow with density stratification. Density stratification decreases energy dissipation and diminishes the interaction with seabed topography. In regions with strong tidal currents there is generally no strong density stratification.
- Neglect of velocity veering in the water column. Earth rotation produces a variable cross-flow acceleration along the water column. The resulting velocity veering affects momentum dissipation and invalidates the quadratic relationship between momentum dissipation and depth-averaged current velocity. This effect is of minor importance in shallow water (water depths of approximately 20 m or less).

- Neglect of horizontal momentum exchange. The horizontal scale of velocity gradients is much larger than the vertical scale; momentum exchange by horizontal velocity fluctuations is therefore much smaller than momentum exchange by vertical velocity fluctuations.
- Neglect of interaction between different tidal constituents. Different tidal constituents may interact through the nonlinear terms in the tidal equations. If one tidal constituent dominates the others (generally the semi-diurnal tidal component) the tides generated by this interaction can be ignored relative to the higher harmonic tides generated by the main tidal constituent.
- Neglect of small-scale topographic structures. We may average the tidal equations over spatial intervals which are small compared to the tidal wavelength but large compared to topographic features without significant influence on tidal propagation characteristics.
- Linearisation of frictional momentum dissipation. The quadratic nature of frictional momentum dissipation generates higher harmonic tides, in particular a tidal component with a frequency three times the frequency of the main tidal constituent. Ignoring this locally generated tidal component has only a minor influence on ebb-flood asymmetry.
- Neglect of nonlinear terms in the tidal equations is equivalent to ignoring tidal distortion and excludes ebb-flood asymmetry. The magnitude of the nonlinear terms relative to the linear terms is of the order of the amplitude-to-depth ratio. If this ratio is small, tidal propagation can be described by linearised tidal equations. Analysis of tidal asymmetry requires consideration of the nonlinear terms.

Linear tidal equations

Considering these simplifications we may formulate tidal equations for the depth-averaged flow components u, v (we have dropped overbars), which are temporally averaged over turbulent fluctuations and spatially averaged over small scale topographic features. Pressure gradients are related to the water surface slope and frictional momentum dissipation is linearly related to the depth-averaged velocity. Momentum advection terms are neglected. The tidal equations then read

$$u_t - fv + g\eta_x + ru/h = 0, \tag{C.1}$$

$$v_t + fu + g\eta_y + rv/h = 0, \tag{C.2}$$

$$\eta_t + (hu)_x + (hv)_y = 0. \tag{C.3}$$

These equations describe several types of wave motion, depending on topography and boundary conditions. Two-dimensional tide propagation is strongly influenced by earth rotation, as the tidal radial frequency ω and the Coriolis parameter f have similar order of magnitude.

Kelvin wave along an infinite uniform coastline

In the vicinity of a coastline (at short distance compared to the tidal wavelength) the cross-shore flow component must vanish. The tidal wave that follows the shoreline with zero cross-shore flow is called Kelvin wave. We assume a straight infinite coastline and choose the x-axis shore-parallel and the y-axis shore-perpendicular. The alongshore tidal velocity u of the Kelvin wave is geostrophically balanced by a cross-shore water level slope

$$fu = -g\eta_y, \quad v = 0. \tag{C.4}$$

The tidal amplitude is therefore largest at the coast. For uniform depth h and negligible friction the Kelvin wave takes the form

$$\eta = ae^{-\frac{fy}{c}} \cos(kx - \omega t), \quad u = c\eta/h, \tag{C.5}$$

where $c = \sqrt{gh}$. The assumption of uniform depth implies that depth variations on the scale of the tidal wave length should be small. Because of the exponential decrease of the tidal amplitude in cross-shore direction, the Kelvin wave can be considered as a coastally trapped wave. An important property of the Kelvin tidal wave is its propagation direction: It follows the coast at the right-hand-side on the northern hemisphere and at the left-hand side on the southern hemisphere.

Kelvin wave in a semi-enclosed coastal sea

We consider a rectangular bay or an approximately rectangular coastal sea of uniform depth, with a single open ocean boundary and characteristic length scale comparable to the tidal wavelength. The North Sea, the English Channel, the Yellow Sea and the Bering Sea are examples of coastal seas that can be schematised to a first approximation as semi-enclosed coastal seas. The incoming tide propagates from the ocean boundary ($x = 0$) into the basin as a damped Kelvin wave, following the right-hand side coastline and, after reflection from the landward boundary ($x = l$), returns along the left-hand side coastline (seen from the ocean) back towards the ocean

$$\eta = \frac{1}{2}a_l\, e^{-\frac{fy}{c_1}} e^{-\mu(x-l)} \cos\left(k(x-l) + \frac{fy}{c_2} - \omega t\right)$$
$$+ \frac{1}{2}a_l\, e^{\frac{fy}{c_1}} e^{\mu(x-l)} \cos\left(k(x-l) - \frac{fy}{c_2} + \omega t\right), \qquad (C.6)$$

where $y = 0$ is chosen as basin axis ($-b/2 \le y \le b/2$) and where a_l is the tidal amplitude at the landward end of the basin axis ($x = l$, $y = 0$). Furthermore we have

$$c_1 = \frac{\omega}{k}(1 + \beta^{-2})^{1/2}, \quad c_2 = \frac{\omega}{\mu}(1 + \beta^{-2})^{1/2}, \quad \beta = \omega h/r,$$

$$k = \frac{\omega}{\sqrt{2gh}}\left(1 + (1 + \beta^{-2})^{1/2}\right)^{1/2}, \quad \mu = \frac{\omega}{\sqrt{2gh}}\left(-1 + (1 + \beta^{-2})^{1/2}\right)^{1/2}.$$

The Kelvin wave does not satisfy the boundary condition $u = 0$ at the landward boundary $x = l$. Other solutions of the tidal equations (C.1) have to be considered as well (Poincaré waves); however, at sufficient distance from the boundary the Kelvin wave is the major tidal wave component.

Amphidromic point

At a quarter wavelength from the landward boundary, at $x = l - \frac{1}{4}L$, we find the amphidromic point, around which the tide rotates in cyclonic direction. Here the tidal amplitude is zero

$$\eta = \frac{1}{2}a_l\left[e^{-\frac{fy}{c_1}}e^{\frac{\mu L}{4}} - e^{\frac{fy}{c_1}}e^{\frac{-\mu L}{4}}\right]\sin\left(\frac{fy}{c_2} - \omega t\right) = 0. \qquad (C.7)$$

The amphidromic point is not situated on the basin axis, but it is shifted in cyclonic direction (seen from the ocean) due to bottom friction. From (C.7) we find

$$y_{amphi} = \frac{c_1\mu}{4f}L. \qquad (C.8)$$

Figure 4.39 shows the phase contours of the semidiurnal tidal wave in the North Sea. The tidal wave in the central North Sea is mainly driven by the Atlantic ocean tide at the northern boundary; the amphidromic point is shifted cyclonically to the eastern part of the basin. The tidal wave in the Southern

Bight is mainly driven by the tidal wave in the central North Sea. However, it is also influenced by the tidal wave in the English Channel, which causes a westward displacement of the amphidromic point.

C.2. Tidal Distortion Along the Coast

Distortion Kelvin wave is similar to 1D wave

If the tidal amplitude is not very small compared to water depth the nonlinear terms in the tidal equations cannot be neglected. While propagating along the coast the tidal wave will be distorted. For a Kelvin wave ($v = 0$) this distortion follows from the equations of motion

$$u_t + uu_x + g\eta_x + c_D \frac{|u|u}{h + \eta} = 0,$$
$$\eta_t + [(h + \eta)u]_x = 0. \tag{C.9}$$

The form of these equations is identical to the 1D tidal equations studied in Appendix B.5. A longshore propagating Kelvin wave is distorted in the same way as a tidal wave in a tidal basin without tidal flats: During propagation, tidal rise speeds up, while tidal fall slows down and a phase difference between flow velocity and water level variation close to $\frac{1}{8}T$ develops.

Field observations

Figure 4.39 shows the distortion of the tidal wave during its northward propagation along the Dutch coast. The strongest tidal asymmetry in the southern North Sea occurs along the coast of central and northern Holland. This is also the shallowest part of the Southern Bight, with relatively strong influence of friction on tidal propagation. The tidal amplitude is rather small along the coast of Holland; this suggests that the tidal depth variation in the friction term contributes more to tidal distortion than the other nonlinear terms. Along the north-western coast of France the tidal wave is distorted in a similar way, see Fig. C.1.

Consequences of tidal distortion in coastal seas

In Sec. 4.6 the influence of offshore tidal asymmetry on the morphology of tidal basins was discussed in detail. This asymmetry corresponds to a faster

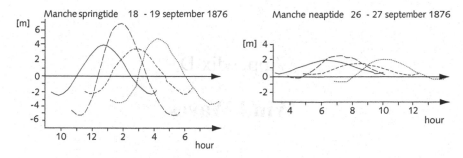

Fig. C.1. Distortion of the tidal wave while propagating along the northwestern French coast at spring tide and neap tide, after [84]. The different curves are for the stations: Brest (solid line), St. Malo (long dashes), Cherbourg (medium dashes), Le Havre (short dashes) and Boulogne (dots). Tidal asymmetry is much stronger for spring tide than for neap tide. Brest is close to the ocean boundary and the tide is almost sinusoidal; tidal rise is getting faster and tidal fall slower upon propagation into the shallow Bay of St. Malo. The tide also propagates from Brest more northward to Cherbourg through deep water; here the tide interferes with the reflected tidal wave from the Bay of St. Malo. The resulting tide is less asymmetric than at St. Malo. From Cherbourg the tide propagates westward towards Le Havre and the mouth of the Seine River; the coastal zone is rather shallow and the tidal distortion is very important. At the Seine mouth tidal rise is much faster than tidal fall. Further to the north, at Boulogne, the tide is influenced by the tidal wave coming from the North Sea. The resulting tide is less asymmetric than at Le Havre.

tidal rise and slower tidal fall and produces flood-dominant sediment transport at the inlet. Morphologic equilibrium requires neutralisation of this flood dominance; this can be achieved by return flow compensating for Stokes drift and by large intertidal flats slowing down HW propagation. The morphologic characteristics of tidal basins on wide-shelf coasts are therefore different from the characteristics of basins which are under direct influence of symmetric ocean tides, see Sec. 4.6.1.

Distortion of the tidal wave in coastal seas also influences longshore residual sediment transport. A shorter tidal rise and longer tidal fall correspond in most cases to shorter flood and longer ebb periods, in particular in the case of important friction. Observations and numerical tidal models show, for instance, that along the Dutch coast net sand transport is directed to the north [350].

Appendix D

Wind Waves

D.1. Introduction

In this appendix we will reproduce some results of classical wave theory, which are relevant for the chapter on wave-topography interaction. A more extensive treatment of wind wave theory can be found in many standard textbooks, e.g. [426, 253].

Monochromatic approximation

Waves incident on the coast are usually part of a wave field consisting of a broad spectrum of waves with different period and wavelength and with different shape and propagation direction. This wave field exerts a force on the seabed and brings sediment in motion. Changes of morphology result from the average wave action over long periods. To avoid the difficulty of determining the action of many different individual waves it is postulated that the average sediment transport produced by the real wavefield is identical to the sediment transport produced by a hypothetical wavefield consisting of single-frequency (monochromatic) waves with the same energy as the average of the real wavefield and with a frequency and direction corresponding to the peak of the energy spectrum. We will adopt this assumption and concentrate on unidirectional single-frequency waves. We ignore the interference of waves with different periods and directions and we will not consider the resulting large-scale patterns in the wave field and the eventual morphologic impact of these large-scale wave patterns.

Difference between wind wave and tidal wave dynamics

Propagation of wind waves differs in at least three important respects from propagation of tidal waves. The first difference pertains to the role of turbulence.

In tidal flow, turbulent shear stresses produce strong dissipation of momentum as the turbulent bottom boundary layer extends over a large part of the water column (in coastal waters often the entire water column). Propagation of wind waves, on the contrary, is hardly influenced by shear stresses generated through bottom friction. The wave period is too short to build up a turbulent boundary layer; the frictional layer at the bottom is often no thicker than approximately 1 cm. Absence of shear stress implies that wave-orbital motion can be considered as irrotational and that we may use the potential flow approximation. The second difference is the importance of vertical accelerations in wave-orbital motion. In tidal flow, vertical accelerations can be ignored due to the large wavelength; for the pressure we may use the hydrostatic approximation. For wave-orbital motion, in contrast, the hydrostatic approximation is not applicable. The third difference relates to the surface inclination and its influence on the development of wave asymmetry. In tidal flow the surface inclination is very small and plays a minor role in producing wave asymmetry. In wave-orbital motion the inclination of the surface imposes a relationship between the vertical and horizontal components of wave-orbital motion; the nonlinear nature of this relationship is responsible for the development of asymmetry both in wave surface and wave-orbital motion.

D.2. Linear Waves

With these considerations in mind, we will formulate equations for the description of wind wave propagation. The solution of these equations yields the major characteristics of wave-orbital motion which are relevant for wave-topography interaction. We consider a single sinusoidal incoming wave, propagating onshore (in negative x-direction) in water of depth $h(x)$ relative to still water. The wave has amplitude a, wave number $k = 2\pi/\lambda$ and period $T = 2\pi/\omega$. The vertical surface displacement caused by the incident wave is described by

$$\eta(x, t) = a \cos(kx + \omega t). \tag{D.1}$$

The wave-orbital motion being irrotational ($u_z = w_x$), we may represent the horizontal and vertical flow components by a potential $\Phi(x, z, t)$,

$$u = \Phi_x, \quad w = \Phi_z. \tag{D.2}$$

The mass balance equation $u_x + w_z = 0$ may now be written as

$$\Delta\Phi = 0. \tag{D.3}$$

The boundary conditions at the bottom (taken as horizontal) and surface read

$$w|_{z=-h} = \Phi_z|_{z=-h} = 0, \tag{D.4}$$

$$w|_{z=\eta} = \Phi_z|_{z=\eta} = \frac{d\eta}{dt} = \eta_t + u\eta_x. \tag{D.5}$$

We will first concentrate on linear wave theory, and neglect the last term on the right-hand side of (D.5). This is equivalent to neglecting terms on the order of $ka = 2\pi a/\lambda$ relative to terms of order 1, as shown by substitution of the linear solution. The wave equations then become linear and the solution reads

$$\Phi(x, z, t) = -a\frac{\omega}{k}\frac{\cosh[k(z+h)]}{\sinh(kh)}\sin(kx + \omega t). \tag{D.6}$$

For the horizontal wave-orbital velocity we find

$$u = -a\omega\frac{\cosh[k(z+h)]}{\sinh(kh)}\cos(kx + \omega t). \tag{D.7}$$

The wave-orbital motion decreases with increasing depth, although bottom friction has been ignored. In order to derive an expression for the wavenumber k as a function of ω we need an additional equation, relating surface elevation to wave-orbital motion. For this we use the momentum balance equations

$$u_t + uu_x + wu_z + \frac{1}{\rho}p_x = 0,$$

$$w_t + uw_x + ww_z + \frac{1}{\rho}p_z = -g, \tag{D.8}$$

which will be integrated over depth. This yields

$$\Phi_t + \frac{1}{2}(u^2 + w^2) + gz - \frac{1}{\rho}p(x, z, t) = f(t). \tag{D.9}$$

Assuming zero pressure at the water surface we find the Bernouilly equation

$$\Phi_t + \frac{1}{2}(u^2 + w^2) + g\eta = 0 \quad \text{at} \quad z = \eta. \tag{D.10}$$

From this equation follows a relation between k and ω, the so-called dispersion relation. Therefore we substitute (D.6) and neglect the quadratic terms (which are of order ka) and find

$$c = \frac{\omega}{k} = \sqrt{\frac{g}{k} \tanh(kh)}. \tag{D.11}$$

Wave energy

The wave energy per unit length is given by the sum of potential and kinetic energy, averaged over the wavelength,

$$E = \frac{1}{\lambda} \int_0^\lambda \left(\frac{1}{2} g \rho \eta^2 + \int_{-h}^\eta \frac{1}{2} \rho (u^2 + w^2) dz \right) dx. \tag{D.12}$$

For linear waves the result is

$$E = \frac{1}{2} \rho \, g \, a^2. \tag{D.13}$$

Wave energy propagates with a different speed than the wave speed, as will be explained below.

Wave groups and energy propagation

Within the real, irregular wave field interference may take place between waves with different periods and wavelengths [284]. This interference produces modulations of the wave amplitude, known as wave groups (see Fig. D.1). These wave groups propagate in the same direction as the constituent waves, but with

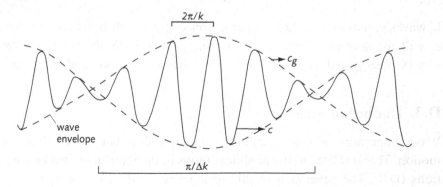

Fig. D.1. Wave train with wave groups.

a lower wave speed. This can be demonstrated by considering the superposition of two waves with close wavenumbers k and $k + \Delta k$. The sum of these two waves can be written

$$\eta_{sum} = \eta(k) + \eta(k + \Delta k) = a_{env} \sin[k(x + ct)], \tag{D.14}$$

where a_{env} is the wave envelope given by

$$a_{env} = 2a \cos[\Delta k(x + c_g t)].$$

The wave group thus travels with the group velocity c_g, which is given by

$$c_g = c + k dc/dk = nc, \tag{D.15}$$

where

$$n = \frac{1}{2} + \frac{kh}{\sin(2kh)}. \tag{D.16}$$

Wave energy also propagates with the group velocity c_g. If the wave energy would propagate at a different speed it would have to travel through the nodes of the wave group, where the wave energy is nil; obviously this is not possible. The wave energy propagates in the same direction as the individual waves, which is the direction of the wave number vector $\vec{k}$. We write $\vec{c}_g = c_g \vec{k}/k$. The energy flux is given by

$$\vec{F} = \vec{c}_g E. \tag{D.17}$$

In sufficiently deep water ($kh \geq 1$) waves dissipate so little energy that

$$\vec{\nabla}.\vec{F} \approx 0. \tag{D.18}$$

If waves propagate in a direction of decreasing depth then the group velocity c_g will also decrease. Continuity of the energy flux (D.18) then implies growth of wave energy and wave amplitude. This phenomenon is called 'shoaling'.

D.3. Radiation Stress

Wave propagation generates not only oscillating flow, but also residual water motion. This is related to the nonlinear terms in the momentum balance equations (D.8). The generation of this residual water motion can be described mathematically by means of so-called 'radiation stresses'. These stresses, first introduced by Longuet–Higgins [289], appear when the momentum balance equations are averaged over depth and wave period,

$$S^{(xx)} = \left\langle \int_{-h}^{\eta} \left(p + \rho u^2\right) dz \right\rangle - \int_{-h}^{\langle \eta \rangle} p_{hydrostat} dz,$$

$$S^{(yy)} = \left\langle \int_{-h}^{\eta} \left(p + \rho v^2\right) dz \right\rangle - \int_{-h}^{\langle \eta \rangle} p_{hydrostat} dz,$$

$$S^{(xy)} = \left\langle \int_{-h}^{\eta} \rho u v dz \right\rangle. \tag{D.19}$$

Upon substitution of the linear wave solution and assuming a uniform wave field propagating in a horizontal x-y-plain with an angle θ to the cross-shore x-axis, we find

$$S^{(xx)} \approx \left(n \left(1 + \cos^2 \theta\right) - \frac{1}{2}\right) E,$$

$$S^{(yy)} \approx \left(n \left(1 + \sin^2 \theta\right) - \frac{1}{2}\right) E,$$

$$S^{(xy)} \approx n E \sin \theta \cos \theta. \tag{D.20}$$

Wave-induced residual flow field

If the incident wave field is not spatially uniform the radiation stresses provide additional momentum to the wave-averaged flow field, both in x and y-direction. After a few algebraic manipulations, using the mass balance equation and the boundary conditions at the bottom and surface, the depth-integrated momentum balance equations become

$$\frac{\partial}{\partial t} \int_{-h}^{\eta} \rho u dz + \frac{\partial}{\partial x} \int_{-h}^{\eta} \rho(u^2 + p) dz + \frac{\partial}{\partial y} \int_{-h}^{\eta} \rho u v dz - p_b \frac{\partial h}{\partial x} + \tau_b^{(x)} = 0,$$

$$\frac{\partial}{\partial t} \int_{-h}^{\eta} \rho v dz + \frac{\partial}{\partial y} \int_{-h}^{\eta} \rho(v^2 + p) dz + \frac{\partial}{\partial x} \int_{-h}^{\eta} \rho u v dz - p_b \frac{\partial h}{\partial y} + \tau_b^{(y)} = 0. \tag{D.21}$$

Here p_b is the pressure at the bottom, for which we substitute the first-order hydrostatic approximation $p_b \approx g\rho h$. These equations describe the depth-averaged residual flow field $(u_0(x, y), v_0(x, y))$ and water surface inclination $(\eta_{0x}(x, y), \eta_{0y}(x, y))$ induced by gradients in the incident wave field. The bottom stress terms $\tau_b^{(x)}$, $\tau_b^{(y)}$ take into account the residual flow momentum dissipated by bottom friction. After averaging the equations over the wave period and introducing the radiation stresses (D.19) we find

$$(u_0{}^2)_x + (v_0 u_0)_y + g\eta_{0x} + [\tau^{(x)}{}_{0b} + S^{(xx)}{}_x + S^{(xy)}{}_y]/\rho D = 0, \quad \text{(D.22)}$$

$$(v_0{}^2)_y + (u_0 v_0)_x + g\eta_{0y} + [\tau^{(y)}{}_{0b} + S^{(yy)}{}_y + S^{(xy)}{}_x]/\rho D = 0, \quad \text{(D.23)}$$

$$(Du_0)_x + (Dv_0)_y = 0, \quad\quad\quad\quad\quad\quad\quad\quad\quad\quad\quad\quad \text{(D.24)}$$

where $D = h + \eta_0$ is the total water depth after averaging over the wave period. The variation of the residual flow over the water column has been ignored, which is a reasonable approximation, considering the small water depth landward of the breaker line.

If we assume that the coastal morphology and the incident wave field are uniform in longshore direction, all terms involving longshore gradients are zero. Because there is no net cross-shore velocity at the shoreline, we also have $u_0 = 0$. The equations then simplify to

$$g\rho D\eta_{0x} + S^{(xx)}{}_x = 0, \quad\quad\quad\quad\quad\quad\quad\quad \text{(D.25)}$$

$$\tau^{(y)}{}_{0b} + S^{(xy)}{}_x = 0. \quad\quad\quad\quad\quad\quad\quad\quad \text{(D.26)}$$

Shoaling and set-down

We first consider waves propagating in shore-normal direction ($\theta = 0$), seaward of the zone where waves are breaking. We assume that outside the surf zone energy dissipation is so small that the energy flux $F = ncE$ is approximately constant (not dependent on x or h). Substitution in (D.20) yields an expression for the radiation stress, which is used in equation (D.25). We then find the following expression for the mean water level inclination

$$\frac{d\eta_0}{dh} = -\frac{F}{g\rho h}\frac{d}{dh}\left(\frac{2n - \frac{1}{2}}{nc}\right). \quad\quad\quad\quad \text{(D.27)}$$

Here the depth D has been approximated by h and differentiation by x has been replaced by equivalent differentiation by h (we assume that h is a monotonously decreasing function of x). The right hand side may be integrated to yield (far offshore $\eta_0 = 0$)

$$\eta_0 = -\frac{1}{2}\frac{ka^2}{\sinh(2kh)} \approx -\frac{1}{4}\frac{a^2}{h}. \tag{D.28}$$

The last approximation is only valid for shallow water where $kh \ll 1$. Upon approaching shallow water — but before wave breaking — the wave amplitude increases and a radiation stress develops which produces a decrease of mean water level. This set-down becomes more important when approaching the shore, since on the one hand h decreases and, on the other hand, the wave amplitude a increases (due to shoaling), see Fig. D.2.

Wave breaking and set-up

Wave breaking results in a decrease of wave amplitude, the shallower the water, the smaller the wave amplitude after breaking. The saturation hypothesis states

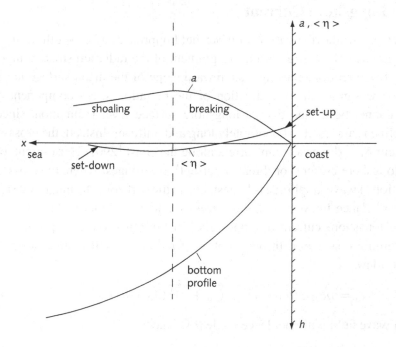

Fig. D.2. Set-down and set-up of mean water level produced by shoaling and wave breaking.

that wave breaking reduces the wave height $H = 2a$ to an approximately fixed ratio of water depth over the entire surf zone,

$$H(x) = \gamma_{br} D(x). \tag{D.29}$$

Observations indicate values of the proportionality constant γ_{br} between 0.4 and 1.3. Substitution in the radiation stress formula (D.20), together with the approximation $kh \ll 1$, yields

$$E = \frac{1}{8} g \rho \gamma_{br}^2 D^2, \quad S^{(xx)} = \frac{3}{16} g \rho \gamma_{br}^2 D^2. \tag{D.30}$$

We use Eq. (D.25) for evaluating the influence of radiation stress on the water level inclination. Assuming shore-normal wave-incidence ($\theta = 0$) we find the wave set-up

$$\eta_{0_x} = -\frac{\frac{3}{8} \gamma_{br}^2}{1 + \frac{3}{8} \gamma_{br}^2} h_x. \tag{D.31}$$

D.4. Longshore Current

Now we consider the case of a wave field approaching a longshore uniform coast at an angle θ. In addition to a gradient of the radiation stress component $S^{(xx)}$ in x-direction, causing a shoreward slope of the mean surface level, we also have a gradient in x-direction of the radiation stress component $S^{(xy)}$. This cannot be balanced by a longshore surface level inclination, since the shoreline is assumed to be infinitely long and uniform. Instead, the cross-shore gradient $S_x^{(xy)}$ drives a mean current along the coast [290]. This current, called the 'longshore current' or 'breaker current', is strongest where the cross-shore variation of wave amplitude is largest, i.e., in the surf zone. Its magnitude results from a balance between radiation stress gradient and momentum dissipation of the longshore current through turbulent bottom friction (Eq. (D.26)). For momentum dissipation through bottom friction we will assume a quadratic friction law,

$$\vec{\tau}_b = \rho c_D \sqrt{(u_0 + u)^2 + (v_0 + v)^2} \, (u_0 + u, v_0 + v). \tag{D.32}$$

For a wave field with incidence angle θ we have

$$u = U \cos \theta \cos \omega t, \quad v = U \sin \theta \cos \omega t.$$

The friction coefficient c_D for a steady flow with superposed wave field is different from the friction coefficient for steady flow; in both cases reliable estimates need to be determined experimentally. Wave-orbital velocities are in general substantially larger in the surf zone than the steady flow components; in this case we may neglect u_0^2, v_0^2 relative to U^2. The wave-incidence angle is generally so small that to a first approximation $\sin \theta$ can be neglected relative to $\cos \theta$. With these approximations, and after averaging over the wave period we find

$$\tau^{(x)}{}_{0b} = 2ru_0, \quad \tau^{(y)}{}_{0b} = rv_0, \quad r \approx 2\rho c_D U / \pi. \tag{D.33}$$

The magnitude of the longshore current V follows from (D.26)

$$V = -v_0 = -\tau^{(y)}{}_{0b}/r = S^{(xy)}{}_x/r. \tag{D.34}$$

The radiation stress $S^{(xy)}$ may be written as

$$S^{(xy)} = nE \cos \theta \sin \theta = F^{(x)} \frac{\sin \theta}{c}, \tag{D.35}$$

where $F^{(x)} = ncE \cos \theta$ is the energy flux perpendicular to the coast.

Outside the breaker zone the shoreward energy flux is approximately constant, $F_x^{(x)} \approx 0$. According to Snells' wave refraction law $\sin \theta / c = $ constant. Hence, outside the breaker zone $S_x^{(xy)} \approx 0$ and no longshore current is generated.

In the surf zone a simplified expression for $S^{(xy)}$ may be obtained if we use for the propagation speed and the wave amplitude the approximations $c_g \approx c \approx \sqrt{gh}$ and $a \approx \frac{1}{2}\gamma_{br}h$, yielding

$$S^{(xy)} \approx S_{br} = \frac{1}{8}\rho g^{3/2} \gamma_{br}^2 h^{5/2} \cos \theta_{br} \left(\frac{\sin \theta_{br}}{c_{br}} \right). \tag{D.36}$$

The index br designates values evaluated at the breaker line x_{br}, the most offshore location of wave breaking (x_{br} is the width of the surf zone). Substitution in the longshore momentum balance (D.34) yields the following expression for the longshore current V

$$V = \frac{5\pi}{16} \frac{\beta \gamma_{br}}{c_D} \sqrt{gh} \cos \theta_{br} \sin \theta_{br}, \tag{D.37}$$

where $\tan \beta \approx \beta$ represents the underwater shoreface slope.

Bibliography

[1] Alldridge, A.L. (1979) The chemical composition of macroscopic aggregation in two neretic seas. *Limnol. Oceanogr.* **24**: 855–866.

[2] Allen, J.L.R. (1968) *Current Ripples.* N. Holland Publ. Comp., Amsterdam, 433 pp.

[3] Allen, J.R.L. (1980) Sandwaves: A model of origin and internal structure. *Sed. Geol.* **26**: 281–328.

[4] Allen, J.R.L. (1982) Sedimentary structures — Their characteristics and physical basis. Vol. 1. In: *Developments in Sedimentology*, Vol. 30A, Elsevier, New York, 593 p.

[5] Allen, G.P., Sauzay, G., Castaing, S.P. and Jouanneau, J.M. (1977) Transport and deposition of suspended sediment in the Gironde estuary, France. In: *Estuarine Processes*, Vol. II. Ed., M. Wiley, Academic Press, New York, pp. 63–81.

[6] Allen, G.P., Salomon, J.C., Bassoulet, P., Du Penhoat, Y. and De Grandpre, C. (1980) Effects of tides on mixing and suspended sediment transport in macrotidal estuaries. *Sediment. Geol.* **26**: 69–90.

[7] Amos, C.L., Daborn, G.L., Christian, H.A., Atkinson, A. and Robertson, A. (1992) In situ erosion measurements on fine-grained sediments from the Bay of Fundy. *Mar. Geol.* **108**: 175–196.

[8] Amos, C.L. (1995) Siliciclastic tidal flats. In: Geomorphology and Sedimentology of Estuaries. Ed., G.M.E. Perillo, Elsevier, New York, pp. 273–306.

[9] Andersen, T.J. and Pejrup, M. (2001) Suspended sediment transport on a temperate, microtidal mudflat, the Danish Wadden Sea. *Mar. Geol.* **173**: 69–85.

[10] Anderson, T.J. (2001) Seasonal variability in erodibility of two temperate microtidal mudflats. *Est. Coast. Shelf Sci.* **45**: 507–524.

[11] Antia, E.E. (1996) Rates and patterns of migration of shoreface-connected sandy ridges along the Southern North Sea Coast. *J. Coast. Res.* **12**: 38–46.

[12] Antia, E.E. (1996) Shoreface-connected ridges in German and US Middle Atlantic Bights: Similarities and contrasts. *J. Coast. Res.* **12**: 141–146.

[13] Ashley, G.M. (1990) Classification of large-scale subaqeous bedforms: A new look at an old problem. *J. Sed. Petrol.* **60**: 160–172.

[14] Ashton, A., Murray, A.B. and Arnault, O. (2001) Formation of coastline features by large-scale instabilities induced by high-angle waves. *Nature* **414**: 296–300.

[15] Aubrey, D. (1979) Seasonal patterns of onshore/offshore sediment movement. *J. Geophys. Res.* **84**: 6347–6354.

[16] Aubrey, D.G. and Speer, P.E. (1985) A study of non-linear tidal propagation in shallow inlet/estuarine systems. Part I: Observations. *Est. Coast. Shelf Sci.* **21**: 185–205.

[17] Austin, J.A. (2004) Estimating effective longitudinal dispersion in the Chesapeake Bay. *Est. Coast. Shelf Sci.* **60**: 359–368.

[18] Avoine, J., Allen, G.P., Nichols, M., Salomon, J.C. and Larsonneur, C. (1981) Suspended-sediment transport in the Seine estuary, France: Effect of man-made modifications on estuary-shelf sedimentology. *Mar. Geol.* **40**: 119–137.

[19] Bagnold, R.A. (1963) Mechanics of marine sedimentation. In: *The Sea*, Vol. 3. Ed., M.N. Hill, Wiley-Interscience, pp. 507–528.

[20] Bailard, J.A. (1981) An energetics total load sediment transport model for a plane sloping beach. *J. Geophys. Res.* **86**: 10938–10954.

[21] Baquerizo, A., Losada, M.A. and Smith, J.M. (1998) Wave reflection from beaches: A predictive model. *J. Coast. Res.* **14**: 291–298.

[22] Barwis, J.H. (1978) Sedimentology of some South Carolina tidal-creek point bars and a comparison with their fluvial counterparts. In: *Fluvial Sedimentology*. Ed., A.D. Miall. *Can. Soc. Petr. Geol. Mem.* **5**: 129–160.

[23] Bass, S.J., Aldridge, J.N., McCave, I.N. and Vincent, C.E. (2002) Phase relationship between fine sediment suspensions and tidal currents in coastal seas. *J. Geophys. Res.* **107**: 10-1–10-6.

[24] Battjes, J.A. (1988) Surf-zone dynamics. *Ann. Rev. Fluid Mech.* **20**: 257–293.

[25] Bauer, B.O., Davidson-Arnott, R.G.D., Nordstrom, K.E., Ollerhead, J. and Jackson, N.L. (1996) Indeterminacy in aeolian sediment transport across beaches. *J. Coast. Res.* **12**: 641–653.

[26] Bearman, G. (Ed.) (1991) *Waves, Tides and Shallow-Water Processes*. The Open University, Pergamon Press, Oxford.

[27] Beets, D.J., Roep, Th.B. and de Jong, J. (1981) Sedimentary sequences of the sub-recent North Sea coast of the Western Netherlands near Alkmaar. In: *Holocene Marine Sedimentation in the North Sea Basin*. Eds., S.D. Nio, R.T.E. Schuttenhelm and Tj.C.E. van Weering. *I. A. S. Spec. Publ.* **5**: 133–145.

[28] Beets, J.D., van der Valk, L. and Stive, M.J.F. (1992) Holocene evolution of the coast of Holland. *Mar. Geol.* **103**: 423–443.

[29] Beets, J.D. and van der Spek, A.J.F. (2000) The holocene evolution of the barrier and the back-barrier basins of Belgium and the Netherlands as a function of late Weichselian morphology, relative sea level rise and sediment supply. *Neth. J. Geosci.* **79**: 3–16.

[30] Bell, P.S. (1999) Shallow water bathymetry derived from an analysis of X-band marine radar images of waves. *Coast. Eng.* **37**: 513–527.

[31] Berendsen, H.J.A. (1998) Birds-eye view of the Rhine-Meuse Delta (The Netherlands). *J. Coast. Res.* **14**: 740–752.

[32] Bernabeu, A. M., Medina, R. and Vidal, C. (2003) A morphological model of the beach profile integrating wave and tidal influences. *Mar. Geol.* **197**: 95–116.

[33] Berné, S., Lericolais, G., Marsset, T., Bourillet, J.F. and De Batist, M. (1998) Erosional offshore sand ridges and lowstand shorefaces: Examples from tide- and wave-dominated environments in France. *J. Sed. Res.* **68**: 540–555.

[34] Berné, S., Vagner, P., Guichard, F., Lericolais, G., Liu, Z., Trentesaux, A., Yin, P. and Yi, H.I. (2002) Pleistocene forced regressions and tidal sand ridges in the East China Sea. *Mar. Geol.* **188**: 293–315.

[35] Bijker, E.W. (1967) Some considerations about scales for coastal models with moveable bed. Thesis, Delft Technological University.

[36] Bird, E.C.F. (1996) Coastal erosion and rising sea-level. In: *Sea-Level Rise and Coastal Subsidence*. Ed., J.D. Milliman and B.U. Haq. Kluwer Ac. Publ., Dordrecht.

[37] Bijker, E.W., Kalkwijk, J.P.Th. and Pieters, T. (1974) Mass transport in gravity waves on a sloping bottom. In: *Procs. 14th Int. Coast. Eng. Conf.*, ASCE, pp. 447–465.

[38] Black, K.P., Gorman, R.M. and Symonds, G. (1995) Sediment transport near the break-point associated with cross-shore gradients in vertical eddy diffusivity. *Coastal Eng.* **26**: 153–175.

[39] Black, K.P., Gorman, R.M. and Byran, K.R. (2002) Bars formed by horizontal diffusion of suspended sediment. *Coastal Eng.* **47**: 53–75.

[40] Blanton, J.O., Lin, G. and Elston, S.A. (2002) Tidal current asymmetry in shallow estuaries and tidal creeks. *Cont. Shelf Res.* **22**: 1731–1743.

[41] Blanton, J.O., Seim, H., Alexander, C., Amft, J. and Kineke, G. (2003) Transport of salt and suspended sediments in a curving channel of a coastal plain estuary: Satilla River, GA. *Est. Coast. Shelf Sci.* **57**: 993–1006.

[42] Blondeaux, P. and Seminara, G. (1985) A unified bar-bend theory of rivers. *J. Fluid Mech.* **157**: 449–470.

[43] Blondeaux, P. (1990) Sand ripples under sea waves. Part 1: Ripple formation. *J. Fluid Mech.* **218**: 1–17.

[44] Blondeaux, P. (2001) Mechanics of coastal forms. *Ann. Rev. Fluid Mech.* **33**: 339–369.

[45] Bokuniewicz, H. (1995) Sedimentary systems of coastal-plain estuaries. In: *Geomorphology and Sedimentology of Estuaries*. Ed., G.M.E. Perillo, Elsevier, Amsterdam, pp. 49–67.

[46] Bolla Pittaluga, M., Repetto, R. and Tubino, M. (2001) Channel bifurcation in one-dimensional models: A physically based nodal point condition. In: *2nd IAHR Symp. Riv. Coas. Est. Morph.* Ed., S. Ikada. Obihiro, Japan, pp. 305–314.

[47] Boothroyd, J.C. (1985) Tidal inlets and tidal deltas. In: *Coastal Sedimentary Environments*. Ed., R.A. Davis, Springer-Verlag, New York, pp. 445–532.

[48] Bowen, A.J. and Inman, D.L. (1971) Edge waves and crescentic bars. *J. Geophys. Res.* **76**: 8662–8671.

[49] Bowen, A.J. (1980) Simple models of nearshore sedimentation; beach profiles and longshore bars. In: *The Coastline of Canada*, Ed., S.B. McCann, Geological Survey of Canada, Ottawa, pp. 1–11.

[50] Bowen, A.J. and Holman, R.A. (1989) Shear instabilities of the mean longshore current, 1, *Theory J. Geophys. Res.* **94**: 18023–18030.

[51] Bowen, M.M. and Geyer, W.R. (2003) Salt transport and the time-dependent salt balance of a partially stratified estuary. *J. Geophys. Res.* **108**: 27.1–27.15.

[52] Boyd, R., Forbes, D.L. and Heffler, D.E. (1988) Time-sequence observations of wave-formed sand ripples on an ocean shoreface. *Sedimentol.* **35**: 449–464.

[53] Boyd, R., Dalrymple, R.W. and Zaitlin, B.A. (1992) Classification of clastic coastal depositional environments. *Sed. Geol.* **80**: 139–150.

[54] Brander, R.W. (1999) Field observations on the morphodynamic evolution of a low-energy rip current system. *Mar. Geol.* **157**: 199–217.

[55] Brenon, I. and Le Hir, P. (1999) Modelling the turbidity maximum in the Seine estuary: Identification of formation processes. *Est. Coast Shelf Sci.* **49**: 525–544.

[56] Bruun, P. (1954) Coast erosion and the development of beach profiles. Beach Erosion Board, US Army Corps of Eng., *Tech. Mem.* **44**: 1–79.

[57] Bruun, P. (1962) Sea-level rise as a cause of shore erosion. *Proc. Am. Soc. Civ. Eng., J. Water Harbors Div.* **88**: 117–130.

[58] Bruun, P. (1963) Longshore currents and longshore throughs. *J. Geophys. Res.* **68**: 1065–1078.

[59] Bruun, P. (1988) The Bruun rule of erosion by sea-level rise: A discussion on large-scale two- and three-dimensional usages. *J. Coast. Res.* **4**: 627–648.

[60] Byrne, R.J., Gammisch, R.A. and Thomas, G.R. (1981) Tidal prism-inlet area relationships for small tidal inlets. In: *Procs. 17th Int. Coast. Eng. Conf., ASCE,* New York, pp. 2517–2533.

[61] Caballeria, M., Coco, G., Falqués, A. and Huntley, A.D. (2002) *J. Fluid Mech.* **465**: 379–410.

[62] Callender, R.A. (1969) Instability and river channels. *J. Fluid Mech.* **36**: 465–480.

[63] Calvete, D., Falqués, A., De Swart, H.E. and Walgreen, M. (2001) Modelling the formation of shoreface-connected sand ridges on storm-dominated inner shelves. *J. Fluid Mech.* **441**: 169–193.

[64] Calvete, D., De Swart, H.E. and Falqués, A. (2002) Effect of depth-dependent wave stirring on the final amplitude of shoreface-connected ridges. *Cont. Shelf Res.* **22**: 2763–2776.

[65] Camenen, B. and Larroude, P. (1999) Nearshore and transport modelling: Application to Trucvert Beach. In: *Proc. IAHR Symp. on River, Coastal and Estuarine Morphodynamics,* Vol. II: 31–40.

[66] Carbajal, N. and Montano, Y. (2001) Comparison between predicted and observed physical features of sandbanks. *Est. Coast. Shelf Sci.* **52**: 435–443.

[67] Carter, R.W.G. (1988) *Coastal Environments.* Academic Press, London, 617 pp.

[68] Cartwright, D.E. (1999) *Tides, A Scientific History.* Cambridge University Press, UK, 292 pp.

[69] Castaing, P. (1989) Co-oscillating tide controls long-term sedimentation on the Gironde estuary, France. *Mar. Geol.* **89**: 1–9.

[70] CERC Shore Protection Manual, I-III. Army Corps of Engineers, US Govt. Printing Office.

[71] Chang, J.H. and Choi, J.Y. (2001) Tidal-flat sequence controlled by holocene sea-level rise in Gomso Bay, West Coast of Korea. *Est. Coast. Shelf Science* **52**: 391–399.

[72] Chang, H., Simons, D.B. and Woolisher, D.A. (1971) Flume experiments on alternate bar formation. *Proc. ASCE, Waterways, Harbors Coast. Eng. Div.* **97**: 155–165.

[73] Chantler, A.G. (1971) The applicability of regime theory to tidal water courses. *J. Hydraul. Res.* **12**: 181–191.

[74] Chen, J., Liu, C., Zhang, C. and Walker, H.J. (1990) Geomorphological development and sedimentation in Qiantang estuary and Hangzou bay. *J. Coast. Res.* **6**: 559–572.

[75] Chung, D.H. and Van Rijn, L.C. (2003) Diffusion approach for suspended sediment transport. *J. Coast. Res.* **19**: 1–11.

[76] Clarke, L.B. and Werner, B.T. (2004) Tidally modulated occurrence of megaripples in a saturated surf zone. *J. Geophys. Res.* **109**(C01012): 1–15.

[77] Cleveringa, J. (2000) Reconstruction and modelling of Holocene coastal evolution of the western Netherlands. Thesis, Utrecht University, Geologica Ultraiectina **200**.

[78] Coco, G., O'Hare, T.J. and Huntley, D.A. (1999) Beach cusps: A comparison of data and theories for their formation. *J. Coast. Res.* **15**: 741–749.

[79] Coco, G., Huntley, D.A. and O'Hare, T.J. (2000) Investigation of a self-organization model for beach cusp formation and development. *J. Geophys. Res.* **105**: 219991–22002.

[80] Coco, G., Burnet, T.K. and Werner, B.T. (2003) Test of self-organisation in beach cusp formation. *J. Geophys. Res.* **108** (C3): 46.

[81] Coleman, S.E. and Melville, B.W. (1996) Initiation of bed forms on a flat sand bed. *J. Hydr. Eng.* **122**: 301–310.

[82] Collins, M.B., Amos, C.L. and Evans, G. (1981) Observations of some sediment-transport processes over intertidal flats, the Wash, UK. In: *Int. Ass. Sediment Spec. Publ.* **5**: 81–98.

[83] Collins, M.B. (1983) Supply, distribution and transport of suspended sediment in a macrotidal environment: Bristol Channel, UK *Can. J. Fish. Aquat. Sci.* **40**(suppl.): 44–59.

[84] Comoy, M. (1881) Etude Pratique sur les Marées Fluviales. Gauthiers-Villars, Paris.

[85] Cornaglia, P. (1889) Delle Spiaggie. Accademia Nazionale dei Lincei, *Atti. Cl. Sci. Fis., Mat. e Nat. Mem.* **5**: 284–304.

[86] Cornish, V. (1898) On sea beaches and sandbanks. *Geograph. J.* **11**: 528–559, 628–647.

[87] Cowell, P.J., Stive, M.J.F., Niedoroda, A.W., de Vriend, H.J., Swift, D.P.J., Kaminsky, G.M. and Capobianco, M. (2003) The coastal-tract: A conceptual approach to aggregate modelling of low-order coastal change. *J. Coast. Res.* **19**: 812–827.

[88] Cox, D.T. and Kobayashi, N. (1998) Application of an undertow model to irregular waves on plane and barred beaches. *J. Coast. Res.* **14**: 1314–1324.

[89] Dalrymple, R.W. (1984) Morphology and internal structure of sand waves in the Bay of Fundy. *Sedimentol.* **31**: 365–382.

[90] Dalrymple, R.W., Zaitlin, B.A. and Boyd, R. (1992) Estuarine facies models: Conceptual basis and stratigraphic implications. *J. Sed. Petrol.* **62**: 1130–1146.

[91] Dalrymple, R.W. and Rhodes, R.N. (1995) Estuarine Dunes and Bars. In: *Geomorphology and Sedimentology of Estuaries.* Ed., G.M.E. Perillo. *Developments in Sedimentology* **53**, Elsevier, Amsterdam: pp. 359–422.

[92] Damgaard, J.S., Van Rijn, L.C., Hall, L.J. and Soulsby, R. (2001) Intercomparison of engineering methods for sand transport. In: *Sediment Transport Modelling in Marine Coastal Environments.* Eds, L.C. van Rijn, A.G. Davies, J. Van de Graaff and J.S. Ribberink. Aqua Publications, Amsterdam, CJ1–CJ12.

[93] Davies, A.G. (1982) On the interaction between surface waves and undulations of the sea bed. *J. Mar. Res.* **40**: 331–368.

[94] Davies, A.G., Ribberink, J.S., Temperville, A. and Zyserman, J.A. (1997) Comparisons between sediment transport models and observations made in wave and current flows above plain beds. *Coastal Eng.* **31**: 163–169.

[95] Davies, A.G. and Villaret, C. (1998) Wave-induced currents above rippled beds and their effects on sediment transport. In: *Physics of Estuaries and Coastal Seas.* Eds., J. Dronkers and M. Scheffers. Balkema, Rotterdam, pp. 187–199.

[96] Davies, J.L. (1980) *Geographical Variation in Coastal Development.* Longman, New York, 212 pp.

[97] Davis, R.A. (1985) Beach and nearshore zone. In: *Coastal Sedimentary Environments.* Ed., R.A. Davis, Springer-Verlag, pp. 379–444.

[98] Dean, R.G. (1973) Heuristic models of sand transport in the surf zone. *Proc. Conf. Eng. Dynamics in the Surf Zone*, Sydney, pp. 208–214.

[99] Dean, R.G. (1977) Equilibrium beach profiles: US Atlantic coast and Gulf coasts. *Ocean Eng. Tech. Rep.* **12**, Univ. of Delaware, Newark, 45 pp.

[100] Dean, R.G. and Maurmeyer, E.M. (1980) Beach cusps at Point Reyes and Drakes Bay beaches, California. In: *Procs. Int. Conf. Coast. Eng. ASCE*, New York, pp. 863–884.

[101] Dean, R.G. (1991) Equilibrium beach profiles: Characteristics and applications. *J. Coast. Res.* **7**: 53–84.

[102] Dean, R.G. and Dalrymple, R.A. (2002) *Coastal Processes with Engineering Applications*. Cambridge Univ. Press, 475 pp.

[103] De Bok, C. and Stam, J.M. (2002) *Long-Term Morphology of the Eastern Scheldt*. Report Rijkswatertstaat, RIKZ 2002/108x.

[104] Defant, A. (1961) *Physical Oceanography*. Vol. II. Permanon Press, Oxford.

[105] De Haas, H. and Eisma, D. (1993) Suspended-sediment transport in the Dollard estuary. *Neth. J. Sea Res.* **31**: 37–42.

[106] Deigaard, R., Drønen, N., Fredsøe, J., Jensen, J.H. and Jørgensen, M.P. (1999) A morphology stability analysis for a long straight barred coast. *Coast. Eng.* **36**: 171–195.

[107] Dette, H.H. (2002) Sandbewegung im Küstenbereich. *Die Küste* **65**: 215–256.

[108] De Vriend, H.J., Bakker, W.T. and Bilse, D.P. (1994) A morphological behaviour model for the outer delta of mixed-energy tidal inlets. *Coast. Eng.* **23**: 305–327.

[109] De Vriend, H. (2001) Long-term morphological prediction. In: *River, Coastal and Estuarine Morphodynamics*. Eds., G. Seminara and P. Blondeaux. Springer, Berlin, pp. 163–190.

[110] Dibbits, H.A.M.C. (1950) Nederland Waterland, a historical-technical perspective. Oosthoek, Utrecht, 286 pp. (In Dutch).

[111] Dibajnia, M. and Watanabe, A. (1996) A transport rate formula for mixed-size sands. In: *Proc. Int. Conf. Coast. Eng.*, Orlando, Florida, ASCE, 3791–3804.

[112] Di Silvio, G. (1991) Averaging operations in sediment transport modelling: Short-step versus long-step morphological simulations. In: *Int. Symp. Transp. Susp. Sed. Mod.* Ed., L. Montefusco. Univ. Florence, pp. 723–739.

[113] Dolan, R. (1971) Coastal landforms: Crescentic and rhythmic. *Geol. Soc. Am. Bul.* **82**: 177–180.

[114] Doodson, A.T. (1921) The harmonic development of the tide-generating potential. *Proc. R. Soc. London, Ser. A* **100**: 305–329.

[115] Dronkers, J.J. (1964) *Tidal Computations in Rivers and Coastal Waters*. North-Holland Publ. Co., Amsterdam, 518 pp.

[116] Dronkers, J.J. (1970) Research for the coastal area of the delta region of the Netherlands. *Proc. 12th Int. Coast. Eng. Conf. Washington, ASCE*, Ch.108.

[117] Dronkers, J. and Zimmerman, J.T.F. (1982) Some principles of mixing in coastal lagoons. *Oceanologica Acta* SP, pp. 107–117.

[118] Dronkers, J. (1982) Conditions for gradient-type dispersive transport in one-dimensional tidally averaged transport models. *Est. Coast. Shelf Sci.* **14**: 599–621.

[119] Dronkers, J. (1984) Import of fine marine sediment in tidal basins. In: *Procs. Int. Wadden Sea Symp. Neth. Inst. for Sea Res. Publ. Series* **10**: 83–105.

[120] Dronkers, J. (1986) Tidal asymmetry and estuarine morphology. *Neth. J. Sea Res.* **20**: 117–131.

[121] Dronkers, J. (1998) Morphodynamics of the Dutch Delta. In: *Physics of Estuaries and Coastal Seas*. Ed., J. Dronkers and M.B.A.M. Scheffers, Balkema, Rotterdam, pp. 297–304.

[122] Dupont, J-P., Lafite, R., Huault, F., Hommeril, P. and Meyer, R. (1994) Continental/marine ratio changes in suspended and settled matter across macrotidal estuary (the Seine estuary, northwestern France). *Mar. Geol.* **120**: 27–40.

[123] Dyer, K.R. (1986) *Coastal and Estuarine Sediment Dynamics*. John Wiley, Chichester, p. 342.

[124] Dyer, K.R. and Huntley, D.A. (1999) The origin, classification and modelling of sand banks and ridges. *Cont. Shelf Res.* **19**: 1285–1330 *Int. Coast. Eng. Conf. Washington, ASCE*, Ch.108.

[125] Dyer, K.R. and Manning, A.J. (1999) Observation of the size, settling velocity and effective density of flocs and their fractal dimension. *J. Sea Res.* **41**: 87–95.

[126] Dyer, K.R., Christie, M.C. and Wright, E.W. (2000) The classification of intertidal mud-flats. *Cont. Shelf Res.* **20**: 1039–1060.

[127] Dyer, K.R., Christie, M.C. and Manning, A.J. (2004) The effect of suspended sediment on turbulence within an estuarine turbidity maximum. *Est. Coast. Shelf Sci.* **59**: 237–248.

[128] Dyer, K.R. Personal communication.

[129] Eidsvik, K.J. (2004) Some contributions to the uncertainty of sediment transport predictions. *Cont. Shelf Res.* **24**: 739–754.

[130] Einstein, H.A. and Krone, R.B. (1962) Experiments to determine modes of cohesive sediment transport in salt water. *J. Geophys. Res.* **67**: 1451–1461.

[131] Eisma, D., Bernard, P., Cadee, G.C., Ittekot, V., Kalf, J., Laane R., Martin, J.M., Mook, W.G., Van Put, A. and Schuhmacher, T. (1983) Suspended-matter particle size in some West-European estuaries; Part I: *Particle Size Distribution. Neth. J. Sea Res.* **28**: 193–214.

[132] Eitner, V. (1996) Morphological and sedimentological development of a tidal inlet and its catchment area (Otzumer Balje, Southern North Sea). *J. Coast. Res.* **12**: 271–293.

[133] Ehlers, J. (1988) The morphodynamics of the Waddensea. Balkema, Rotterdam.

[134] El Ganaoui, O., Schaaff, E., Boyer, P., Amielh, M., Anselmet, F. and Grenz, C. (2004) The deposition and erosion of cohesive sediment determined by a multi-class model. *Est. Coast. Shelf Sci.* **60**: 457–475.

[135] Elias, E., Stive, M., Bonekamp, H. and Cleveringa, J. (2003) Tidal inlet dynamics in response to human intervention. *J. Coast. Eng.* **45**: 629–658.

[136] Elliott, A.J. (1987) Observations of meteorologically induced circulation in the Potomac estuary. *Est. Coast. Mar. Sci.* **6**: 285–299.

[137] Elgar, S., Gallagher, E.L. and Guza, R.T. (2001) Nearshore sandbar migration. *J. Geophys. Res.* **106**: 11623–11727.

[138] Engelund, F. (1970) Instability of erodible beds. *J. Fluid Mech.* **42**: 225–244.

[139] Engelund, F. and Hansen, E. (1972) *A Monograph on Sediment Transport in Alluvial Streams*, 3rd Edn. Technical Press, Copenhagen.

[140] EUROSION: Living with coastal erosion in Europe (2004) Ed., P. Doody. Off. Publ. European Communities, Luxembourg, ISBN 9289474963.

[141] Falqués, A., Calvete, D. and Montoto, A. (1998) Bed-flow instabilities of coastal currents. In: *Physics of Estuaries and Coastal Seas*. Eds., J. Dronkers and M.B.A.M. Scheffers. Balkema, Rotterdam, pp. 417–424.

[142] Falqués, A. and Iranzo, I. (1994) Numerical simulation of vorticity waves in the nearshore. *J. Geophys. Res.* **99**: 835–841.

[143] Falqués, A., Coco, G. and Huntley, D.A. (2000) A mechanism for the generation of wave-driven rhythmic patterns in the surf zone. *J. Geophys. Res.* **105**(C10): 24071–24087.

[144] Falqués, A. (2003) On the diffusivity in coastline dynamics. *Geophys. Res. Letters* **30**(21): OCE 4.

[145] Falqués, A. and Calvete, D. (2005) Large-scale dynamics of sandy coastlines: *Diffusivity and Instability. J. Geophys. Res.* **110**, C03007.

[146] Feddersen, F. (2004) Effect of wave directional spread on the radiation stress: Comparing theory and observations. *Coast. Eng.* **51**: 473–481.

[147] Field, M.E., Nelson, C.H., Cacchione, D.A. and Drake, D.E. (1981) Sand waves on an epicontinental shelf: Northern Bering Sea. *Mar. Geol.* **42**: 233–258.

[148] Figueiredo, A.G., Swift, D.J.P., Stubblefield, W.L. and Clarke, T.L. (1981) Sand ridges on the inner Atlantic shelf of North America: Morphometric comparisons with Huthnance stability model. *Geomarine Letters* **1**: 187–191.

[149] Figueiredo, A.G., Sanders, J.E. and Swift, D.J.P. (1982) Storm-graded layers on inner continental shelves: Examples from Southern Brazil and the Atlantic coast of the central United States. *Sed. Geol.* **31**: 171–190.

[150] Fischer, H.B., List, E.J., Koh, R.C.Y., Imberger, J. and Brooks, N.H. (1979) *Mixing in Inland and Coastal Waters*. Academic Press, New York.

[151] FitzGerald, D.M. (1988) Shoreline erosional-depositional processes associated with tidal inlets. In: *Hydrodynamics and Sediment Dynamics of Tidal Inlets*. Eds., D.G. Aubrey and L. Weishar. Springer-Verlag, New York, pp. 186–225.

[152] Flemming, B.W. (1988) Zur klassifikation subaquatischer, strömungstransversaler transportkörper. *Bochumer Geologische und Geotechnische Arbeiten*, **29**: 44–47.

[153] Foda, M.A. (2003) Role of wave pressure in bedload sediment transport. *J. Waterway, Port, Coastal and Ocean Eng.* **129**: 243–249.

[154] Fredsøe, J. (1974) On the development of dunes in erodible channels. *J. Fluid Mech.* **64**: 1–16.

[155] Fredsøe, J. (1978) Meandering and braiding of rivers. *J. Fluid Mech.* **84**: 609–624.

[156] Fredsøe, J. (1982) Shape and dimensions of stationary dunes in rivers. *J. Hydr. Div., ASCE* **111**: 1041–1059.

[157] Fredsøe, J. and Deigaard R. (1992) *Mechanics of Coastal Sediment Transport*. World Scientific Publishing, Singapore.

[158] Friedrichs, C.T., Aubrey, D.G. and Speer, P.E. (1990) Impacts of relative sea-level rise on evolution of shallow estuaries. In: *Coastal and Estuarine Studies 38, Residual Currents and Long-Term Transport*. Ed., R.T. Cheng, Springer-Verlag, New York, pp. 105–122.

[159] Friedrichs C.T. and Madsen, O.S. (1992) Non-linear diffusion of the tidal signal in frictionally dominated embayments. *J. Geophys. Res.* **97**: 5637–5650.

[160] Friedrichs, C.T., Lynch, D.R. and Aubrey, D.G. (1992) Velocity asymmetries in frictionally-dominated tidal embayments: Longitudinal and lateral variability. In:

Dynamics and Exchanges in Estuaries and the Coastal Zone. Ed., D. Prandle. Springer-Verlag, New York, pp. 277–312.

[161]   Friedrichs C.T. and Aubrey, D.G. (1994) Tidal propagation in strongly convergent channels. *J. Geophys. Res.* **99**: 3321–3336.

[162]   Friedrichs, C.T. (1995) Stability shear stress and equilibrium cross-sectional geometry of sheltered tidal channels. *J. Coast. Res.* **11**: 1062–1074.

[163]   Friedrichs, C.T. and Aubrey, D.G. (1996) Uniform bottom shear stress and equilibrium hypsometry of intertidal flats. In: *Mixing in Estuaries and Coastal Seas, Coastal Estuarine Stud.* 50. Ed., C. Pattiaratchi. AGU, Washington D.C., pp. 405–429.

[164]   Friedrichs, C.T., Armbrust, B.D. and De Swart, H.E. (1998) Hydrodynamics and equilibrium sediment dynamics of shallow funnel-shaped tidal estuaries. In: *Physics of Estuaries and Coastal Seas.* Ed., J. Dronkers and M.B.A.M. Scheffers, Balkema, Rotterdam, pp. 315–328.

[165]   Gallagher, B. (1971) Generation of surfbeat by nonlinear wave interactions. *J. Fluid Mech.* **49**: 1–20.

[166]   Gallagher, E.L., Guza, T. and Elgar, S. (1998) Observations of sandbar evolution on a natural beach. *J. Geophys. Res.* **103**: 3203–3215.

[167]   Gallagher, E.L., Elgar, S. and Thornton, E.B. (1998) Observations and predictions of megaripple migration in a natural surf zone. *Nature* **394**: 165–168.

[168]   Gao, S. and Collins, M. (1994) Tidal inlet equilibrium in relation to cross-sectional area and sediment transport patterns. *Est. Coast. Shelf Science* **38**: 157–172.

[169]   Gerkema, T. (2000) A linear analysis of tidally generated sand waves. *J. Fluid Mech.* **417**: 303–322.

[170]   Gerritsen, F., Dunsbergen, D.W. and Israel, C.G. (2003) A rational stability approach for tidal inlets, including analysis of the effect of wave action. *J. Coast. Res.* **19**: 1066–1081.

[171]   Geyer, W.R. and Farmer, D.M. (1989) Tide-induced variation of the dynamics of a salt wedge estuary. *J. Phys. Ocean.* **19**: 1060–1072.

[172]   Glenn, S.M. and Grant, W.D. (1987) A suspended sediment stratification correction for combined wave and current flows. *J. Geophys. Res.* **92**: 8244–8264.

[173]   Godin, G. (1991) Frictional effects in river tides. In: *Tidal Hydrodynamics.* Ed., B.B. Parker, Wiley, New York, p. 19.

[174]   Goff, J.A., Swift, D.J.P., Duncan, C.S., Mayer, L.A. and Hughes-Clarke, J. (1999) High-resolution swath sonar investigation of sand ridge, dune and ribbon morphology in the offshore environment of the New Jersey margin. *Mar. Geol.* **161**: 307–337.

[175]   Gourlay, M.R. (1968) Beach and dune erosion tests, Rep. m935/m936, Delft Hydraul. Lab., Delft.

[176]   Grabemann, I., Uncles, R.J., Krause, G. and Stephens, J.A. (1997) Behaviour of turbidity maxima in the Tamar and Weser estuaries. *Est. Coast. Shelf Science* **45**: 235–246.

[177]   Graf, W.H. (1971) *Hydraulics of Sediment Transport.* McGraw-Hill, NY, 513 pp.

[178]   Grant, W.D. and Madsen, O.S. (1979) Combined wave and current interaction with a rough bottom. *J. Geophys. Res.* **84**: 1797–1808.

[179]   Grant, W.D. and Madsen, O.S. (1982) Movable bed roughness in unsteady oscillatory flow. *J. Geophys. Res.* **87**: 469–481.

[180]   Groen, P. (1967) On the residual transport of suspended matter by an alternating tidal current. *Neth. J. Sea Res.* **3**: 564–574.

[181] Grunnet, N.M., Walstra, D-J.R. and Ruessink, B.G. (2004) Process-based modelling of a shoreface nourishment. *Coast. Eng.* **51**: 581–607.

[182] Guézennec, L., Lafite, R., Dupont, J-P., Meyer, R. and Boust, D. (1999) Hydrodynamics of suspended particulate matter in the tidal freshwater zone of a microtidal estuary. *Estuaries* **22**: 717–727.

[183] Gust, G. and Walger, E. (1976) The influence of suspended cohesive sediments on boundary-layer structure and erosive activity of turbulent seawater flow. *Mar. Geol.* **22**: 189–206.

[184] Guza, R.T. and Inman, D.L. (1975) Edge waves and beach cusps. *J. Geophys. Res.* **80**: 2997–3012.

[185] Haas, K.A., Svendse, I.A., Haller, M.C. and Zhao, Q. (2003) Quasi-three-dimensional modeling of rip current systems. *J. Geophys. Res.* **108**: 10-1–10-21.

[186] Haller, M.C., Dalrymple, R.A. and Svendsen, I.A. (2002) Experimental study of nearshore dynamics on a barred beach with rip channels. *J. Geophys. Res.* **107**(C6): 14-1-21.

[187] Hallermeyer, R.J. (1981) A profile zonation for seasonal sand beaches from wave climate. *Coast. Eng.* **4**: 253–277.

[188] Hands, E.W. and Shepsis, V. (1999) Cyclic movement at the entrance to Willapy Bay, Washington, USA. In: *Coastal Sediments*. Ed., N.C. Kraus and W.G. McDougal. ASCE, pp. 1522–1536.

[189] Hanes, D. and Huntley, D. (1986) Continuous measurements of suspended sand concentration in a wave dominated nearshore environment. *Cont. Shelf Res.* **6**: 585–596.

[190] Hanes, D.M., Alymov, V. and Chang, Y.S. (2001) Wave-formed sand ripples at Duck, North Carolina. *J. Geophys. Res.* **106**: 22575–22592.

[191] Hansen, D.V. (1965) Currents and mixing in the Columbia river estuary. In: *Ocean Science and Ocean Engineering*, Vol. 2. The Marine Technology Society, Wahington D.C., pp. 943–955.

[192] Haring, J. (1970) Historische ontwikkeling in het Noordelijk Deltabekken 1879–1966. Nota W-70.060, Deltadienst, Rijkswaterstaat (in Dutch).

[193] Harms, J.C. (1969) Hydraulic significance of some sand ripples. *Geol. Soc. Amer. Bull.* **80**: 363–396.

[194] Harris, P.T. (1988) Large-scale bedforms as indicators of mutually evasive sand transport and the sequential infilling of wide-mouthed estuaries. *Sediment. Geol.* **57**: 273–298.

[195] Harris, P.T., Baker, E.K., Cole, A.R. and Short, S.A. (1993) A preliminary study of sedimentation in the tidally dominated Fly River delta, Gulf of Papua. *Cont. Shelf Res.* **13**: 441–472.

[196] Haslett, S.K., Cundy, A.B., Davies, C.F.C., Powell, E.S. and Croudace, I.W. (2003) Salt marsh sedimentation over the past c. 120 years along the West Cotentin coast of Normandy (France): Relationship to sea-level rise and sediment supply. *J. Coast. Res.* **19**: 609–620.

[197] Heathershaw, A.D. and Davies, A.G. (1985) Resonant wave reflection by transverse bedforms and its relation to beaches and offshore bars. *Mar. Geol.* **62**: 321–338.

[198] Hedegaard, I.B., Deigaard, R. and Fredsøe, J. (1991) Onshore/offshore sediment transport and morphological modelling of coastal profiles. *Procs. Coast. Sed.* **91**: 643–654.

[199] Hennings, I., Lurin, B., Vernemmen, C. and Vanhessche, U. (2000) On the behaviour of tidal currents due to the presence of submarine sand waves. *Mar. Geol.* **169**: 57–68.

[200] Hibma, A., de Vriend, H.J. and Stive, M.J.F. (2003) Numerical modelling of shoal pattern formation in well-mixed elongated estuaries. *Est. Coast. Shelf Sci.* **57**: 981–991.

[201] Hino, M. (1974) Theory on formation of rip current and cuspidal coast. *Procs. 14th Int. Coast. Eng. Conf., ASCE*, pp. 901–919.

[202] Hitching, E. and Lewis, A.W. (1999) Bed roughness over vortex ripples. In: *Proc. 4th Int. Symp. Coast. Eng. and Coast. Sed. Processes.* Long Island, ASCE, pp. 18–30.

[203] Hoefel, F. and Elgar, S. (2003) Wave-induced sediment transport and sandbar migration. *Science* **299**: 1885–1887.

[204] Hoitink, A.F.J., Hoekstra, P. and van Mare, D.S. (2003) Flow asymmetry associated with astronomical tides: Implications for residual transport of sediment. *J. Geophys. Res.* **108**: 13-1–13-8.

[205] Holland, K.T. and Holman, R.A. (1996) Field observations of beach cusps and swash motions. *Mar. Geol.* **134**: 77–93.

[206] Holloway, P.E. (1981) Longitudinal mixing in the upper reaches of the Bay of Fundy. *Est. Coast. Shelf Sci.* **13**: 495–515.

[207] Holman, R.A., Lippmann, T.C., O'Neill, P.V. and Haines, J.W. (1993) The application of video image processing to the study of nearshore processes. *Oceanography* **6**: 78–85.

[208] Holman, R.A. (2001) Pattern formation in the nearshore. In: *River, Coastal and Estuarine Morphodynamics.* Eds., G. Seminara and P. Blondeaux, Springer-Verlag, Berlin, pp. 141–162.

[209] Horel, J.D. (1984) Complex principal component analysis: Theory and examples. *J. Clim. Appl. Meteorol.* **23**: 1660–1673.

[210] Horikawa, K. and Kuo, C.T. (1966) A study on wave transformation in the surf zone. *Proc. 10th Int. Coast. Eng. Conf.*, Tokyo, ASCE, pp. 217–233.

[211] Horton, R.E. (1945) Erosional development of streams and their drainage basins; hydrophysical approach to quantitative morphology. *Geol. Soc. Am. Bull.* **56**: 275–370.

[212] Houbolt, J.J.H.C. (1968) Recent sediments in the Southern Bight of the North Sea. *Geologie en Mijnbouw* **47**: 245–273.

[213] Howarth, M.J. and Huthnance, J.M. (1984) Tidal and residual currents around a Norfolk sandbank. *Est. Coast. Shelf Sci.* **19**: 105–117.

[214] Hsu, J.R.C., Silvester, R. and Xia, Y.M. (1989) Static equilibrium bays — New relationships. *J. Waterway, Port, Coastal and Ocean Eng., ASCE* **115**: 285–298.

[215] Hughes, F.W. and Rattray, M. (1980) Salt flux and mixing in the Columbia river estuary. *Est. Coast. Shelf Sci.* **10**: 470–493.

[216] Hulscher, S.J.M.H., de Swart, H.E. and de Vriend, H.J. (1993) The generation of offshore tidal sand banks and sand waves. *Cont. Shelf Res.* **13**: 1183–1204.

[217] Hulscher, S.J.M.H. (1996) Tidal-induced large-scale regular bed form patterns in a three-dimensional shallow water model. *J. Geophys. Res.* **101**: 20727–20744.

[218] Hulscher, S.J.M.H. (2001) Comparison between predicted and observed sandwaves and sandbanks in the North Sea. *J. Geophys. Res.* **106**: 9327–9338.

[219] Hume, T.M. and Herdendorf, C.E. (1992) Factors controlling tidal inlet characteristics on low drift coasts. *J. Coast. Res.* **8**: 355–375.

[220] Hunkins, K. (1981) Salt dispersion in the Hudson estuary. *J. Phys. Ocean.* **11**: 729–738.

[221] Hunt, I.A. (1959) Design of seawalls and breakwaters. *J. Waterw. Harb. Div., ASCE* **85**: 123–152.

[222] Hunt, J.R. (1986) Particle aggregate break-up by fluid shear. In: *Estuarine Cohesive Sediment Dynamics*. Ed., A.J. Mehta. *Lecture Notes on Coastal and Estuarine Studies*, Vol. 14. Springer-Verlag, Berlin, pp. 85–109.

[223] Hunter, K.A. and Liss, P.S. (1982) Organic matter and the surface charge of suspended particles in estuarine waters. *Limnol. Oceanogr.* **27**: 322–335.

[224] Huntley, D.A. and Bowen, A.J. (1973) Field observations of edge waves. *Nature* **243**: 160–161.

[225] Huntley, J.R., Nicholls, R.J., Liu, C. and Dyer, K.R. (1994) Measurements of the semidiurnal drag coefficient over sand waves. *Cont. Shelf Res.* **14**: 437–456.

[226] Huthnance, J.M. (1973) Tidal current asymmetries over the Norfolk sandbanks. *Est. Coast. Mar. Sci.* **1**: 89–99.

[227] Huthnance, J.M. (1982) On one mechanism forming linear sandbanks. *Est. Coast. Mar. Sci.* **14**: 79–99.

[228] Ikeda, S., Parker, G. and Sawai, K. (1981) Bend theory of river meanders. Part 1. Linear development. *J. Fluid Mech.* **112**: 363–377.

[229] Inman, D.L., Elwany, M.H. and Jenkins, S.A. (1993) Shorerise and bar-berm profiles on ocean beaches. *J. Geophy. Res.* **98**: 18181–18199.

[230] Ippen, A.T. and Goda, Y. (1963) Wave-induced oscillations in harbors: The solution for a rectangular harbor connected to the open sea. *Rep. Hydr. Lab MIT*, p. 59.

[231] Ippen, A.T. (1966) *Estuary and Coastline Hydrodynamics*. McGraw-Hill, New York, 744 pp.

[232] Israel, C.G. and Dunsbergen, D.W. (1999) Cyclic morphological development of the Ameland Inlet, The Netherlands. In: *River, Coastal and Estuarine Morphodynamics. Proc. Conf. IAHR*, pp. 705–715.

[233] Izumi, N. and Parker, G. (1995) Inception of channelization and drainage basin formation. Upstream-driven theory. *J. Fluid Mech.* **283**: 341–363.

[234] Jackson, R.G. (1976) Sedimentological and fluid-dynamic implications of the turbulent bursting phenomenon in geophysical flows. *J. Fluid Mech.* **77**: 531–560.

[235] Jaffee, B. and Rubin, D. (1996) Using non-linear forecasting to determine the magnitude and phasing of time-varying sediment suspension in the surf zone. *J. Geophys. Res.* **101**: 14238–14296.

[236] Janssen-Stelder, B. (2000) The effect of different hydrodynamic conditions on the morphodynamics of a tidal mudflat in the Dutch Wadden Sea. *Cont. Shelf Res.* **20**: 1461–1478.

[237] Jarret, J.T. (1976) *Tidal Prism-Inlet Area Relationships*. GITI, Rep.3, US Army Eng. Waterw. Exp. Station, Vicksburg.

[238] Jay, D.A. and Smith, J.D. (1990) Residual circulation in shallow estuaries. 1. Highly stratified, narrow estuaries. *J. Geophys. Res.* **95**: 711–731.

[239] Jay, D.A. (1991) Green's law revisited: Tidal long-wave propagation in channels with strong topography. *J. Geophys. Res.* **96**: 20585–20598.

[240] Jeuken, M.C.J.L. (2000) *On the Morphologic Behaviour of Tidal Channels in the Westerschelde Estuary*. PhD thesis, Utrecht University.

[241] Ji, Z.G. and Mendoza, C. (1997) Weakly nonlinear stability analysis for dune formation. *J. Hydr. Eng.* **123**: 979–985.

[242] Jiyu, C., Cangzi, L., Chongle, Z. and Walker, H.J. (1990) Geomorphological development and sedimentation in Qiantang estuary and Hangzou Bay. *J. Coast. Res.* **6**: 559–572.

[243] Johnson, D.W. (1919) *Shore Processes and Shoreline Development*. Prentice Hall, NY, 584 pp.

[244] Jones, N.V. and Elliot, M. (2000) Coastal zone topics: Process, ecology and management, 4. The Humber estuary and adjoining Yorkshire and Lincolnshire coasts. *Est. Coast. Sci. Ass.*, Hull, UK.

[245] De Jong, H. and Gerritsen, F. (1985) Stability parameters of the Western Scheldt estuary. In: *Procs. 19th Int. Coast. Eng. Conf., ASCE*, New York, pp. 3079–3093.

[246] Jonsson, I.G. (1966) Wave boundary layers and friction factors. In: *Proc. Int. Conf. Coast. Eng.*, Tokyo, Japan. ASCE, pp. 127–148.

[247] Jouanneau, J.M. and Latouche, C. (1981) The Gironde estuary. In: *Contributions to Sedimentology 10 E. Schweizerbartsche Verlagsbuchhandlung, Nagele und Obermiller*, Stuttgart, 115 pp.

[248] Julien, P.Y. and Wargadalam, J. (1995) Alluvial channel geometry: Theory and applications. *J. Hydr. Eng.* **121**: 312–325.

[249] Kaczmarek, L.M., Biegowski, J. and Ostrowski, R. (2004) Modelling cross-shore intensive sand transport and changes of bed grain size distributions versus field data. *Coast. Eng.* **51**: 501–529.

[250] Kamphuis, J.W. (1991) Alongshore sediment transport rate. *J. Waterway, Port, Coastal and Ocean Eng. Div, ASCE* **117**: 624–640.

[251] Kang, S.K., Lee, S.R. and Lie, H.J. (1998) Fine-grid tidal modelling of the Yellow and East China seas. *Cont. Shelf Res.* **18**: 739–772.

[252] Kennedy, J.F. (1969) The formation of sediment ripples, dunes and antidunes. *Ann. Rev. Fluid Mech.* **1**: 147–168.

[253] Kinsman, B. (1965) *Wind Waves*. Prentice-Hall, Englewood Cliffs, N.J.

[254] Kirby, R. and Parker, W.R. (1983) Distribution and behaviour of fine sediment in the Severn Estuary and Inner Bristol Channel. *UK Can. J. Fish. Aquat. Sci.* **40**(suppl.): 83–95.

[255] Kirby, R. (1992) Effects of sea-level on muddy coastal margins. In: *Dynamics and Exchanges in Estuaries and the Coastal Zone*. Ed., D. Prandle. Springer-Verlag, New York, pp. 313–334.

[256] Kitinades, P.K. and Kennedy, J.F. (1984) Secondary currents and river-meander formation. *J. Fluid Mech.* **144**: 217–229.

[257] Kohsiek, L.H.M. and Terwindt, J.H.J. (1981) Characteristics of foreset and topset bedding in megaripples related to hydrodynamic conditions on an intertidal shoal. In: *Holocene Marine Sedimentation in the North Sea Basin*. Eds., S.D. Nio., R.T.E. Schuttenhelm and Tj.C.E. van Weering. *Int. Ass. Sed. Soc. Publ.* **5**: 27–37.

[258] Kohsiek, L.H.M., Buist, H.J., Bloks, P., Misdorp, R., van der Berg, J.H. and Visser, J. (1988) Sedimentary processes on a sandy shoal in a mesotidal estuary (Oosterschelde, The Netherlands). In: *Tide-Influenced Sedimentary Environments and Facies*. Ed., P.L. de Boer *et al.* Reidel Publ. Co., pp. 210–214.

[259] Komar, P.D. (1998) *Beach Processes and Sedimentation*. Prentice Hall, London, p. 544.

[260] Konicki, K.M. and Holman, R.A. (2000) The statistics and kinematics of transverse bars on an open coast. *Mar. Geol.* **169**: 69–101.

[261] Kraak, A., Balfoort, H.M., Vroon, J. and Hallie, F. (2002) Tradition, trends and tomorrow. *The 3rd Coastal Policy Document of The Netherlands*. RIKZ/Rijkswaterstaat, The Hague.

[262] Krone, R.B. (1986) The significance of aggregate properties to transport processes. In: *Estuarine Cohesive Sediment Dynamics*. Ed., A.J. Mehta. *Lecture Notes on Coastal and Estuarine Studies*, Vol. 14. Springer-Verlag, Berlin, pp. 66–84.

[263] Kroon, A. Personal communication.

[264] Lacey, G. (1929) Stable channels in alluvium. *Proc. Inst. Civ. Eng.*, London, **229**: 259–290.

[265] Lamb, H. (1932) *Hydrodynamics*. Cambridge Univ. Press.

[266] Langhorne, D.N. (1973) A sand wave field in the outer Thames Estuary, Great Britain. *Mar. Geol.* **121**: 1–21.

[267] Lanckneus, J., De Moor, G. and Stolk, A. (1994) Environmental setting, morphology and volumetric evolution of the Middelkerke Bank (southern North Sea). *Mar. Geol.* **121**: 1–21.

[268] Lanzoni, S. and Seminara, G. (1998) On tide propagation in convergent estuaries. *J. Geophys. Res.* **103**: 30793–30812.

[269] Lanzoni, S. and Seminara, G. (2002) Long-term evolution and morphodynamic equilibrium of tidal channels. *J. Geophys. Res.* **107**: 1-1–1-13.

[270] Larras, J. (1963) *Embouchures, Estuaires, Lagunes et Deltas*. Collection Centre de Chatou, Eyrolles, France, 171 pp.

[271] LeBlond, P.H. (1978) On tidal propagation in shallow rivers. *J. Geophys. Res.* **83**: 4717–4721.

[272] Le Hir, P., Roberts, W., Cazaillet, O., Christie, M., Bassoullet, P. and Bacher, C. (2000) Characterization of intertidal flat hydrodynamics. *Cont. Shelf Res.* **20**: 1433–1459.

[273] Le Hir, P., Ficht, A., Silva Jacinto, R., Lesueur, P., Dupont, J.-P., Lafitte, R., Brenon, I., Thouvenin, B. and Cugier, P. (2001) Fine sediment transport and accumulations at the mouth of the Seine estuary (France). *Estuaries* **24**: 950–963.

[274] Leopold, L.B., Wolman, M.G. and Miller, J.P. (1964) *Fluvial Processes in Geomorphology*. Freeman, San Francisco.

[275] Lessa, G. (1996) Tidal dynamics and sediment transport in a shallow macrotidal estuary. In: *Mixing in Estuaries and Coastal Seas, Coastal and Estuarine Studies, Am. Geophys. Un.* **50**: 338–360.

[276] Levoy, F., Anthony, E.J., Monfort, O. and Larsonneur, C. (2000) The morphodynamics of megatidal beaches in Normandy, France. *Mar. Geol.* **171**: 39–59.

[277] Li, M.Z. and Amos, C.L. (1998) Predicting ripple geometry and bed roughness under combined waves and currents in a continental shelf environment. *Cont. Shelf Res.* **18**: 941–947.

[278] Li, M.Z. and Amos, C.L. (1999) Field observations of bedforms and sediment transport thresholds of fine sand under combined waves and currents. *Mar. Geol.* **158**: 147–160.

[279] Li, M.Z. and Gust, G. (2000) Boundary layer dynamics and drag reduction in flows of high cohesive sediment suspensions. *Sedimentol.* **47**: 71–86.

[280] Liang, G. and Seymour, R.J. (1991) Complex principle component analysis of wave-like sand motions. In: *Proc. Coastal Sediments*, New York, ASCE, pp. 2175–2186.

[281] Lincoln, J.M. and Fitzgerald, D.M. (1988) Tidal distortions and flood dominance at five small tidal inlets in Southern Maine. *Mar. Geol.* **82**: 133–148.

[282] Linley, E.A.S. and Field, J.G. (1982) The nature and significance of bacterial aggregation in a nearshore upwelling ecosystem. *Est. Coast. Shelf Sci.* **14**: 1–11.

[283] Lippmann, T.C. and Holman, R.A. (1990) The spatial and temporal variability of sandbar morphology. *J. Geophys. Res.* **95**: 11575–11590.

[284] List, J.H. (1986) Wave groupiness as a source for nearshore long waves. *Proc. Int. Conf. Coast. Eng., ASCE*, New York, pp. 497–511.

[285] Liu, Z. (1985) A preliminary study of tidal current ridges. *Chin. J. Ocean. Limnol.* **3**: 118–133.

[286] Liu, Z., Huang, Y. and Zhang, Q. (1989) Tidal current ridges in the southwestern Yellow Sea. *J. Sed. Petr.* **59**: 432–437.

[287] Longuet-Higgins, M.S. (1953) Mass transport in water waves. Royal Soc. London, *Phil. Trans.* **245A**: 535–581.

[288] Longuet-Higgins, M.S. and Stewart, R.W. (1962) Radiation stress and mass transport in gravity waves, with application to 'surf beats'. *J. Fluid Mech.* **8**: 563–583.

[289] Longuet-Higgins, M.S. and Stewart, R.W. (1964) Radiation stresses in water waves: A physical discussion with applications. *Deap-Sea Res.* **11**: 529–562.

[290] Longuet-Higgins, M.S. (1970) Longshore currents generated by obliquely incident sea waves. *J. Geohys. Res.* **75**: 6778–6801.

[291] Louisse, C.J. and Kuik, T.J. (1990) Coastal defence alternatives in The Netherlands. *Int. Conf. Coast. Eng., ASCE*, 1862–1875.

[292] Louda, J.W., Loitz, J.W., Melisiotis, A. and Orem, W.H. (2004) Potential sources of hydrogel stabilisation of Florida Bay lime mud sediments and implications for organic matter preservation. *J. Coast. Res.* **20**: 448–463.

[293] Louters, T. and Gerritsen, F. (1994) The riddle of the sands. *Min. Publ. Works*, The Netherlands, RIKZ-90.040.

[294] Louters, T., van den Berg, J.H. and Mulder, J.P.M. (1998) Geomorphological changes of the Oosterchelde tidal system during and after the implementation of the Delta project. *J. Coast. Res.* **14**: 1134–1151.

[295] Lueck, R.G. and Lu, Y. (1997) The logarithmic layer in a tidal channel. *Cont. Shelf Res.* **17**: 1785–1801.

[296] Lynch, D.K. (1982) Tidal Bores, Scientific American **247**: 134–143.

[297] Madsen, O.S., Wright, L.D., Boon, J.D. and Chisholm, T.A. (1993) Wind stress, bed roughness and sediment suspension on the inner shelf during an extreme storm event. *Cont. Shelf Res.* **13**: 1303–1324.

[298] Malikides, M., Harris, P.T. and Tate, P.M. (1989) Sediment transport and flow over sand waves in a non-rectilinear tidal environment. *Cont. Shelf Res.* **9**: 203–221.

[299] Mallet, C., Howa, H.L., Garlan, T., Sottolichio, A. and Le Hir, P. (2000) Residual transport model in correlation with sedimentary dynamics over an elongate tidal sandbar in the Gironde estuary (Southwestern France). *J. Sed. Res.* **70**: 1005–1016 In: *Coastal Sedimentary Environments*. Ed., R.A. Davis, Springer-Verlag, New York, pp. 77–186.

[300] Marin, F. (2004) Eddy viscosity and Eulerian drift over rippled beds in waves. *Coast. Eng.* **50**: 139–159.

[301] Masselink, G. and Short, A.D. (1993) The effect of tide range on beach morphodynamics and morphology: A conceptual beach model. *J. Coast. Res.* **9**: 785–800.

[302] Masselink, G. (1995) Group bound long waves as a source of infragravity waves in the surf zone. *Cont. Shelf Res.* **15**: 1525–1547.

[303] Masselink, G. and Pattiaratchi, C.B. (1998) Morphological evolution of beach cusps and associated swash circulation patterns. *Mar. Geol.* **146**: 93–113.

[304] Masselink, G. and Hughes, M. (2003) *Introduction to Coastal Processes and Geomorphology.* Oxford University Press.

[305] McBride, R.A. and Moslow, T.F. (1991) Origin, evolution and distribution of shoreface sand ridges, Atlantic inner shelf. *USA Marine Geol.* **97**: 57–85.

[306] McCave, I.N. (1971) Sand waves in the North Sea off the coast of Holland. *Mar. Geol.* **10**: 199–225.

[307] McCave, I.N. and Langhorne, D.N. (1982) Sand waves and sediment transport around the end of a tidal sand bank. *Sedimentol.* **29**: 95–110.

[308] McDowell, D.M. and O'Connor, B.A. (1977) *Hydraulic Behaviour of Estuaries.* MacMillan Press, London, 292 pp.

[309] McLean, S.R., Wolfe, S.R. and Nelson, J.M. (1999) Predicting boundary shear stress and sediment transport over bed forms. *J. Hydr. Eng.* **125**: 725–736.

[310] Meene, J.W.H. van de, Boersma, J.R. and Terwindt, J.H.J. (1996) Sedimentary structures of combined flow deposits from the shoreface-connected ridges along the central Dutch coast. *Mar. Geol.* **131**: 151–175.

[311] Mehta, A.J. and Partheniades, E. (1975) An investigation of the depositional properties of flocculated fine sediments. *J. Hydr. Res.* **13**: 361–381.

[312] Mehta, A.J. (1986) Characteristics of cohesive sediment properties and transport processes in estuaries. In: *Estuarine Cohesive Sediment Dynamics.* Ed., A.J. Mehta. *Lecture Notes Coastal and Estuarine Studies* 14, Springer-Verlag, Berlin, pp. 427–445.

[313] Mehta, A.J. (1996) Interaction between fluid mud and water waves. In: *Environmental Hydraulics.* Eds., V.P. Singh and W.H. Hager. Kluwer Ac. Publ., Dordrecht, pp. 153–187.

[314] Mei, C.N., Fan, S. and Jin, K. (1997) Resuspension and transport of fine sediments by waves. *J. Geophys. Res.* **102**: 15807–15821.

[315] Migniot, C. (1968) Etude des propriétés physiques de différents sédiments très fins et de leur comportement sous des actions hydrodynamiques. *La Houille Blanche* **7**. 391–620.

[316] Monin, A.S. and Yaglom, A.M. (1971) Statistical fluid mechanics: Mechanics of turbulence. MIT Press, MA, 769 pp.

[317] Miles, J. and Munk, W.H. (1961) The harbor paradox. ASCE *J. Waterw. Harb. Div.* p. 2288.

[318] Munk, W.H. and Anderson, E.R. (1948) Notes on the theory of the thermocline. *J. Mar. Res.* **7**: 276–295.

[319] Murray, A.B., LeBars, M. and Guillon, C. (2003) Tests of a new hypothesis for non-bathymetrically driven rip currents. *J. Coast. Res.* **19**: 269–277.

[320] Neumeier, U. and Ciavola, P. (2004) Flow resistance and associated sedimentary processes in a Spartina maritama salt marsh. *J. Coast. Res.* **20**: 435–447.

[321] Nicholls, R.J., Birkemeyer, W.A. and Lee, G.H. (1998) Evaluation of depth of closure using data from Duck, NC, USA. *Mar. Geol.* **148**: 179–201.

[322] Nichols, M.M. and Biggs, R.B. (1985) Estuaries. In: *Coastal Sedimentary Environments.* Ed., R.A. Davis, Springer-Verlag, New York, pp. 77–186.

[323] Nichols, M.M. (1989) Sediment accumulation rates and relative sea-level rise in lagoons. *Mar. Geol.* **88**: 201–219.

[324] Nichols, M.M. and Boon, J.D. (1994) Sediment transport processes in coastal lagoons. In: *Coastal Lagoon Processes.* Ed., B. Kjerfve, Elsevier, Amsterdam, pp. 157–219.

[325] Nicolis, G. and Prigogine, I. (1989) *Exploring Complexity.* Freeman and Co., New York, 313 pp.

[326] Nielsen, P. (1992) Coastal bottom boundary layers and sediment transport. In: *Advanced Series on Ocean Engineering, IV.* World Scientific.

[327] Nikora, V., Goring, D., McEwan, I. and Griffiths, G. (2001) Spatially averaged open-channel flow over rough bed. *J. Hydr. Eng.* **127**: 123–133.

[328] Nordstrom, C.E., Psuty, N. and Carter, R.W.G. (Eds.) (1990) *Coastal Dunes: Processes and Morphology.* John Wiley and Sons, Chichester, 392 pp.

[329] O'Brien, M.P. (Ed.) (1950) *Procs. 1st Conf. Coast. Eng.* Engineering Foundation, University of California, Berkeley.

[330] O'Brien, M.P. (1969) Equilibrium flow areas of inlets and sandy coasts. *J. Waterw. Harbor Coast. Eng. Div.* **95**: 43–52.

[331] O'Connor, B.A., Nunes, C.R. and Sarmento, A.J.N.A. (1996) Sand wave dimensions and statistics. In: *CSTAB Handbook and Final Report.* Ed., B.A. O'Connor. Univ. Liverpool, pp. 336–353.

[332] Odd, N.V.M. and Owen, M.W. (1972) A two-layer model of mud transport in the Thames estuary. In: *Procs. Instn. Civ. Eng.*, Suppl. paper 75175, pp. 175–205.

[333] O'Donoghue, T. and Wright, S. (2001) Experimental study of graded sediments in sinusoidal oscillatory flow. In: *Coastal Dynamics*, Lund, Sweden pp. 918–927.

[334] Oertel, G.F. (1977) Geomorphic cycles in ebb deltas and related patterns of shore erosion and accretion. *J. Sed. Petr.* **47**: 1121–1131.

[335] Oertel, G.F. (1988) Processes of sediment exchange between tidal inlets, ebb deltas and barrier islands. In: *Hydrodynamics and Sediment Dynamics of Tidal Inlets.* Eds., D.G. Aubrey and L. Weishar. Springer-Verlag, New York, pp. 297–318.

[336] Off, T. (1963) Rhythmic linear sand bodies caused by tidal currents. *AAPG Bull.* **47**: 324–341.

[337] Officer, C.B. (1976) *Physical Oceanography of Estuaries and Associated Coastal Waters.* John Wiley, New York.

[338] Oltman-Shay, J., Howd, P.A. and Birkemeier, W.A. (1989) Shear instabilities in the longshore current, 2. Field observations. *J. Geophys. Res.* **94**: 18031–18042.

[339] Osborne, P. and Greenwood, B. (1993) Sediment suspension under waves and currents: Time scales and vertical structure. *Sedimentology*, **40**: 599–622.

[340] Owen, M.W. (1971) The effect of turbulence on the settling velocities of silt flocs. In: *Proc. 14th Conf. IAHR*, Paris, D4: 1–5.

[341] Parker, B.B. (1991) The relative importance of the various nonlinear mechanisms in a wide range of tidal interactions. In: *Tidal Hydrodynamics.* Ed., B.B. Parker, Wiley, New York, pp. 237–268.

[342] Parker, G. (1976) On the cause and the characteristic scales of meandering and braiding in rivers. *J. Fluid Mech.* **76**: 457–480.

[343] Parker, G., Lanfredi, N.W. and Swift, D.J.P. (1982) Seafloor response to flow in a southern hemisphere sand ridge field: Argentine inner shelf. *Sed. Geol.* **33**: 195–216.

[344] Partheniades, E. (1965) Erosion and deposition of cohesive soils. ASCE *J. Hydr. Div.* **91**: 105–139.

[345] Pattiaratchi, C.B. and Collins, M.B. (1987) Mechanisms for linear sandbank formation and maintenance in relation to dynamical oceanographic observations. *Progr. Oceanogr.* **19**: 117–176.

[346] Pelnard-Considère, J.R. (1954) Essai de théorie de l'évolution des formes de rivages en plages de sable et de galets. Soc. Hydrotechnique de France, IVmes Journées de l'Hydraulique, Les Energies de la Mer, Paris, Question 3, 1953.

[347] Perillo, G.M.E. (1995) Definitions and geomorphologic classifications of estuaries. In: *Geomorphology and Sedimentology of Estuaries.* Ed., G.M.E. Perillo. Elsevier, Amsterdam, pp. 17–47.

[348] Pethick, J.S. (1984) *An Introduction to Coastal Geomorphology.* Arnold, London, 260 pp.

[349] Pethick, J.S. (1992) Saltmarsh geomorphology. In: *Saltmarshes: Morphodynamics, Conservation and Engineering Significance.* Eds., J.R.L. Allen and K. Pye, Cambridge Univ. Press, pp. 41–63.

[350] Pingree, R.D. and Griffiths, D.K. (1979) Sand transport paths around the British isles resulting from M2 and M4 tidal interactions. *J. Mar. Biol Ass. UK* **59**: 497–513.

[351] Pingree, R.D. and Griffiths, D.K. (1987) Tidal friction for semidiurnal tides. Cont. Shelf Res. **7**: 1181–1209.

[352] Plant, N.G., Ruessink, B.G. and Wijnberg, K.M. (2001) Morphologic properties derived from a simple cross-shore sediment transport model. *J. Geophys. Res.* **106**(C1): 945–958.

[353] Pontee, N.I., Whitehead, P.A. and Hayes, C.M. (2004) The effect of freshwater flow on siltation in the Humber estuary, northeast UK. *Est. Coast. Shelf Sci.* **60**: 241–249.

[354] Postma, H. (1954) Hydrography of the Dutch Wadden Sea. *Arch. Néerl. Zool.* **12**: 319–349.

[355] Postma, H. (1961) Transport and accumulation of suspended matter in the Dutch Wadden Sea. *Neth. J. Sea Res.* **1**: 148–190.

[356] Postma, H. (1967) Sediment transport and sedimentation in the estuarine environment. In: *Estuaries.* Ed., G.H. Lauff, *Am. Ass. Adv. Sci.* 83, Washington, D.C., pp. 158–179.

[357] Powell, M.A., Thieke, R.J. and Mehta, A.J. (2004) Ebb and flood delta volumes at Florida's sandy tidal entrances. In: *Proceedings Physics of Estuaries and Coastal Seas, 2004.*

[358] Prandle, D. and Rahman, M. (1980) Tidal response in estuaries. *J. Phys. Ocean.* **10**: 1552–1573.

[359] Prandle, D. (2003) Relationships between tidal dynamics and bathymetry in strongly convergent estuaries. *J. Phys. Ocean.* **33**: 2738–2750.

[360] Prandle, D. (2004) Salt intrusion in partially mixed estuaries. *Est. Coast. Shelf Sci.* **59**: 385–397.

[361] Prandle, D. (2004) Sediment trapping, turbidity maximum and bathymetric stability in macrotidal estuaries. *J. Geophys. Res.* **109**, C09001.

[362] Price, W.A. (1947) Equilibrium of form and forces in tidal basins on coasts of Texas and Louisiana. *Bul. Am. Ass. Petr. Geol.* **31**: 1619–1663.

[363] Puleo, J.A., Holland, K.T., Plant, N.G., Slinn, D.N. and Hanes, D.M. (2003) Fluid acceleration effects on suspended sediment transport in the swash zone. *J. Geophys. Res.* **108**: C11.

[364] Pullen, T. and She, K. (2002) A numerical study of breaking waves and a comparison of breaking criteria. *Proc. 28th Int. Conf. Coast. Eng.*, Cardiff, ASCE, pp. 293–305.

[365] Ranasinghe, R., Symonds, G., Black, K. and Holman, R. (2004) Morphodynamics of intermediate beaches: A video imaging and numerical modelling study. *Coast. Eng.* **51**: 629–665.

[366] Raubenheimer, B., Guza, R.T. and Elgar, S. (1996) Wave transformation across the inner surf zone. *J. Geophys. Res.* **101**: 25589–25597.

[367] Raubenheimer, B., Guza, R.T. and Elgar, S. (2001) Field observations of wave-driven setdown and setup. *J. Geophys. Res.* **106**: 4629–4638.

[368] Raubenheimer, R., Elgar, S. and Guza, T. (2004) Observations of swash zone velocities: A note on friction coefficients. *J. Geophys. Res.* **109**(C01027): 1–8.

[369] Raudkivi, A.J. and Witte, H.H. (1990) Development of bed features. *J. Hydr. Eng.* **116**: 1063–1079.

[370] Raudkivi, A.J. (1997) Ripples on stream bed. *J. Hydr. Eng.* **123**: 58–64.

[371] Reniers, A.J.H.M., Thornton, E.B., Stanton, T.P. and Roelvink, J.A. (2004) Vertical flow structure during Sandy Duck: observations and modelling. *Coast. Eng.* **51**: 237–260.

[372] Reniers, A.J.H.M. (2005) Personal communication.

[373] Ribas, F., Falqués, A., Plant, N. and Hulscher, S. (2001) Self-organization in surf zone morphodynamics: Alongshore uniform instabilities. In: *Procs. 4th Int. Conf. Coastal Dynamics*. Eds., H. Hanson and M. Larson. ASCE, pp. 1068–1077.

[374] Ribas, F., Falqués, A. and Montoto, A. (2003) Nearshore oblique sand bars. *J. Geophys. Res.* **108**: C4.

[375] Richards, K.J. (1980) The formation of ripples and dunes on an erodible bed. *J. Fluid Mech.* **99**: 597–618.

[376] Ridderinkhof, H. (1990) *Residual Currents and Mixing in the Wadden Sea*. Thesis, Utrecht University, p. 91.

[377] Ridderinkhof, H., van der Ham, R. and van der Lee, W. (2000) Temporal variations in concentration and transport of suspended sediments in a channel-flat system in the Ems-Dollard estuary. *Cont. Shelf Res.* **20**: 1479–1493.

[378] Riethmüller, R., Fanger, H.U., Grabemann, I., Krasemann, H.L., Ohm, K., Böning, J., Neumann, L.J.R., Lang, G., Markofsky, M. and Schubert, R. (1988) Hydrographic measurements in the turbidity maximum of the Weser estuary. In: *Physical Processes in Estuaries*. Eds., J. Dronkers and W. van Leussen. Springer-Verlag, Berlin, pp. 332–344.

[379] RIKZ National Institute for Coastal and Marine Management (1994) Average tidal curves for the Dutch tidal waters. 1991.0. (De gemiddelde getijkromme, in Dutch). RIKZ, The Netherlands, ISBN 90-369-0453-6.

[380] Robinson, A.H.W. (1965) Residual currents in relation to shoreline evolution of the east Anglian coast. *Mar. Geol.* **4**: 57–84.

[381] Robinson, I.S. (1983) A tidal model of the Fleet — An English tidal lagoon. *Set. Coast. Shelf Sci.* **16**: 669–688.

[382] Roelvink, D.J.A. and Stive, M.J.F. (1989) Bar-generating cross-shore flow mechanisms on a beach. *J. Geophys. Rev.* **94**: 4785–4800.

[383] Roy, P.S., Cowell, P.J., Ferland, M.A. and Thom, B.G. (1994) Wave dominated coasts. In: *Coastal Evolution: Late Quaternary Shoreline Morphodynamics*. Eds., R.W.G. Carter and C.D. Woodroffe, Cambridge University Press, pp. 121–185.

[384] Rubin, D.M. and Ikeda, H. (1990) Flume experiments on the alignment of transverse, oblique and longitudinal dunes in directionally varying flows. *Sedimentol.* **37**: 673–684.

[385] Ruessink, B.G., van Enckevort, I.M.J., Kingston, K.S. and Davidson, M.A. (2000) Analysis of observed two- and three-dimensional nearshore bar behaviour. *Mar. Geol.* **169**: 161–183.

[386] Ruessink, B.G., van Enckevort, I.M.J., Kingston, K.S. and Davidson, M.A. (2002) In: *Coast3D-Egmond; The Behaviour of a Straight Sandy Coast on the Time Scale of Storms and Seasons*. Eds., L.C. Van Rijn, B.G. Ruessink and J.P.M. Mulder. ISBN 90-800356-5-3, Aqua Publ., Amsterdam, L1–L23.

[387] Russel, R.J. and McIntire, W.G. (1965) Beach cusps. *Geol. Soc. Am. Bull.* **76**: 307–320.

[388] Ryu, S.O. (2003) Seasonal variation of sedimentary processes in a semi-enclosed bay: Hampyong Bay, Korea. *Est. Coast. Shelf Sci.* **56**: 481–492.

[389] Salomons, W. and Mook, W.G. (1981) Field observations of isotopic composition of particulate organic carbon in the Southern North Sea and adjacent estuaries. *Mar. Geol.* **41**: 11–20.

[390] Savenije, H.H.G. (2003) The width of a bankfull channel; Lacey's formula explained. *J. Hydrol.* **276**: 176–183.

[391] Scharp, J.C. (1949) Hydrografie. In: *Handboek der Geografie van Nederland*. Eds., G.J.A. Mulder, J.J. De Erven, Z. Tijl. **1**: 378–529.

[392] Schielen, R., Doelman, A. and De Swart, H.E. (1993) On the nonlinear dynamics of free bars in straight channels. *J. Fluid Mech.* **252**: 325–356.

[393] Schijf, J.B. and Schönfeld, J.C. (1953) Theoretical considerations on the motion of salt and fresh water. In: *Proc. Minn. Int. Hydraul. Conv.*, Mineanopolis, p. 321.

[394] Schramkowski, G.P., Schuttelaars, H.M. and de Swart, H.E. (2002) The effect of geometry and bottom friction on local bed forms in a tidal embayment. *Cont. Shelf Res.* **22**: 1821–1833.

[395] Schröder, M. and Siedler, G. (1989) Turbulent momentum and salt transport in the mixing zone of the Elbe estuary. *Est. Coast. Shelf Sci.* **28**: 615–638.

[396] Schumm, S.A. (1969) River metamorphosis. *J. Hydr. Div. Proc.* ASCE **96**: 201–222.

[397] Schuttelaars, H.M. and De Swart, H.E. (1999) Formation of channels and shoals in a short tidal embayment. *J. Fluid Mech.* **386**: 15–42.

[398] Seminara, G. and Tubino, M. (1997) Bed formation in tidal channels: Analogy with fluvial bars. In: *Morphology of Rivers, Estuaries and Coasts*. Ed., DiSilvio, IAHR, London.

[399] Seminara, G. and Blondeax, P. (Eds.) (2001) *River, Coastal and Estuarine Morphodynamics*. Springer, Berlin, p. 211.

[400] Seminara, G. and Tubino, M. (2001) Sand bars in tidal channels. Part 1. Free bars. *J. Fluid Mech.* **440**: 49–74.

[401] Seminara, G., Lanzoni, S., Bolla Pittaluga, M. and Solari, L. (2001) Estuarine Patterns: An introduction to their morphology and mechanics. In: *Geomorphological Fluid Mechanics*. Eds., M.J. Balmforth and E. Provenxale. Springer, pp. 455–499.

[402] Sha, L.P. (1998) Sand transport patterns in the ebb-tidal delta off Texel Inlet, Wadden Sea, The Netherlands. *Mar. Geol.* **86**: 137–154.

[403] Sha, L.P. and van den Berg, J.H. (1993) Variation in ebb-tidal delta geometry along the coast of the Netherlands and the German Bight. *J. Coast. Res.* **9**: 730–746.

[404] Shi, N.C. and Larsen, L.H. (1984) Reverse transport induced by amplitude-modulated waves. *Mar. Geol.* **54**: 181–200.

[405] Shepard, F.P. (1973) *Submarine Geology*. Harper and Row, New York, 517 pp.

[406] Short, A.D. (1991) Macro-meso tidal beach morphodynamics — An overview. *J. Coast. Res.* **7**: 417–436.

[407] Simpson, J.H., Crawford, W.R., Rippeth, T.P., Campbell, A.R. and Cheok, J.V.S. (1996) The vertical structure of turbulent dissipation in shelf seas. *J. Phys. Ocean.* **26**: 1579–1590.

[408] Sistermans, P.J.G., Van de Graaff, J. and Van Rijn, L.C. (2001) Vertical sorting of graded sediments by waves and currents. In: *Proc. Int. Conf. Coast. Eng.*, Sydney, Australia. ASCE, pp. 2780–2793.

[409] Sleath, J.F.A. (1976) On rolling grain ripples. *J. Hydr. Res.* **14**: 69–80.

[410] Sleath, J.F.A. (1984) *Sea Bed Mechanics*. Wiley, New York.

[411] Sleath, J.F.A. (1991) Velocities and shear stress in wave-current flows. *J. Geophys. Res.* **96**: 15237–15244.

[412] Small, C. and Nicholls, R.J. (2003) A global analysis of human settlement in coastal zones. *J. Coast. Res.* **19**: 584–599.

[413] Smith, J.B. and FitzGerald, D.M. (1994) Sediment transport patterns at the Essex River Inlet ebb-tidal delta, Massachusetts, USA. *J. Coast. Res.* **10**: 752–774.

[414] Smith, J.D. (1969) Geomorphology of a sand ridge. *J. Geol.* **77**: 39–55.

[415] Smith, J.D. and McLean, S.R. (1977) Spatially averaged flow over a wavy surface. *J. Geophys. Res.* **82**: 1735–1746.

[416] Sonu, C.J. (1968) Collective movement of sediment in littoral environment. In: *Proc. Int. Conf. Coast. Eng.*, ASCE, pp. 373–400.

[417] Soulsby, R.L., Atkins, R. and Salkfield, P. (1994) Observations of the turbulent structure of a suspension of sand in a tidal current. *Cont. Shelf Res.* **14**: 429–435.

[418] Soulsby, R.L. (1997) *Dynamics of Marine Sands*. Thomas Telford, pp. 249.

[419] Southard, J.B. and Dingler, J.R. (1971) Flume study of ripple propagation behind mounds on flat sand beds. *Sedimentol.* **16**: 251–263.

[420] Southard, J.B. and Boguchwal, L.A. (1980) Bed configurations in steady inudirectional water flows. Part 2. Synthesis of flume data. *J. Sed. Petrol.* **60**: 658–779.

[421] Spanhoff, R., Biegel, E.J., Burger, M. and Dunsbergen, D.W. (2004) *Shoreface Nourishments in the Netherlands* (In press).

[422] Spanhoff, R. Personal communication.

[423] Speer, P.E. and Aubrey, D.G. (1985) A study of non-linear tidal propagation in shallow inlet/estuarine systems. Part II: Theory. *Estuarine, Coast. Shelf Sci.* **21**: 207–224.

[424] Stive, M.J.F., Roelvink, D.J.A. and De Vriend, H.J. (1991) Large-scale coastal evolution concept. *Procs. 22nd Int. Conf. Coast. Eng. ASCE*, pp. 1962–1974.

[425] Stive, M.J.F., Aarninkhof, S.G.J., Hamm, L., Hanson, H., Larson, M., Wijnberg, K.M., Nicholls, J. and Capobianco, M. (2002) Variability of shore and shoreline evolution. *Coastal Eng.* **47**: 211–235.

[426] Stoker, J.J. (1957) Water waves. *Interscience*, New York.

[427] Straaten, L.M.J.U. and Kuenen, P.H. (1957) Accumulation of fine-grained sediments in the Dutch Wadden Sea. *Geol. Mijnbouw* (N.S.) **19**: 329–354.

[428] Sunamura, T. (2004) A predictive relationship for the spacing of beach cusps in nature. *Coast. Eng.* **51**: 697–711.

[429] Sutherland, A.J. (1967) Proposed mechanism for sediment entrainmment by turbulent flows. *J. Geophys. Res.* **72**: 6183–6194.

[430] Swift, D.J.P., Duane, D.B. and McKinney, T.F. (1973) Ridge and swale topography of the middle Atlantic Bight, North America: Secular response to the Holocene hydraulic regime. *Marine Geology* **15**: 227–247.

[431] Swift, D.J.P., Parker, G., Lanfredi, N.W., Perillo, G. and Figge, K. (1978) Shoreline-connected sand ridges on American and European shelves — A comparison. *Est. Coast. Mar. Sci.* **7**: 227–247.

[432] Swift, D.J.P. and Field, M.E. (1981) Evolution of a classic sand ridge field: Maryland sector, North American inner shelf *Sedimentol.* **28**: 461–482.

[433] Swift, D.J.P. and Thorne, J.A. (1991) Sedimentation on continental margins, I: A general model for shelf sedimentation. In: *Shelf Sand and Sandstone Bodies*. Eds., D.J.P.

Swift, G.F. Oertel, R.W. Tillman and J.A. Thorne. *Int. Ass. Sed.*, Blackwell, Oxford, pp. 3–31.

[434] Swart, D.H. (1776) Coastal sediment transport. *Computation of Longshore Transport.* rep. R968: Part I. W.L. Delft Hydraulics, Delft.

[435] Tang, E.C.S. and Dalrymple, R.A. (1989) Nearshore circulation: B. Rip currents and wave groups. In: *Nearshore Sediment Transport.*

[436] Temmerman, S., Govers, G., Meire, P. and Wartel, S. (2003) Modelling long-term marsh growth under changing tidal conditions and suspended sediment concentrations, Scheldt estuary, Belgium. *Mar. Geol.* **193**: 151–169.

[437] Ten Brinke, W.B.M., Dronkers, J. and Mulder, J.P.M. (1994) Fine sediments in the Eastern-Scheldt tidal basin before and after partial closure. *Hydrobiol.* **282/283**: 41–56.

[438] Terwindt, J.H.J. (1971) Sand waves in the southern North Sea. *Mar. Geol.* **10**: 51–67.

[439] Thompson, C.E.L., Amos, C.L., Lecouturier, M. and Jones, T.E.R. (2004) Flow deceleration as a method of determining drag coefficient over roughened flat beds. *J. Geophys. Res.* **109**: C03001, 1–12.

[440] Thornton, E.B. and Guza, R.T. (1982) Energy saturation and phase speeds measured on a natural beach. *J. Geophys. Res.* **87**: 9499–9508.

[441] Thornton, E.B. and Guza, R.T. (1983) Transformation of wave height distribution. *J. Geophys. Res.* **88**: 5925–5938.

[442] Thornton, E., Humiston, R. and Birkemeyer, W. (1996) Bar-trough generation on a natural beach. *J. Geophys. Res.* **101**: 12097–12110.

[443] Tomasicchio, G.R. and Sancho, F. (2002) On wave induced undertow at a barred beach. *Proc. 28th Int. Conf. Coast. Eng.*, Cardiff, ASCE, pp. 557–569.

[444] Trembanis, A.C., Wright, L.D., Friedrichs, C.T., Green, M.O. and Hume, T. (2004) The effects of spatially complex inner shelf roughness on boundary layer turbulence and current and wave friction: Tairua embayment, New Zealand. *Cont. Shelf Res.* **24**: 1549–1571.

[445] Trowbridge, J.H. and Madsen, O.S. (1984) Turbulent wave boundary layers: 2. Second-order theory and mass transport. *J. Geophys. Res.* **89**: 7999–8007.

[446] Trowbridge, J.H. (1995) A mechanism for the formation and maintenance of the shore-oblique sand ridges on storm-dominated shelves. *J. Geophys. Res.* **100**: 16071–16086.

[447] Turrell, W.R. and Simpson, J.H. (1988) The measurement and modelling of axial convergence in shallow well-mixed estuaries. In: *Physical Processes in Estuaries.* Eds., J. Dronkers and W. van Leussen, Springer-Verlag, Berlin, pp. 130–145.

[448] Uncles, R.J. and Jordan, M.B. (1979) Residual fluxes of water and salt at two stations in the Severn Estuary. *Est. Coast. Mar. Sci.* **9**: 287–302.

[449] Uncles R.J. and Jordan, M.B. (1980) A one-dimensional representation of residual currents in the Severn estuary and associated observations. *Est. Coast. Shelf Sci.* **10**: 39–60.

[450] Uncles, R.J., Elliot, R.C.A. and Weston, S.A. (1985) Observed fluxes of water, salt and suspended sediment in a partly mixed estuary. *Est. Coast. Shelf Sci.* **20**: 147–167.

[451] Uncles, R.J., Elliot, R.C.A. and Weston, S.A. (1986) Observed and computed lateral circulation patterns in a partially mixed estuary. *Est. Coast. Shelf Sci.* **22**: 439–457.

[452] Uncles, R.J. and Stephens, J.A. (1990) Salinity stratification and vertical shear transport in an estuary. In: *Coastal and Estuarine Studies* 38, Residual currents and long-term transport. Ed., R.T. Cheng, Springer-Verlag, New York, pp. 137–150.

[453] Uncles, R.J. and Stephens, J.A. (1993) The nature of the turbidity maximum in the Tamar estuary. *Est. Coast. Shelf Sci.* **36**: 413–431.

[454] Uncles, R.J. (2002) Estuarine physical processes research: Some recent studies and progress. *Est. Coast. Shelf Sci.* **55**: 829–856.

[455] Uncles, R.J., Bale, A.J., Brinsley, M.D., Frickers, P.E., Harris, C., Lewis, R.E., Pope, N.D., Staff, F.J., Stephens, J.A., Turley, C.M. and Widdows, J. (2003) Intertidal mudflat properties, currents and sediment erosion in the partially mixed Tamar Estuary, UK. *Ocean Dynamics* **53**: 239–251.

[456] Van den Berg, J.H. (1987) Bed form migration and bedload transport in some rivers and tidal environments. *Sedimentology* **34**: 681–698.

[457] Van de Kreeke, J. Stability of tidal inlets; Escoffier's analysis. *Shore and Beach* **60**: 9–12.

[458] Van de Kreeke, J. and Iannuzzi, R.A. (1998) Second-order solutions for damped cooscillating tide in narrow canal. *J. Hydr. Eng.* **124**: 1253–1260.

[459] Van der Molen, J. (2002) The influence of tides, wind and waves on the net sand transport in the North Sea. *Cont. Shelf Res.* **22**: 2739–2762.

[460] Van der Spek, A.F.J. (1994) *Large-Scale Evolution of Holocene Tidal Basins in the Netherlands.* Thesis, Utrecht University, 191 pp.

[461] Van der Spek, A.F.J. (1997) Tidal asymmetry and long-term evolution of Holocene tidal basins in The Netherlands: Simulation of paleo-tides in the Schelde estuary. *Mar. Geol.* **141**: 71–90.

[462] Van der Wal, D. (1999) *Aeolian Transport of Nourishment Sand in Beach-Dune Environments.* Thesis, University of Amsterdam.

[463] Van der Wal, D., Pye, K. and Neal, A. (2002) Long-term morphological change in the Ribble estuary, northwest England. *Mar. Geol.* **189**: 249–266.

[464] Van Dongeren, A.R. and De Vriend, H.J. (1994) A model of morphological behaviour of tidal basins. *Coast. Eng.* **22**: 287–310.

[465] Van Duin, M.J.P., Wiersma, N.R., Walstra, D.-J.R., Van Rijn, L.C. and Stive, M.J.F. (2004) Nourishing the shoreface: Observations and hindcasting of the Egmond case, The Netherlands. *Coast. Eng.* **51**: 813–837.

[466] Van Goor, M.A., Zitman, T.J., Wang, Z.B. and Stive, M.J.F. (2003) Impact of sea-level rise on the morphological equilibrium state of tidal inlets. *Mar. Geol.* **202**: 211–227.

[467] Van Heteren, S., Baptist, M.J., Van Bergen Henegouwen, V.N., Van Dalfsen, J.A., Van Dijk, T.A.G.P., Hulscher, N.H.B.M., Knaapen, M.A.F., Lewis, W.E., Morelissen, R., Passchier, S., Penning, W.E., Storbeck, F., Van der Spek, A.F.J., Van het Groenewoud, H. and Weber, A. (2003) *Eco-Morphodynamics of the Seafloor.* Delft Cluster Publ. 03.01.05-04, Univ. Delft, p. 52

[468] Van Lancker, V.R.M. and Jacobs, P. (2000) The dynamical behaviour of shallow-marine dunes. In: *Marine Sandwave Dynamics.* Eds., A. Trentesaux and T. Garlan. *Procs. Int. Workshop,* Univ. Lille, pp. 213–220.

[469] Van Leussen, W. (1994) *Estuarine Macro-Flocs and their Role in Fine-Grained Sediment Transport.* Thesis, Utrecht University, p. 488.

[470] Van Maldegem, D.C., Mulder, H.P.J. and Langerak, A. (1991) A cohesive sediment balance for the Scheldt estuary. *Neth. J. Aq. Ecol.* **27**: 247–256.

[471] Van Rijn, L.C. (1984) Sediment transport, Part II: Suspended load transport. *J. Hydraul. Div. Proc. ASCE* **110**: 1613–1641.

[472] Van Rijn, L.C. (1993) *Handbook Sediment Transport in Rivers, Estuaries and Coastal Seas.* Aqua Publ., Amsterdam.

[473] Van Rijn, L.C., Ruessink, B.G. and Mulder, J.P.M. (2002) Summary of project results. In: *Coast3D-Egmond; The Behaviour of a Straight Sandy Coast on the Time Scale of Storms and Seasons.* ISBN 90-800356-5-3, Aqua Publ., Amsterdam.

[474] Van Rijn, L.C., Caljauw, M. and Kleinhout, K. (2002). Basic features of morphodynamics at the Egmond site on the medium-term time scale of seasons. In: *Coast3D-Egmond; The Behaviour of a Straight Sandy Coast on the Time Scale of Storms and Seasons.* Eds., L.C. Van Rijn, B.G. Ruessink and J.P.M. Mulder. ISBN 90-800356-5-3, Aqua Publ., Amsterdam, J1–J20.

[475] Van Straaten, L.M.J.U. and Kuenen, P.H. (1957) Accumulation of fine grained sands in the Dutch Wadden sea. *Geol. en Mijnbouw* **19**: 406–413.

[476] Van Veen, J. (1950) Eb- en vloedschaar systemen in de Nederlandse getijdewateren. *Tijdschrift Kon. Ned. Aardrijkskundig Genootschap* **67**: 303–325.

[477] Vincent, C.E. and Green, M.O. (1990) Field measurements of the suspended sand concentration profiles and fluxes and of the resuspensionm coefficient over a rippled bed. *J. Geophys. Res.* **95**: 11591–11601.

[478] Vincent, C.E., Stolk, A. and Porteer, C.F.C. (1998) Sand suspension and transport on the Middelkerke Bank (southern North Sea) by storms and tidal currents. *Mar. Geol.* **150**: 113–129.

[479] Vittori, G. (2003) Sediment suspension due to waves. *J. Geophys. Res.* **108**(C6), 4: 1–17.

[480] Walton, T.L. and Adams, W.D. (1976) Capacity of inlet outer bars to store sand. *Procs. 15th Int. Coast. Eng. Conf., ASCE,* New York, pp. 1919–1937.

[481] Wang, Z.D., Fokkink, R.J., de Vries, M. and Langerak, A. (1995) Stability of river bifurcations in 1D morphological models. *J. Hydraul. Res.* **33**: 739–750.

[482] Wang, P., Ebersole, B.A. and Smith, E.R. (2003) Beach-profile evolution under spilling and plunging breakers. *J. Waterway, Port, Coast. Ocean Eng., ASCE* **129**: 41–46.

[483] Wells, J.T. (1997) Tide-dominated estuaries and tidal rivers. In: *Geomorphology and Sedimentology of Estuaries.* Ed., G.M.E. Perillo, Elsevier, Amsterdam, pp. 179–205.

[484] Warrick, R.A., Barrow, E.M. and Wigley, T.M.I. (Eds.) (1993) Climate and sea level change: Observations, projections and implications. Cambridge Univ. Press, 424 pp.

[485] Werner, B.T. and Fink, T.M. (1993) Beach cusps as self-organized patterns. *Science,* **260**: 968–971.

[486] Werner, S.R., Beardsley, R.C. and Williams, A.J. (2003) Bottom friction and bed forms on the southern flank of Georges Bank. *J. Geophys. Res.* **108**(C11), GLO 5: 1–21.

[487] West, J.R. and Mangat, J.S. (1986) The determination and prediction of longitudinal dispersion coefficients in a narrow, shallow estuary. *Est. Coast. Shelf Sci.,* pp. 161–181.

[488] Whitehouse, R.J.S. and Mitchener, H.J. (1998) Observations of the morphodynamic behaviour of an intertidal mudflat at different timescales. In: *Sedimentary Processes in the Intertidal Zone.* Eds., K.S. Black, D.M. Paterson and A. Cramp, Geological Society, London, Special Publ. **139**: 255–271.

[489] Wijnberg, K. (1995) *Morphologic Behaviour of a Barred Coast over a Period of Decades.*
       Thesis, Utrecht University, ISBN 90-6266-125-4, 245 pp.

[490] Wijnberg, K. and Terwindt, J.H.J. (1995) Extracting decadal morphological behaviour
       from high-resolution long-term bathymetric surveys along the Holland coast using eigen-
       function analysis. *Marine Geol.* **126**: 301–330.

[491] Williams, P.B. and Kemp, P.H. (1971) Initiation of ripples on flat sediment beds. *J. Hydr.
       Div. ASCE* **97**: 505–522.

[492] Winkelmolen, A.H. and Veenstra, H.J. (1980) The effect of a storm surge on
       nearshore sediments in the Ameland-Schiermonnikoog area. *Geologie Mijnbouw* **59**:
       97–111.

[493] Winterwerp, J.C. (2001) Stratification effects by cohesive and non-cohesive sediments.
       J. Geophys. Res. **106**: 22559–22574.

[494] Wolanski, E., King, B. and Galloway, D. (1997) Salinity intrusion in the Fly River estuary,
       Papua New Guinea. *J. Coast. Res.* **13**: 983–994.

[495] Wood, R. and Widdows, J. (2002) A model of sediment transport over an intertidal
       transect, comparing the influence of biological and physical factors. *Limnol. Ocean.* **47**:
       848–855.

[496] Woodroffe, C.D. (2002) Coasts, form, processes and evolution. Cambridge Univ. Press,
       623 pp.

[497] Wright, L.D., Coleman, J.M. and Thom, B.G. (1973) Processes of channel development
       in a high tide-range environment: Cambridge Gulf-Ord River Delta, Western Australia.
       J. Geol. **81**: 15–41.

[498] Wright, L.D., Chappel, J., Thom, B.G., Bradshow, M.P. and Cowell, P. (1979) Morphody-
       namics of reflective and dissipative beach and inshore systems: Southeastern Australia.
       Mar. Geol. **32**: 105–140.

[499] Wright, L.D. and Short, A.D. (1984) Morphodynamic variability of surf zones and
       beaches: A synthesis. *Mar. Geol.* **56**: 93–118.

[500] Wright, L.D. (1985) River Deltas. In: *Coastal Sedimentary Environments.* Ed., R.A.
       Davis. Springer-Verlag, New York, pp. 1–76.

[501] Yalin, M.S. (1964) Geometrical properties of sand waves. *Proc. Am. Soc. Civil Eng.* **90**:
       105–119.

[502] Yalin, M.S. and da Silva, A.M.F. (1992) Horizontal turbulence and alternating bars. *J.
       Hydrosci. Hydraul. Eng.* **9**: 47–58.

[503] Yu, J. and Slinn, D.N. (2003) Effects of wave-current interaction on rip currents. *J.
       Geophys. Res.* **108**: 33-1–33-19.

[504] Zagwijn, W.H. (1983) Sea level changes in the Netherlands during the Eemian. *Geol.
       Mijnb.* **62**: 437–450.

[505] Zagwijn, W.H. (1986) Nederland in het Holoceen. Rijks Geologische Dienst. *Haarlem*,
       46 pp.

[506] Zang, D.P. and Sunamura, T. (1994) Multiple bar formation by breaker induced vor-
       tices: A laboratory approach. In: *Proc. 24th Int. Coast. Eng.*, Kobe, Japan. ASCE, pp.
       2856–2870.

[507] Zedler, E.A. and Street, R.L. (2001) Large-eddy simulation of sediment transport: Current
       over ripples. *J. Hydr. Eng.* **127**: 444–452.

[508] Zimmerman, J.T.F. (1973) The influence of the subaqeous profile on wave-induced bot-
       tom stress. *Neth. J. Sea Res.* **6**: 542–549.

[509] Zimmerman, J.T.F. (1976) Mixing and flushing of tidal embayments in the western Wadden Sea, I: Distribution of salinity and calculation of mixing time scales. *Neth. J. Sea Res.* **10**: 149–191.

[510] Zimmerman, J.T.F. (1978) Topographic generation of residual circulation by oscillatory (tidal) currents. *Geophys. Astrophys. Fluid Dyn.* **11**: 35–47.

[511] Zimmerman, J.T.F. (1981) Dynamics, diffusion and geomorphological significance of tidal residual eddies. *Nature* **290**: 549–555.

[512] Zimmerman, J.T.F. (1986) The tidal whirlpool: A review of horizontal dispersion by tidal and residual currents. *Neth. Journal of Sea Research* **20**: 133–154.

Index

512